**TETRAHEDRON ORGANIC**

*Series Editors:*

**J E Baldwin, FRS**
*University of Oxford,*
*Oxford, OX1 3QY, UK*

**R M Williams**
*Colorado State University,*
*Fort Collins, CO 80523, USA*

**VOLUME 22**

# Organic Syntheses Based on Name Reactions

## SECOND EDITION

**Related Pergamon Titles of Interest**

**BOOKS**

*Tetrahedron Organic Chemistry Series:*
CARRUTHERS: Cycloaddition Reactions in Organic Synthesis
CLARIDGE: High-Resolution NMR Techniques in Organic Chemistry
FINET: Ligand Coupling Reactions with Heteroatomic Compounds
GAWLEY & AUBÉ: Principles of Asymmetric Synthesis
HASSNER & STUMER: Organic Syntheses Based on Name Reactions and Unnamed Reactions
McKILLOP: Advanced Problems in Organic Reaction Mechanisms
OBRECHT & VILLALGORDO: Solid-Supported Combinatorial and Parallel Synthesis of Small-Molecular-Weight Compound Libraries
PERLMUTTER: Conjugate Addition Reactions in Organic Synthesis
SESSLER & WEGHORN: Expanded, Contracted & Isomeric Porphyrins
TANG & LEVY: Chemistry of C-Glycosides
WONG & WHITESIDES: Enzymes in Synthetic Organic Chemistry
LI & GRIBBLE: Palladium in Heterocyclic Chemistry
PIETRA: Biodiversity and Natural Product Diversity

*Other Series:*
RAHMAN: Studies in Natural Products Chemistry *(series)*

**JOURNALS**

BIOORGANIC & MEDICINAL CHEMISTRY
BIOORGANIC & MEDICINAL CHEMISTRY LETTERS
TETRAHEDRON
TETRAHEDRON LETTERS
TETRAHEDRON: ASYMMETRY

*Full details of all Elsevier Science publications are available on www.elsevier.com or from your neare Elsevier Science office*

# Organic Syntheses Based on Name Reactions

## SECOND EDITION

A. HASSNER

and

C. STUMER

*Department of Chemistry*
*Bar-Ilan University*
*Ramat-Gan, Israel*

2002

PERGAMON

An imprint of Elsevier Science

Amsterdam - Boston - London - New York - Oxford - Paris
San Diego - San Francisco - Singapore - Sydney - Tokyo

ELSEVIER SCIENCE Ltd
The Boulevard, Langford Lane
Kidlington, Oxford OX5 1GB, UK

First edition 2002

Library of Congress Cataloging-in-Publication Data

Hassner, Alfred, 1930-
    Organic syntheses based on name reactions/A. Hassner and C. Stumer.--2nd ed.
      p.cm. -- (Tetrahedron organic chemistry series ; v. 22)
    Rev. ed. of: Organic syntheses based on name reactions and unnamed reactions, 1994.
    Includes indexes.
    ISBN 0-08-043260-3 (h) -- ISBN 0-08-043259-X (p)
      1. Organic compounds--Synthesis. I. Stumer, C. II. Title. III. Series.

    QD262 .H324 2002
    547'.2--dc21

                                                    2002024278
British Library Cataloguing in Publication Data

Hassner, Alfred, 1930-
    Organic syntheses based on name reactions. - 2nd ed. -
    (Tetrahedron organic chemistry series ; v. 22)
    1.Organic compounds - Synthesis
    I.Title II.Stumer, C.
    547.2

ISBN:   0 08 043260 3 (hardbound)
ISBN:   0 08 043259 X (paperback)

Transferred to digital print 2008
Printed and bound by CPI Antony Rowe, Eastbourne

# Contents

Foreword to first edition      vii

Foreword to second edition      ix

Name Reactions (arranged alphabetically)      1

Names Index      425

Reagents Index      435

Reactions Index      441

Functional Group Transformations Index      445

*Non nova sed nove*

## FOREWORD to FIRST EDITION

### And these are the names...

The above are the opening words of Exodus, the second book of the Pentateuch. Already in ancient times, names were important in association with events. As organic chemistry developed during the 20th century, researchers started associating synthetically useful reactions with the names of discovers or developers of these reactions. In many cases such names serve merely as a mnemonic, to remember a reaction more easily; there are few chemistry undergraduates who do not know what the Friedel-Crafts reaction is.

In recent years there has been a proliferation of new reactions and reagents that have been so useful in organic synthesis that often people refer to them by name. Many of these are stereoselective or regioselective methods. While the expert many know exactly what the Makosza vicarious nucleophilic substitution, or the Meyers asymmetric synthesis refers to, many students as well as researchers would appreciate guidance regarding such "Name Reactions".

It is in this context that we perceived the necessity to incorporate the older name reactions with some newer name reactions or "unnamed reactions", that are often associated with a name but for which details, references and experimental details are not at everyone's fingertips. This was our inspiration for the current monograph *"Organic Syntheses Based on Name Reactions and Unnamed Reactions"*.

In particular, we thought it would be useful to include cross-references of functional group transformations and an experimental procedure, so that the reader will be able to evaluate the reaction conditions at a glance; for instance, is this reaction carried out at room temperature or at 200 °C? For 1 h or 5 days? Are special catalysts required? How is the reaction worked up, what yield can be expected?

The choice of which reactions to include is not an easy one. First there are the well known "Name Reactions", that have appeared in various monographs or in the old Merck index. Some of these are so obvious mechanistically to the modern organic chemistry practitioner that we have in fact omitted them; for instance, esterification of alcohols with acid chlorides – the Schötten-Baumann procedure. Others are so important and so well entrenched by name, like the Baeyer-Villiger ketone oxidation, that it is impossible to ignore them. In general, we have kept older name reactions that are not obvious at first glance.

In some cases we have combined similar reactions under one heading, for instance, the Hunsdiecker-Borodin-Cristol-Firth decarboxylative bromination. It is not a simple task to decide whether credit is due to the first discoverer of a reaction or to is developer. Often an improvement on a method is more useful than the original discovery, and usually one reaction owes its inception to some previous discovery; *non nova sed nove*.

Except in the case of reactions that have been known for a long time under shared names, we often took the liberty to include in the title, as well as in the references (here to save space), only the name of the major author; for this we apologize to the co-authors, whose contributions are often seminal. For reactions named after contemporary authors, we have tried to consult the authors about choice of examples, etc. This led, for instance, to the Mannich-Eschenmoser methylination.

Among the newer reactions, we have chosen those that are not only synthetically useful, but, at first glance, not immediately obvious transformations. Another criterion was the stereochemical implication of the process. Yet, we admit our own bias in choosing from the plethora of novel transformations that have appeared in the literature over the past 30 years or so. Space limitation was by necessity a criterion. Nevertheless, we have included approximately 450 name reactions and 2100 references. We sincerely apologize if we have inadvertently omitted important reactions.

In all cases we have tried to include the first reported reference, a reference to an experimental procedure, and whenever possible, a review reference (journal or *Organic Reactions*). In general, we did not include references to books, series of monographs, or to *Organic Syntheses*; chemists will of course consult these where available.

Furthermore, we have compiled four indices, which should be helpful to the reader:

1. **A names index** with cross references to multiple names;

2. **A reagents index**;

3. **An index to types of reactions,** e.g. alkylations, stereoselective reductions, cyclizations, etc.; and

4. Most important for the synthetic chemist is **an index to the synthesis of functional groups**, e.g., *synthesis* of alkenes *from* ketones, as well as *conversion* of ketones *to* alkenes.

We thank our families for their support and understanding during the travail on this book. Special thanks are due to my son, Lawrence Hassner, for constructive suggestions and invaluable help.

We are grateful to the TEVA Pharmaceutical Co. for support of this project.

*Alfred Hassner*

*Carol Stumer*

# Foreword to Second Edition

The success of the first edition of "*Organic Syntheses Based on Name Reactions and Unnamed Reactions*" and the proliferation of new Name Reactions are the reason for this new revised edition. It became obvious that many new reagents and reactions are being referred to in the organic chemistry research community by their names. Hence, in addition to over 170 new reactions (previously referred to as Unnamed Reactions) in the first edition, we have included in the second edition 157 new Name Reactions bringing the total to 545. However, we have eliminated the term "Unnamed Reactions" from the title of the monograph, since these reactions are now no longer unnamed. Furthermore, we omitted some older and less utilized Name Reactions that appeared in the first edition but have included them in the Name Index, by providing reference to the page number in the first edition (e.g. Baudisch I-27, refers to first edition, p.27).

The new additions are all synthetically useful or not immediately obvious transformations. In choosing them, emphasis was placed on stereoselective or regioselective reagents or reactions including asymmetric syntheses. The latter are particularly timely with the recent Nobel Prize in Chemistry awarded in this area.

Again we admit our own bias in choosing from the many interesting newer transformations reported in the literature. Where possible we have tried to consult with the Name Reaction major author. We apologize if inadvertently important reactions were omitted.

We have maintained the useful format of providing important references (over 3,300); in each case this includes one of the first references to the reaction and a review reference where available. Furthermore, an example of an experimental procedure is provided.

Important features of this monograph remain the indexes, which should be helpful to the reader:

**A names index** with cross references to multiple names;

**A reagent index;**

**A reaction index**, e.g. acylations, asymmetric synthesis; epoxidation, heteroannulations, rearrangements, etc.; as well as

**A functional group transformation index**, which allows one to search for conversions of one functional group to another. The latter has proved valuable to the synthetic chemist searching for pathways to perform such synthetic procedures.

Hence, the monograph should be of interest to chemists in industry and academia. In fact this format has led to the monograph being adopted as a text in advanced organic chemistry courses.

We thank our families for their understanding during the travail on this book and are grateful to TEVA Pharmaceutical Co. for their support.

This monograph is dedicated to the memory of my dear wife Cyd (A.H.).

*Alfred Hassner*
*Carol Stumer*

# A B R A M O V Phosphonylation

Stereoselective phosphonylation of aldehydes by means of phosphorodiamidates

| 1 | Abramov, V. S. | *Dokl. Akad. NauKSSSR* | **1954** | 95 | 991 |
|---|---|---|---|---|---|
| 2 | Kee, T. P. | *J. Chem. Soc. Perkin 1* | **1994** | | 3183 |
| 3 | Evans, D. A. | *J. Am. Chem. Soc.* | **1978** | *100* | 3467 |
| 4 | Devitt, P. G. | *J. Chem. Soc. Perkin 1* | **1993** | | 2701 |

**(1R, 2S)-O, N-Ephedrine P-N (SiMe₃)₂ (2).**[2] To a stirred solution of (1R, 2S)-0, N-ephedrine PCl **1** (240 mg, 1.07 mmol) in THF (20 mL) at −78 °C was added a solution of LiN (SiMe₃)₂ in THF (1.07 mL of 1 M, 1.07 mmol). After allowing the mixture to warm to 20 °C, it was stirred for another hour. The solvent was removed under vacuum and the residue was extracted with pentane. Evaporation of the pentane under reduced pressure gave 290 mg of **2** (83%) of 96-98% epimeric purity. Flash filtration of the pentane solution through basic alumina improved the epimeric purity to 98%.
**(1R, 2S)-O, N-Ephedrine P(NSiMe₃)CHPh(OSiMe₃) (4).** To a solution of **2** (410 mg, 1.15 mmol) in pentane (15 mL) was added at 20 °C a solution of benzaldehyde **3** (120 mg, 1.15 mmol) under stirring. After 3 h the solution was filtered and the volatile components were removed in vacuum, to afford 440 mg of **4** (83%), 92% diastereoselectivity.

**A D L E R**  Phenol Oxidation

Oxidation of o-alkoxyphenols with sodium metaperiodate to afford 6,6-spiro-2,4-cyclohexadienones which dimerize spontaneously to a Diels-Alder adduct.

**1** n = 1; 2                              **2**                    **3** (74%)

**4**                          **5**                    **6**

| 1 | Adler, E. | *Acta Chem. Scand.* | **1959** | 13 | 1959 |
|---|---|---|---|---|---|
| 2 | Adler, E. | *Acta Chem. Scand.* | **1960** | 14 | 1261; 1580 |
| 3 | Adler, E. | *Acta Chem. Scand.* | **1962** | 16 | 529 |
| 4 | Adler, E. | *Acta Chem. Scand.* | **1971** | 25 | 2055 |

**Spirooxirane 3.**[4] NaIO$_4$ (47 g; 0.22 mol) in water (1000 mL) was added to a stirred solution of 2-hydroxybenzyl alcohol **1** (24.83 g; 0.2 mol) in water (1500 mL). After 10 min, colorless crystals appear. The mixture was kept for 24 h at 4°C in the dark. The crystalline product was filtered, washed (water) and dried in vacuum over P$_2$O$_5$ to afford 18.05 g of **3** (74%), mp 194-195°C.

# ALDER (Ene) Reaction

Thermal or catalytic sigmatropic rearrangement with H-transfer and C-C bond formation either inter or intramolecular and with chiral induction (see 1st edition).

| 1 | Alder, K. | *Chem.Ber* | **1943** | *76* | 27 |
|---|---|---|---|---|---|
| 2 | Usieli, V. | *J.Org.Chem.* | **1973** | *38* | 1703 |
| 3 | Achmatowicz, O. | *J.Org.Chem.* | **1980** | *45* | 1228 |
| 4 | Snider, B. B. | *J.Org.Chem.* | **1982** | *47* | 745 |
| 5 | Hill, R. | *J.Am.Chem.Soc.* | **1964** | *86* | 965 |
| 6 | Oppolzer, W. | *Angew.Chem.Int.Ed.* | **1978** | *17* | 476 |
| 7 | Sarkar, T.K. | *Synlett.* | **1996** | | 97 |

**Methyl 2-hydroxy-2-carbomethoxy-4-heptenoate 3.**[3] A solution of dimethyl mesoxalate **2** (1.46 g, 10 mmol) and 1-pentene **1** (0.7 g, 10 mmol) in $CH_2Cl_2$ was heated at 140°C for 16 h. The solvent was removed and the residue distilled under reduced pressure. The fraction collected between 90 and 105°C (0.5 torr) was diluted with $Et_2O$ (20 mL), washed with water and dried. The residue after evaporation of the solvent, gave on distillation 1.55 g of **3** (62 %), bp 89-90°C (0.2 torr).

**Diethyl (2-isopropenyl-4,4-dimethyl cyclopentyl)-1-malonate 5.**[7]
The catalyst: $LiClO_4$ (2.0 g) in $Et_2O$ (10 mL) was stirred with silica gel for 30 min. After evaporation of the solvent in vacuum the catalyst was dried for 24 h at 150°C and 0.1 torr.
The catalyst (50 mg) and **4** (298 mg, 1 mmol) in $CH_2Cl_2$ (2 mL) was stirred at 20°C for 5 h under Ar. After removal of the catalyst and evaporation of the solvent, **5** was obtained in quantitative yield.

## **A L D E R – R I C K E R T** Cycloaddition

Synthesis of polysubstituted benzenes by a Diels-Alder reaction of cyclohexadienes with acetylenes, via bicyclooctadienes.

| 1 | Alder, K., Richert, H. F. | *Liebigs Ann.* | **1936** | *524* | 180 |
| 2 | Birch, A. J. | *Aust. J. Chem.* | **1969** | *22* | 2635 |
| 3 | Danishefsky, S. | *J. Am. Chem. Soc.* | **1674** | *96* | 7807 |
| 4 | Patterson, J. W. | *J. Org. Chem.* | **1995** | *60* | 560 |

**1-Chloro-4-methyl-2-(2-propenyl)-3-(trimethylsiloxy)-1,3-cyclo-hexadiene (2).[4]** A solution of LDA was prepared by adding n-BuLi (40.6 mL of 1.6 N hexane solution) to iPr$_2$NH (9.1 mL, 65 mmol) in THF (110 mL) at –40 °C. After cooling to –70 °C, the reaction mixture was treated with TMS-Cl (12 mL) added over 10 min, followed by 3-chloro-6-methyl-2-(2-propenyl)cyclohex-2-enone **1** (11 g, 59 mmol). After 30 min stirring at –70 °C, Et$_3$N (20 mL) was added and the mixture was poured into ice water and hexane. The organic layer was washed, dried (K$_2$CO$_3$) and distilled (Kugelrohr) to afford 12.02 g of **2** (79%), bp 80 °C/0.2 mm.

**Dimethyl 3-chloro-5-hydroxy-6-methyl-4-(2-propenyl)-phthalate (4).** A solution of **2** (12 g, 47 mol) and DMAD (dimethyl acetylenedicarboxylate) (9 mL, 73 mmol) in xylene (45 mL) was heated at 70 °C for 2 h and then at 145 °C for 4 h. Evaporation of the solvent in vacuum followed by routine work up and chromatography (silica gel, ethyl acetate: hexane) afforded 9.48 g of **4** (53%) as an oil.

# ALLEN - MILLAR - TRIPPETT Phosphonium Rearrangement

Ring enlargement via hydrolysis of cyclic phosphonium salts obtained by alkylation (acylation) of cyclic phosphines.

| 1 | Allen, D.W., Millar, I.T. | *Chem.Ind.* | **1967** | | 2178 |
|---|---|---|---|---|---|
| 2 | Allen, D.W., Millar, I.T. | *J.Chem.Soc. C* | **1969** | | 252 |
| 3 | Trippett, S. | *Chem.Comm.* | **1967** | | 1113 |
| 4 | Tebby, J.C. | *J.Chem.Soc. C* | **1971** | | 1064 |
| 5 | Mathey, F. | *Tetrahedron* | **1972** | *28* | 4171 |
| 6 | Mathey, F. | *Tetrahedron* | **1973** | *29* | 707 |
| 7 | Allen, D.W. | *J.Chem.Soc. Perkin 1* | **1976** | | 2050 |
| 8 | Markl, G. | *Angew.Chem.Int.Ed.* | **1987** | *26* | 1134 |
| 9 | Keglevich, Gy. | *J.Org.Chem.* | **1990** | *55* | 6361 |
| 10 | Keglevich, Gy. | *Synthesis* | **1993** | | 931 |

**9-Methyl-9,10-dihydro-9-phosphaphenanthrene-9-oxide (3).[1]** The phosphonium salt **2** (R = Me) (0.7 g, 1.5 mmol) in aqueous acetone containing KOH solution was heated to reflux for 2 h. Extraction of the cold mixture with $CHCl_3$, evaporation of the solvent and chromatography (silica gel, EtOAc : EtOH 7:3) afforded 0.24 g of **3** (71%). Purification by preparative TLC with EtOAc.

**2-Hydroxy-1,2-dihydroxyphosphinine oxide 6.[6]** Benzoyl chloride (10 g, 71.1 mmol) was added to **4** (7.52 g, 40 mmol) and Et₃N (20 mL) in $Et_2O$ (300 mL). After 3 h stirring under reflux **5** was hydrolyzed with water (150 mL) (reflux 2 h). The next day the precipitate was removed by filtration and the filtrate dried (MgSO₄). Evaporation of the solvent and recrystalization from PhMe afforded 10.8 g of **6** (87%).

## A L P E R Carbonylation

Carbonylation of cyclic amines, hydroformylation (CO-$H_2$) of amino olefins catalyzed by metal (Pd, Ru, Rh) complexes (see 1st edition).

| # | Author | Journal | Year | Vol | Page |
|---|--------|---------|------|-----|------|
| 1 | Alper, H. | *J. Chem. Soc. Chem. Commun.* | **1983** | *102* | 1270 |
| 2 | Alper, H. | *Tetrahedron, Lett.* | **1987** | *28* | 3237 |
| 3 | Alper, H. | *J. Org. Chem.* | **1992** | *57* | 3328 |
| 4 | Alper, H. | *J. Am. Chem. Soc.* | **1990** | *112* | 2803 |
| 5 | Alper, H. | *Aldrichimica Acta* | **1991** | *24* | 3 |
| 6 | Alper, H. | *J. Am. Chem. Soc.* | **1992** | *114* | 7018 |

**N-(n-Butyl)-α-methlene-β-lactam (2).[2]** CO was bubbled through Pd(OAc)$_2$ or Pd(Ph$_3$P)$_4$ (0.136 mmol) in CH$_2$Cl$_2$ (4 mL). After 2 min Ph$_3$P (0.54 mmol) in CH$_2$Cl$_2$ (2 mL) was added followed by aziridine **1** in CH$_2$Cl$_2$. After 40 h evaporation and preparative TLC (silica gel hexane :EtOAc 8:1) yielded **2** (79%).

**Perhydroquinolinone (4).[6]** Perhydroindole **3** (311 mg, 1.32 mmol), a mixture of Co$_2$(CO)$_8$ and Ru$_3$(CO)$_{12}$ in PhH (10 mL) in a glass lined autoclave purged and pressurized with 54 atm of CO was heated to 200-220 °C for 3 days. Work up and preparative TLC gave 249 mg of **4** (79%).

**N-Cyclohexyl-2-pyrrolidone (6).[3]** **5** (278 mg, 2 mmol), NaBH$_4$ (75 mg, 2.25 mmol) and HRh(CO)(Ph$_3$P)$_3$ (18.36 mg, 0.02 mmol) in i-PrOH (0.5 mL) and CH$_2$Cl$_2$ (5 mL) was treated with CO at 34.5 atm, with heating and stirring for 30 h at 100 °C. Work up and chromatography afforded 260 mg of **6** (78%).

## A M A D O R I   Glucosamine Rearrangement

Rearrangement of N-glucosides of aldoses to glucosides of ketoses (see 1st edition).

| | | | | | |
|---|---|---|---|---|---|
| 1 | Amadori, M. | *Atti. Accad. Lincei* | **1925** | *2* | 337 (6) |
| 2 | Weygand, F. | *Chem. Ber.* | **1940** | *73* | 1259 |
| 3 | Hixon, R.M. | *J. Am. Chem. Soc.* | **1944** | *66* | 483 |
| 4 | Ames, G.R. | *J. Org. Chem.* | **1962** | *27* | 390 |
| 5 | Gomez-Sanchez, A. | *Carbohydrate Res.* | **1992** | *229* | 302 |
| 6 | Winckel, D. | *Rec. Trav. Chim.* | **1995** | *114* | 321 |
| 7 | Horvat, S. | *J. Chem. Soc. Perkin 1* | **1998** | | 909 |

**1-Deoxy-1-*p*-tolylamino-D-fructose 2.**[3] A mixture of $\alpha$-D-glucose **1** (100 g; 555 mmol), *p*-toluidine (80 g; 533 mmol), water (25 mL) and 2N AcOH (5 mL) was heated to 100°C for 30 min. To the cooled mixture was added anh. EtOH (100 mL) and after 24 h the precipitate was filtered, washed with EtOH:Et$_2$O (2:3), to give 94 g of **2** (60%), m.p. 152-153°C.

## A N G E L I – R I M I N I   Hydroxamic Acid Synthesis

Synthesis of hydroxamic acids from aldehydes and N-sulfonylhydroxylamine; also used as a color test for aldehydes (see 1st edition).

| | | | | | |
|---|---|---|---|---|---|
| 1 | Angeli, A. | *Gazz. Chim. Ital.* | **1896** | *26* | 17 (II) |
| 2 | Rimini, E. | *Gazz. Chim. Ital.* | **1901** | *31* | 84 (I) |
| 3 | Hassner, A. | *J. Org. Chem.* | **1970** | *35* | 1952 |
| 4 | Lwowsky, W. | *Angew. Chem. Int. Ed.* | **1967** | *6* | 897 |

## **APPEL** Halogenation Reagent

Triphenyl phosphine and carbon tetrachloride (or tetrabromide), a reagent for chlorine (bromine) substitution, dehydration.

$$Ph_3P \ + \ CCl_4 \longrightarrow [Ph_3PCl]^+ CCl_3^-$$
$$\mathbf{3}$$

$$\underset{R_2}{\overset{R_1}{>}}C{=}N{-}OH \ \xrightarrow{\ \mathbf{3}\ } \ R_1{-}\underset{Cl}{\overset{|}{C}}{=}N{-}R_2 \ + \ Ph_3PO \ + \ CHCl_3$$

| 1 | Rabinowitz, R. Marcus, R. | *J.Am.Chem.Soc.* | **1962** | *84* | 1312 |
| 2 | Appel, R. | *Chem.Ber* | **1971** | *104* | 1030 |
| 3 | Appel, R. | *Chem.Ber* | **1975** | *108* | 2680 |
| 4 | Evans, S.A. Jr. | *J.Org.Chem.* | **1981** | *46* | 3361 |
| 5 | Appel, R. | *Angew.Chem.Int.Ed.* | **1975** | *14* | 801 |
| 6 | Brinkman, H.R. | *Synthesis* | **1992** | | 1093 |

**Trans-2-chlorocyclohexanol (2).**[4] Trans-1,2-cyclohexandiol **1** (1.91 g, 16.5 mmol) was added to a solution of **3** (triphenylphosphine 4.93 g, 16.5 mmol in anh. $CCl_4$ 30 mL) and MeCN (10 mL). After 24 h reflux and work up, there was obtained 1.95 g of **2** (88%).

**2-Cyano-adamantan-4,8-dione (5).**[2] To 4,4,8,8 - tetramethoxy - 2 - carboxamido - adamantane **4** (300 mg, 1,0 mmol), $Ph_3P$ (393 mg, 1.5 mmol) and $Et_3N$ anh. (101 mg, 1mmol) in anh. $CH_2Cl_2$ (30 mL), was added $CCl_4$ (154 mg, 1mmol). After 15 h reflux, the solvent was removed by distillation and the residue chromatographed (silica gel, petroleum ether / $Me_2CO$). The product in water : $Me_2CO$ (1:1, 40 mL) and conc $H_2SO_4$ (5 drops) was refluxed for 3 h. Evaporation of the solvent and recrystalization from petroleum ether afforded 168 mg of **5** (89%), mp 255-257°C.

## ARNDT-EISTERT Homologation

Homologation of carboxylic acids or ketones via diazocompounds (see 1st edition).

| | | | | | |
|---|---|---|---|---|---|
| 1 | Eistert, B.; Arndt, F. | *Chem. Ber.* | **1927** | *60* | 1364 |
| 2 | Barbier, F. | *Helv. Chim. Acta* | **1940** | *23* | 523 |
| 3 | Gokel, G. | *Synthesis* | **1976** | | 181 |
| 4 | Aryama, T. | *Chem. Pharm. Bull.* | **1981** | *29* | 3249 |
| 5 | Smith, A.B. | *J. Am. Chem. Soc.* | **1986** | *108* | 3110 |
| 6 | Weigand, F. | *Angew. Chem.* | **1960** | *72* | 535 |
| 7 | Bachmann, W.E. | *Org. React.* | **1942** | *1* | 38 |

**Ketones 2 and 3.**[2] To cooled **1** (100 g; 0.71 mol) in MeOH (225 mL) and 50% KOH was added slowly nitrosomethylurea (74 g; 0.68 mol) at 0°C so that the solution became colorless before the next portion was added. After several hours filtration neutralization with AcOH and distillation afforded a mixture of **2** and **3**, bp 70-95°C/11 mm, see also ref 4.

**A S I N G E R** Thiazoline Synthesis

Synthesis of thiazolines from ketones, sulfur and $NH_3$ with the possibility to obtain thioketones.

|   |              |               |          |     |     |
|---|--------------|---------------|----------|-----|-----|
| 1 | Asinger, F.  | *Liebigs Ann.* | **1957** | 602 | 37  |
| 2 | Asinger, F.  | *Liebigs Ann.* | **1957** | 606 | 67  |
| 3 | Asinger, F.  | *Liebigs Ann.* | **1964** | 674 | 57  |
| 4 | Asinger, F.  | *Angew. Chem.* | **1958** | 70  | 372 |
| 5 | Lyle, E. R.  | *J. org. Chem.* | **1965** | 30  | 293 |
| 6 | Domling, A.  | *Tetrahedron*  | **1995** | 51  | 755 |

**1, 6-Dimethyl-2',4',5',6',7',7'a-hexahydrospiro piperidine-4,2-thiazolo [5,4-c] pyridine. 2HCl (2).**[5] A stirred and ice cooled suspension of sulfur (6.0 g, 187 mmol) in 1-methyl-4-piperidone **1** (40 g, 354 mmol) was treated with a flow of $NH_3$ maintaining the temperature between 40-50 °C. The bubbling of $NH_3$ was continued until all traces of sulfur disappeared (ca 2). The excess of $NH_3$ was removed in vacuum, the mixture was diluted with 50% $K_2CO_3$ solution (200 mL) and extracted with $Et_2O$ (5×100 mL). The dried solution ($K_2CO_3$ anh) was treated with dry HCl, The solid was filtered , washed ($Et_2O$) and dried (vacuum) to give 53.5 g of **2**. HCl (97.8%), mp 200-205 °C. After recrystallization, mp 240-241 °C(EtOH:iPrOH).

**2,2-Pentamethylene-4,5-tetramethylene-3-thiazoline (4).**[2] Into a mixture of sulfur (32 g, 1 mol) in cyclohexanone **3** (196 g, 2 mol) was bubbled a stream of $NH_3$ at 40-50 °C for 1-2 h. After another 30 min bubbling of $NH_3$ under gentle heating at the same temperature, usual work up and vacuum distillation afforded 170 g of **4** (80%), mp 81.5-82 °C, bp 156-157 °C/11 mm.

## ARENS – VAN DORP Cinnamaldehyde Synthesis

Synthesis of cinnamaldehydes from aryl ketones, and ethoxyacetylene (see 1st edition).

$$HC \equiv C - OC_2H_5 \xrightarrow[0°]{EtMgBr} \xrightarrow[0° \quad \mathbf{1}]{Ph-CO-Me} \underset{Me \quad OH}{\overset{Ph \quad C \equiv C - OC_2H_5}{C}} \quad \mathbf{3}\ (48\ \%)$$

**2**

$$\mathbf{3} \xrightarrow[BaSO_4]{H_2 / Pd} \underset{Me \quad OH}{\overset{Ph \quad CH = CH - OC_2H_5}{C}} \xrightarrow[refl., 15\ min]{0.1\ N\ HCl} \underset{Me}{\overset{Ph - C = CH - CHO}{|}}$$

**4 (98 %)**      **5**

| | | | | | |
|---|---|---|---|---|---|
| 1 | Arens, J.F., v. Dorp, A.D. | *Nature* | **1947** | *160* | 189 |
| 2 | Isler, O. | *Helv.Chim.Acta* | **1956** | *39* | 259 |
| 3 | Kell, P.S. | *J.Am.Chem.Soc.* | **1959** | *81* | 4117 |

## ATHERTON – TODD Phosphoramidate Synthesis

Synthesis phosphoramidates from formamides and dialkyl phosphite (see 1st edition).

$$C_6H_5NH = CHO \quad + \quad \underset{O}{\overset{HP(OC_2H_5)_2}{\|}} \xrightarrow[R_4N^+Br^-,\ CCl_4]{30\%\ NaOH} \quad \underset{O}{\overset{C_6H_5-NH-P(OC_2H_5)_2}{\|}}$$

**1**      **2**      **3**

| | | | | | |
|---|---|---|---|---|---|
| 1 | Atherton, F.R., Todd, A.R. | *J.Chem.Soc.* | **1945** | | 660 |
| 2 | Wadsworth, W.S. | *J.Am.Chem.Soc.* | **1962** | *84* | 1316 |
| 3 | Zwierzak, A. | *Synthesis* | **1982** | | 922 |
| 4 | Lukanow, L.K. | *Synthesis* | **1985** | | 671 |
| 5 | Hovalla, D. | *Tetrahedron Lett.* | **1992** | *33* | 2817 |
| 6 | Garrigue, B. | *Synth.Commun.* | **1995** | *25* | 871 |
| 7 | Liu, L.Z. | *Org.Prep.Proced.Int.* | **1996** | *28* | 490 |

**Diethyl N-phenylphosphoramide (3).[4]** To an ice cooled stirred suspension of formylanilide **1** (605 mg, 5mmol) in CHCl$_4$ (25 mL) was added 30 % NaOH (10 mL) and benzyltriethylammonium bromide (0.2 g). Diethyl phosphite **2** (828 mg, 6 mmol) in CCl$_4$ (5 mL) was added dropwise. After 1 h at 0° and 4 h at 20°C, the organic layer gave **3**, after crystallization, 0.687 g (60 %), mp 96-97°C.

## AUWERS Flavone Synthesis

Synthesis of benzopyran-4-ones (flavones) from o-hydroxychalcones or from benzofuran-3-ones (see 1st edition).

| 1 | Auwers, K.   | *Chem.Ber.*  | **1908** | *41*  | 4233 |
| 2 | Minton, T.H. | *J.Chem.Soc.*| **1922** | *121* | 1598 |
| 3 | Ingham, B.H. | *J.Chem.Soc.*| **1931** |       |  895 |
| 4 | Acharya, B.C.| *J.Chem.Soc.*| **1940** |       |  817 |

## AUWERS – INHOFFEN Dienone–Phenol Rearrangement

Rearrangement of dienones to phenols catalyzed by acids.

| 1 | Auwers, K.    | *Liebigs Ann.*  | **1921** | 425 |  217 |
| 2 | Inhoffen, C.  | *Angew.Chem.*   | **1940** |  53 |  473 |
| 3 | Djerassi, C.  | *J.Am.Chem.Soc.*| **1951** |  73 |  990 |
| 4 | Winstein, S.  | *J.Am.Chem.Soc.*| **1957** |  79 | 3109 |
| 5 | Eneyama, K.   | *J.Org.Chem.*   | **1995** |  60 | 6402 |

## B A E Y E R  Pyridine Synthesis

Synthesis of pyridines from pyrones (see 1st edition).

| 1 | Baeyer, A. | *Chem. Ber.* | **1910** | *43* | 2337 |
| 2 | Nenitzescu, C.D. | *Liebigs Ann.* | **1959** | *625* | 74 |
| 3 | Cavallieri, L.F. | *Chem. Rev.* | **1947** | *41* | 525 |
| 4 | Dimroth, K. | *Angew. Chem.* | **1960** | *72* | 331 |
| 5 | Balaban, A.T. | *Liebigs Ann.* | **1992** | | 173 |

## B A E R - F I S C H E R  Amino Sugar Synthesis

Synthesis of 3-nitro and derived 3-amino sugars by aldol condensation of sugar-derived dialdehydes with nitroalkanes (see 1st edition).

| 1 | Baer, H.H; Fischer, H.O.L. | *Proc. Nat. Acad. Sci. USA* | **1958** | *44* | 991 |
| 2 | Baer, H.H. | *Adv. Carbohydr. Chem.* | **1969** | *24* | 67 |
| 3 | Brimacombe, J.S. | *J. Chem. Soc. Perkin I* | **1974** | | 62 |
| 4 | Santoyo-Gonzales, F. | *Synlett* | **1990** | | 715 |

**Nitrosugar 4.**[3] Methyl-*L*-rhamnoside **3** (100 g; 0.55 mol) in 1000 mL water was treated with $NaIO_4$ (200 g; 0.83 mol) at 20°C. After 3 h $NaHCO_3$ was added, the mixture poured into EtOH (4000 mL) and filtered. The filtrate was concentrated and extracted with hot EtOH. The extract was cooled, filtered and treated with nitroethane (104.5 g; 1.4 mol) followed by a solution of Na (12 g; 0.52 at.g.) in EtOH (750 mL). After 4 h at 20°C the solution was treated with $CO_2$, filtered and concentrated. The mixture was treated with pyridine (400 mL) and $Ac_2O$ (300 mL) at 20°C for 12 h. Work up left a residue which dissolved in $Et_2O$:petroleum ether (1:1) (500 mL) and cooled afforded 36 g of **4** (19%), mp 137-138°C, $[\alpha]_D$= -130° (c 1).

## B A E Y E R - V I L L I G E R Ketone Oxidation

Regioselective peroxide oxidation of ketones to esters or lactones with retention of configuration (see 1st edition).

| 1 | Bayer, A.; Villiger, V. | *Chem.Ber.* | **1899** | *32* | 3625 |
|---|---|---|---|---|---|
| 2 | Hassner, A. | *J.Org.Chem.* | **1978** | *43* | 1774 |
| 3 | Sarapanami, C.R. | *J.Org.Chem.* | **1986** | *51* | 2322 |
| 4 | Johnson, C.R. | *J.Am.Chem.Soc.* | **1990** | *112* | 6729 |
| 5 | Morimoto, T. | *Synth.Commun.* | **1995** | *25* | 3765 |
| 6 | Yamashita, M. | *J.Org.Chem.* | **1997** | *62* | 2633 |
| 7 | Hassal, C.H. | *Org.React.* | **1957** | *9* | 73 |
| 8 | Krow, G.R. | *Org.React.* | **1993** | *43* | 251 |

**Bicyclic lactone (2).**[2] To a solution of **1** (790 mg, 5 mmol) in 90% HOAc (5 mL) at 0°C, was added 30% $H_2O_2$ (2.5 mL) in 90% HOAc (3 mL). The mixture was kept at 0°C for 24 h, poured into water and extracted with hexane. The organic layer after washing ($NaHSO_3$ and $H_2O$) was evaporated to give 570 mg of **2** (65%).

$\varepsilon$ **-Caprolactone (4).** Cyclohexanone **1** (196 mg, 2 mmol) and moist bentonite clay (2 g) in MeCN (10 mL) was heated to 80°C with stirring and magnesium monoperoxyphthalate (MMPP) (3 mmol) was added in six portions at ten minute intervals. After additional 1 h stirring, followed by cooling, filtering and washing the precipitate with $CHCl_3$ (100 mL), evaporation of the solvent afforded 200 mg of **4** (88%).

## BAILEY Crisscross Cycloaddition

A bis 3+2 cycloaddition between aromatic aldazines and olefins or acetylenes, called "crisscross" cycloaddition.

| 1 | Bailey, J.R. | *J.Am.Chem.Soc.* | 1917 | *39* | 279; 1322 |
|---|--------------|------------------|------|------|-----------|
| 2 | Forshaw, T.P. | *J.Chem.Soc. (C)* | 1971 | | 2404 |
| 3 | Forshaw, T.P. | *J.Chem.Soc.Perkin 1* | 1972 | | 1059 |
| 4 | Shimizu, T. | *J.Org.Chem.* | 1987 | *52* | 2277 |
| 5 | Burger, K. | *Liebgs Ann.* | 1982 | | 853 |
| 6 | Matur, S.S. | *J.Chem.Soc.Pekin 1* | 1975 | | 2479 |
| 7 | Radl, S. | *Aldrichimica Acta* | 1997 | *30* | 97 |

**Diazabicyclooctadiene 2.**[4] A mixture of acetylenic aldehyde **1** (1.56 g, 5 mmol) and hydrazine.2HCl (260 mg, 2.5 mmol) in EtOH (80 mL) was refluxed for 4 h under stirring. To the cooled mixture (20°C) was added triethylamine (0.5 g, 5 mmol) and the mixture was stirred for 1 h at the same temperature. The crystals were filtered. Recrystalization afforded 1.042 g of **2** (69%), mp 276-278°C.

## BAKER-VENKATARAMAN Flavone Synthesis

Rearrangement of aromatic o-keto esters of phenols to o-hydroxy-1,3-diketones followed by cyclization to flavones (see 1st edition).

| | | | | | |
|---|---|---|---|---|---|
| 1 | Baker, W. | *J. Chem. Soc.* | **1938** | | 1381 |
| 2 | Venkataraman, K. | *J. Chem. Soc.* | **1939** | | 1767 |
| 3 | Kramm, E. | *J. Org. Chem.* | **1984** | 49 | 3212 |
| 4 | Krupadavam, G.L.D. | *J. Heterocycl. Chem.* | **1996** | 33 | 1561 |
| 5 | Levine, E. | *Chem. Rev.* | **1954** | 54 | 493 |

## B A L A B A N - N E N I T Z E S C U - P R A I L L  Pyrylium Salt Synthesis

Synthesis of pyrylium salts by acylation of unsaturated ketones or by diacylation of alkenes.

| | | | | | |
|---|---|---|---|---|---|
| 1 | Balaban, A.T.; Nenitzescu, C.D. | *Liebigs Ann.* | **1959** | *625* | 66; 74 |
| 2 | Balaban, A.T.; Nenitzescu, C.D. | *J. Chem. Soc.* | **1961** | | 3553; 3561 |
| 3 | Balaban, A.T.; Nenitzescu, C.D. | *J. Chem. Soc.* | **1961** | | 3564; 3566 |
| 4 | Praill, P.F.G.; Whitear, A.L. | *J. Chem. Soc.* | **1961** | | 3573 |
| 5 | Balaban, A.T.; Nenitzescu, C.D. | *Org.Synth.Coll.* | | 5 | 1106 |
| 6 | Balaban, A.T.; Boulton, A.J. | *Org.Synth.Coll.* | | 5 | 1112; 1114 |

**2,4,6-Trimethylpyrylium perchlorate 2.**[2,5] Anh. *t*-BuOH **1** (148 g; 2 mol) and Ac$_2$O (10 mL) at -10°C were cautiously treated with 70% HClO$_4$ (1.75 mol) and the temperature was controlled at 90-100°C. The mixture was heated at 100°C for 2 h. After cooling **2** was filtered and washed (AcOH, Et$_2$O) to give 205-215 g of **2** (53-57%), explosive when dry. The tetrafluoroborate or triflate[5] are not explosive.

## B A M B E R G E R  Benzotriazine Synthesis

Synthesis of benzotriazines from pyruvic acid hydrazone **2** and aryldiazonium salts **1** (see 1st edition).

| | | | | | |
|---|---|---|---|---|---|
| 1 | Bamberger, E. | *Chem. Ber.* | **1892** | 25 | 3201 |
| 2 | Abramovitch, R.A. | *J. Chem. Soc.* | **1955** | | 2326 |

**B A M B E R G E R**  Imidazole cleavage

Synthesis of 2-substituted imidazoles from imidazoles via cleavage with acid chlorides to enediamides (see 1st edition).

| 1 | Bamberger, E. | *Liebigs Ann.* | **1893** | *273* | 342 |
|---|---------------|----------------|----------|-------|------|
| 2 | Babad, E. | *J. Heterocycl. Chem.* | **1969** | *6* | 235 |
| 3 | Grace, M.E. | *J. Am. Chem. Soc.* | **1980** | *102* | 6784 |
| 4 | Kimoto, H. | *J. Org. Chem.* | **1978** | *43* | 3403 |
| 5 | Altman, J. | *J. Chem. Soc. Perkin I* | **1984** | | 59 |

**Imidazole 2.**[5] Imidazole **1** (9.2 g; 54 mmol) in EtOAc (140 mL) was treated with benzoyl chloride (15.7 g; 112 mmol) in EtOAc (40 mL) and 1M NaHCO$_3$ (380 mL) added simultaneously in 1 h under ice-cooling. The mixture was stirred for 1 h, then a further portion of benzoyl chloride (15.7 g; 112 mmol) in EtOAc and 1M NaHCO$_3$ (280 mL) was added followed by more 1M NaHCO$_3$ (200 mL). After 24 h the organic layer was concentrated and the residue dissolved in THF (300 mL). The THF solution was stirred with 10% NaHCO$_3$ (600 mL) for 24 h to decompose any N-formyl intermediate and to remove benzoic acid. Extraction with EtOAc, drying (Na$_2$SO$_4$), solvent evaporation and recrystallization from EtOAc:hexane afforded 16.24 g of **2** (84%), mp 128-129°C.

## BAMFORD-STEVENS-CAGLIOTI-SHAPIRO Olefination

Conversion of ketones to olefins via tosylhydrazones with NaOR, LAH, LDA or BuLi. But 2-naphthaldehyde tosylhydrazone is reduced by LAH to 2-methylnaphthalene (see 1st edition).

| 1 | Bamford, W.;Stevens, T. | *J.Chem.Soc.* | **1952** | | 4735 |
| 2 | Farnum, D.G. | *J.Org.Chem.* | **1963** | *28* | 870 |
| 3 | Nikon, A. | *J.Org.Chem.* | **1981** | *46* | 4692 |
| 4 | Stadler, H. | *Helv.Chim.Acta.* | **1984** | *67* | 1379 |
| 5 | Caglioti, R. | *Tetrahedron Lett.* | **1962** | | 1261 |
| 6 | Caglioti, R. | *Tetrahedron* | **1963** | *19* | 1127 |
| 7 | Shapiro, R.H. | *J.Am.Chem.Soc.* | **1967** | *89* | 1442; 5734 |
| 8 | Siemeling, E. | *J.Org.Chem.* | **1997** | *62* | 3407 |
| 9 | Shapiro, R.H. | *Org.React.* | **1976** | *23* | 405 |

$\beta$-**Methylnaphthalene 5.**[7] To a solution of **4** (2.0 g, 6.17 mmol) in THF (50 mL) was added LiAlH$_4$ (3.0 g, 78.9 mmol) and the mixture refluxed for 18 h. After careful decomposition of excess hydride with moist Et$_2$O and water, the organic phase was washed with dil. H$_2$SO$_4$ and water, dried and evaporated, to yield 620 mg of **5** (70.7 %).

**1,3-Diphenyl-4,5-di(2-pyridyl)cyclopentene 8.**[8] A solution of **7** (30.2 g, 54 mmol) in THF (300 mL) was treated with LDA at 0°C. After 14 h stirring at 20°C, the mixture was quenched with brine at 0°C. Workup gave 16.2 g of **8** (80 %).

**B A R B I E R**   Reaction

In situ Grignard generation in the presence of an electrophile (see 1st edition).

**1**                    **2** (70%)

**3**              **4**                        **5** (93%)

| 1 | Barbier, P. | *C. R.* | **1899** | *128* | 110 |
|---|---|---|---|---|---|
| 2 | Grignard, V. | *C. R.* | **1900** | *130* | 1322 |
| 3 | Ashby, R. | *Pure & Appl. Chem.* | **1980** | *52* | 545 |
| 4 | Huang, X.Z. | *Tetrahedron Lett.* | **1988** | *29* | 1395 |
| 5 | Blomberg, C. | *Synthesis* | **1977** | | 18 |
| 6 | Hassner, A. | *J. Organomet. Chem.* | **1978** | *156* | 227 |
| 7 | Imai, T. | *Synthesis* | **1993** | | 395 |
| 8 | Banik, Bak. | *Tetrahedron Lett.* | **2001** | *42* | 187 |

**2-Chloro-1-nonen-4-ol 5.**[7] To **3** (500 mg; 5 mmol) and **4** (611 mg; 5.5 mmol) was added successively $ScCl_2 \cdot 2H_2O$ (1.7 g; 7.5 mmol) and NaI (1.1 g; 7.5 mmol). After 20 h stirring at 20°C, 30% $NH_4F$ (10 mL) and $Et_2O$ (20 mL) were added. Usual work up and chromatography followed by distillation gave 820 mg of **5** (93%).

## B A R B I E R - W I E L A N D   Degradation

A multi-step (Grignard reaction, elimination, oxidative cleavage) procedure for chain degradation of carboxylic acids (esters) (see 1st edition).

$$H_3C-\underset{\underset{\textstyle 1}{|}}{\overset{\overset{\textstyle OH}{|}}{CH}}-(CH_2)_8-CO_2Me \xrightarrow[\text{Et}_2\text{O, }\Delta]{PhMgBr} H_3C-\underset{\underset{\textstyle 2}{|}}{\overset{\overset{\textstyle OH}{|}}{CH}}-(CH_2)_8-\overset{\overset{\textstyle OH}{|}}{C}Ph_2 \xrightarrow[20\%H_2SO_4]{\text{refl.}}$$

$$H_3C-\underset{\underset{\textstyle 3 \,(80\%)}{|}}{\overset{\overset{\textstyle OH}{|}}{CH}}-(CH_2)_7-CH{=}CPh_2 \xrightarrow[\text{AcOH}]{CrO_3} H_3C-\overset{\overset{\textstyle OH}{|}}{CH}-(CH_2)_7-CO_2H$$

| 1 | Barbier, P. | C. R. | 1913 | 156 | 1443 |
|---|---|---|---|---|---|
| 2 | Wieland. E. | Chem. Ber. | 1912 | 45 | 484 |
| 3 | Sarel, S. | J. Org. Chem. | 1959 | 24 | 2081 |
| 4 | Fetisson, M. | C. R. | 1961 | 252 | 139 |
| 5 | Djerassi, C. | Chem. Rev. | 1946 | 38 | 526 |
| 6 | Chadha. M.S. | Synthesis | 1978 | | 468 |

**9-Oxodecanoic acid 4.**[6] To PhMgBr (from PhBr; 29.8 g, and Mg 4.6 g in Et$_2$O 100 mL) was added the hydroxy ester **1** (7 g; 32 mmol) in Et$_2$O (25 mL) over 1 h and refluxed for 2.5 h. Aq. NH$_4$Cl was added and the etheric extracts were concentrated to give diol **2** which was refluxed with 20% H$_2$SO$_4$ (100 mL) for 1 h. Extraction (Et$_2$O), washing and evaporation afforded 7.3 g of **3** (80%), distilled at 180°C (Bath) / 0.5 torr. CrO$_3$ (6 g; 60 mmol) in water (8 mL) was added to crude **3** (6.4 g) in AcOH (75 mL) over 1.5 h. After stirring at 35°C for 1 h, work up gave 2.2 g of **4** (60%), mp 48°C.

## BARLUENGA Iodination Reagent

Bis(pyridine)iodonium(I) tetrafluoroborate reagent for 1,2-iodofunctionalization of isolated or conjugated olefins, or cyclization of alkynyl sulfides.

$$HgO.HBF_4.silica\ gel\ +\ I_2 \xrightarrow[CH_2Cl_2]{pyridine} I(py)_2BF_4 \quad \mathbf{1}$$

$$\mathbf{1}\ +\ \ce{C=C}\ +\ HBF_4\ +\ Nu^-\ \longrightarrow\ \ce{-C-C-}\ \underset{Nu}{} \quad Nu = F,\ Cl,\ Cr,\ AcOH,\ MeOH$$

**4 (92%)**

| 1 | Barluenga, J. | *J.Chem.Soc.Perkin 1* | **1984** | | 2623 |
|----|---------------|------------------------|----------|-----|------|
| 2 | Barluenga, J. | *Angew.Chem.Int.Ed.* | **1985** | 24 | 319 |
| 3 | Barluenga, J. | *Tetrahedron Lett.* | **1986** | 27 | 1715 |
| 4 | Barluenga, J. | *J.Org.Chem.* | **1990** | 55 | 3104 |
| 5 | Barluenga, J. | *J.Org.Chem.* | **1993** | 58 | 2058 |
| 6 | Barluenga, J. | *Angew.Chem.Int.Ed.* | **1993** | 32 | 893 |
| 7 | Goldfinger, M.B. | *J.Am.Chem.Soc.* | **1994** | 116 | 7895 |
| 8 | Barluenga, J. | *J.Am.Chem.Soc.* | **1997** | 119 | 6933 |
| 9 | Barluenga, J. | *Tetrahedron Lett.* | **1998** | 39 | 7393 |
| 10 | Barluenga, J. | *Angew.Chem.Int.Ed.* | **1998** | 37 | 3136 |
| 11 | Barluenga, J. | *Pure Appl.Chem.* | **1999** | 71 | 431 |
| 12 | Barluenga, J. | *Angew.Chem.Int.Ed.* | **2001** | 40 | 3389 |

**Thiaanthracene 4.**[10] To a solution of IPy$_2$BF$_4$ **1** (3.72 g, 10 mmol) in CH$_2$Cl$_2$ (100 mL) cooled to −80°C was added HBF$_4$ (1.36 mL, 54% in Et$_2$O, 10 mmol). After 10 min a solution of diyne **3** (3.08 g, 10 mmol) in CH$_2$Cl$_2$ was added and the reaction mixture was stirred. Quenching with 10% Na$_2$S$_2$O$_3$, washing, drying and filtration through Al$_2$O$_3$ (elution with EtOAc : hexane) afforded 3.99 g of **4** (92%), mp 102-103°C.

## BARTON Nitrite Photolysis

Long range functionalization of alcohols via nitrites leading to $\gamma$-hydroxy oximes (see 1st edition).

| 1 | Barton, D.H.R. | *J.Am.Chem.Soc.* | **1960** | *82* | 2640 |
|---|---|---|---|---|---|
| 2 | Barton, D.H.R. | *J.Am.Chem.Soc.* | **1961** | *83* | 4076 |
| 3 | Barton, D.H.R. | *Pure Appl.Chem.* | **1968** | *16* | 1 |
| 4 | Baldwin, S.W. | *J.Am.Chem.Soc.* | **1982** | *104* | 4990 |
| 5 | Barton, D.H.R. | *Aldrichimica Acta* | **1990** | *23* | 3 |

## BARTON Deamination

Free radical deamination of primary amines via isocyanides (see 1st edition).

| 1 | Barton, D.H.R. | *J.Chem.Soc.Perkin I* | **1980** | | 2657 |
|---|---|---|---|---|---|
| 2 | Swindell, C.S. | *J.Org.Chem.* | **1990** | *55* | 3 |
| 3 | Barton, D.H.R. | *Aldrichimica Acta* | **1990** | *23* | 3 |

**Octadecane (4).**[1] A solution of **3** (0.279 g, 1 mmol) and azoisobutyronitrile (AIBN) (0.1 g) in dry xylene (50 mL) was added dropwise to a solution of tri-n-butyl stannane (0.64 g, 2.2 mol equiv). A solution of AIBN (0.1 g) in xylene (50 mL) was slowly added at 80°C over 5 h. The solvent was removed in vacuum, the residue dissolved in pentane and iodine in pentane was added until the iodine color persisted. The solvent was evaporated and **4** was isolated by preparative TLC (silica gel, pentane). Sublimation in vacuum gave 0.205 g of **4** (81%), mp 29°C.

## BARTON Phenylation of Phenols, Enols

Phenylation of phenols, enols and other anions by a pentavalent organo-bismuth reagent under neutral, acidic or basic conditions.

| 1 | Barton, D.H.R. | *J.Chem.Soc.Chem.Commun.* | **1980** |    | 246, 827 |
|---|---|---|---|---|---|
| 2 | Barton, D.H.R. | *J.Chem.Soc.Chem.Commun.* | **1981** |    | 503 |
| 3 | Barton, D.H.R. | *Tetrahedron Lett.* | **1982** | 23 | 3365 |
| 4 | Barton, D.H.R. | *J.Chem.Soc.Perkin Trans* | **1985** |    | 2657, 2667 |
| 5 | Barton, D.H.R. | *Tetrahedron* | **1988** | 44 | 3039 |
| 6 | Barton, D.H.R. | *Aldrichim Acta* | **1990** | 23 | 3 |

**1-Phenyl-2-naphthol (2).**[4] To a stirred solution of Ph$_3$BiCl$_2$ (550 mg, 1.07 mmol) and 2-naphthol **1** (144 mg, 1 mmol) in THF (1mL) at 20°C under an Ar atmosphere was added tetramethyl-2-t-butylguanidine (TMBG) (500 mg, 0.11 mmol). After 5 h stirring, usual work up and chromatography (silica gel, Et$_2$O:hexane 1:4) afforded 198 mg of **2** (90%).

**1,3,5-Trihydroxy-2,4,6-triphenylbenzene 4 and 2,2,4,5-tetraphenyl cyclopent-4-ene-1,3-dione (5).**[4] A mixture of phloroglucinol **3** (300 mg, 3.9 mmol) and Ph$_3$BiCO$_2$ (3.0 g, 6 mmol) in dioxane (10 mL) was heated to reflux under Ar for 11 h. After removing insoluble material by filtration, the solvent was evaporated and the residue chromatographed (hexane:EtOAc 7:3) to give 195 mg of **4** (24%) and 368 mg of **5** (40%). The same reaction but using a molar ratio of **3**:Ph$_3$BiCO$_3$=1:5.7 and heating for 24 h at 80°C afforded **4** in 60% yield.

## B A R T O N  Decarboxylation

Decarboxylation of a mixed anhydride (thiohydroxamic-carboxylic) and interception of radicals as a sulfide, selenide or bromo derivative (see 1st edition).

| 1 | Barton, D.H.R. | *J. Chem. Soc. Chem. Commun.* | **1983** | | 939 |
|---|---|---|---|---|---|
| 2 | Barton, D.H.R. | *Tetrahedron Lett.* | **1984** | 25 | 5777 |
| 3 | Barton, D.H.R. | *Tetrahedron Lett.* | **1985** | 26 | 5939 |
| 4 | Tamm, Ch. | *Helv. Chim. Acta* | **1995** | 78 | 403 |
| 5 | Renault, P. | *Synlett* | **1997** | | 181 |
| 6 | Barton, D.H.R. | *Aldrichimica Acta* | **1990** | 23 | 3 |

## B A R T O N - K E L L O G  Olefination

Olefin synthesis (especially tetrasubstituted) from hydrazones and thioketones via $\Delta^3$-1,3,4-thiazolidines (see 1st edition).

| 1 | Barton, D.H.R. | *J. Chem. Soc. Perkin I* | **1972** | | 305 |
|---|---|---|---|---|---|
| 2 | Barton, D.H.R. | *Chem. Soc.* | **1970** | | 1225 |
| 3 | Kellog, R.M. | *Tetrahedron Lett.* | **1970** | | 1987 |
| 4 | Kellog, R.M. | *J. Org. Chem.* | **1972** | 37 | 4045 |
| 5 | Barton, D.H.R. | *J. Chem. Soc. Perkin I* | **1974** | | 1794 |

**(-)-2-Diphenylethylenecamphane 5. 2** (585 mg; 3 mmol) (from **1**, lead tetraacetate and TEA in $CH_2Cl_2$ at -20°C)[5] and **4** (505 mg; 3 mmol) in THF (5 mL) were heated to reflux under $N_2$ for 3 h. After chromatography, the product was refluxed with $Ph_3P$ (870 mg) in THF (5 mL) for 16 h and evaporated. The residue in petroleum ether was treated with 1 mL of MeI (exothermic) and stirred 2 h. Chromatography (silica) afforded 545 mg of **6** (90%), mp 69.5-72.5°C (EtOH).

## BARTON-MCCOMBIE Alcohol Deoxygenation

Deoxygenation of secondary alcohols to hydrocarbons via xantates (see 1st edition).

| 1 | Barton, D.H.R.; McCombie, S.W. | *J. Chem. Soc. Perkin I* | **1975** | | 1574 |
|---|---|---|---|---|---|
| 2 | Cristol, S.J. | *J. Org. Chem.* | **1982** | 47 | 132 |
| 3 | Barton, D.H.R. | *Tetrahedron* | **1986** | 42 | 2329 |
| 4 | McClure, C.K. | *J. Org. Chem.* | **1991** | 56 | 2326 |
| 5 | Chatgilialoglu, C. | *Tetrahedron Lett.* | **1995** | 36 | 3897 |
| 6 | Crich, D. | *Aldrichimica Acta* | **1987** | 20 | 36 |

**Cholestane 3.**[5] To a stirred solution of **1** (100 mg; 0.19 mmol) and 5,10-dihydrosilanthrene **2** (67 mg; 0.32 mmol) in cyclohexane (20 mL) was added AIBN (5 mg) and the mixture was heated for 1 h at 80°C. Evaporation of the solvent and chromatography (hexane) gave 95 mg of **3** (85%).

## BENARY Conjugated Aldehyde Synthesis

Formation of polyunsaturated aldehydes from vinyl halides and enaminoaldehydes (see 1st edition).

| 1 | Benary, E. | *Chem. Ber.* | **1930** | 63 | 1573 |
|---|---|---|---|---|---|
| 2 | Normant, H. | *C. R.* | **1958** | 247 | 1744 |
| 3 | Schiess, P. | *Helv. Chim. Acta* | **1972** | 55 | 2363 |
| 4 | Näff, F. | *Helv. Chim. Acta* | **1974** | 57 | 1317 |

**3 (via Grignard reagent). 1** (4.42 g; 25 mmol) and Mg (0.6 g; 25 mat) in THF followed by **2** (4.02 g; 25 mmol) and usual work up gave 1.32 g of **3** (33%), bp 95-103°C, as a mixture of 12% (*E,Z*) and 88% (*E,E*).

## B E C K M A N N   Rearrangement or Fragmentation
Acid catalyzed rearrangement of oximes to amides or cleavage of oximes to nitriles.

**1**          **2** (96%)      **3**   SnBu$_3$     **4** (73%)

$$Ph-\underset{Tol}{\overset{NOH}{\underset{|}{CH-\overset{||}{C}-Me}}} \xrightarrow{PCl_5} Ph-\underset{Tol}{\underset{|}{CH-Cl}}$$

$$Ph-\underset{NOH}{\overset{|}{\underset{||}{C}-Ph}} \xrightarrow[Clay]{FeCl_3} Ph-CONH-Ph$$

**5**          **6 + MeCN**       **7**        **8** (100%)

| | | | | | |
|---|---|---|---|---|---|
| 1 | Beckmann, E. | *Chem. Ber.* | **1886** | *19* | 988 |
| 2 | Conley, R.T. | *J. Org. Chem.* | **1963** | *28* | 210 |
| 3 | Hassner, A. | *Tetrahedron Lett.* | **1965** | | 525 |
| 4 | Eaton, P.E. | *J. Org. Chem.* | **1973** | *38* | 4071 |
| 5 | Nishiyama, H. | *Tetrahedron* | **1988** | *44* | 2413 |
| 6 | Johnson, C.R. | *J. Am. Chem. Soc.* | **1990** | *112* | 6729 |
| 7 | Samant, G.D. | *Synth. Commun.* | **1997** | *27* | 379 |
| 8 | Popp, I. | *Chem. Rev.* | **1958** | *58* | 370 |
| 9 | Heldt, W.Z. | *Org. React.* | **1960** | *11* | 1 |
| 10. | Denz, y. | *Tetrahedron Lett.* | **2001** | *42* | 403 |

ε-**Caprolactam 2**.[4] To a solution of P$_2$O$_5$ (36 g) in MeSO$_3$H (360 g) was added **1** (2 g; 20 mmol) under stirring. After 1 h at 100°C quenching with NaHCO$_3$, extraction (CHCl$_3$), evaporation of the solvent and recrystallization from hexane gave 1.92 g of **2** (96%), mp 65-68°C.
ω-**Hexenenitrile 4**.[6] To **3** (99 mg; 0.5 mmol) in CH$_2$Cl$_2$ (2 mL) was added P$_2$O$_5$ (70 mg; 0.5 mmol). After 24 h at 20°C Et$_2$O (2 mL) and Et$_3$N (0.12 mL) were added followed by chromatography to afford 43 g of **4** (73%).
**N-Phenylbenzamide 8**.[7] FeCl$_3$ (15 g) was dissolved in MeCN (60 mL) and Montmorillonit K-10 (10 g) was added. After 5 h stirring the clay was filtered, washed and dried (5 h at 280°C). Ketoxime **7** (400 mg; 2 mmol), clay catalyst (1 g) in PhMe was refluxed (TLC monitoring). Filtration and concentration in vacuum followed by chromatography (EtOAc:hexane) gave 400 mg of **8** (100%).

## BERCHTOLD Enamine Homologation

Addition of acetylenic esters to cyclic enamines leading by rearrangement ring expansion to cyclic ketones with two more carbon atoms.

**3 (48%)**

**4 (90%)**

| 1 | Brannock, K.C. | *J.Org.Chem.* | **1961** | *26* | 625 |
|---|---|---|---|---|---|
| 2 | Berchtold, G.A. | *J.Org.Chem.* | **1961** | *26* | 3043 |
| 3 | Berchtold, G.A. | *J.Org.Chem.* | **1963** | *28* | 1459 |

**1-(N-Morpholino)-2,3-dicarbomethoxy-1,3-cycloheptadiene (3).**[3] Dimethyl acetylene dicarboxylate **1** (16.2 g, 77.4 mmol) was added to morpholinocyclopentene **2** (11 g, 77.4 mmol) in PhMe (40 mL) under $N_2$ with ice cooling and stirring at such a rate that the temperature never rose above 50°C. After a short supplementary stirring, the mixture was heated to reflux for 12 h. The solution was treated with excess of $Et_2O$ under cooling and the precipitate was filtered off. Recrystallization from $Me_2CO$ afforded 11.4 g of **3** (48%), mp 167-168°C.

**2,3-Dicarbomethoxy-3-cycloheptenone (4).** A solution of **3** (1 g, 3.25 mmol) in MeOH (5 mL) and 32% HCl (1 mL) was heated to reflux. Water (2 mL) was added and the mixture was heated for another 10 min to reflux. After cooling the precipitate was crystallized from MeOH:$H_2O$ 2:1 to give 610.5 mg of **4** (90%), mp 63.5-64°C.

## BERGMAN Cycloaromatization

Ring annulation by radical cyclization of ene-diynes and (Z)-allene-ene-ynes in a thermal reaction to give aromatics (electrocyclization).

| solvent | X | Y |
|---|---|---|
| hydrocarbon | H | H |
| $CCl_4$ | Cl | Cl |
| $CH_3OH$ | $CH_2OH$ | H |

| | | | | | |
|---|---|---|---|---|---|
| 1 | Bergman, R.G. | *J. Am. Chem. Soc.* | **1972** | *94* | 660 |
| 2 | Bergman, R.G. | *J. Am. Chem. Soc.* | **1981** | *103* | 4082; 4091 |
| 3 | Schreiber, S.L. | *J. Am. Chem. Soc.* | **1988** | *110* | 631 |
| 4 | Maier, M.E. | *Liebigs Ann.* | **1992** | | 855 |
| 5 | Grissom, J.W. | *Tetrahedron Lett.* | **1992** | *33* | 2315 |
| 6 | Bergman, R.G. | *Acc. Chem. Res.* | **1973** | *6* | 25 |
| 7 | Myers, A.G. | *J. Am. Chem. Soc.* | **1989** | *111* | 8057 |
| 8 | Myers, A.G. | *J. Am. Chem. Soc.* | **1992** | *114* | 9369 |
| 9 | Ming-Jung Wu | *Tetrahedron Lett.* | **1994** | *35* | 1879 |
| 10 | Cramer, C.J. | *J. Am. Chem. Soc.* | **1998** | *120* | 6269 |
| 11 | Grissom, J.W. | *Tetrahedron* | **1996** | *52* | 6453 |

**3,4-Dihydrobenz-[e]-indene 2.**[5] A mixture of ene-diyne **1** (39.9 mg; 0.17 mmol), PhCl (1.8 mL) and 1,4-cyclohexadiene **3** (0.4 mL; 4.2 mmol) under $N_2$ was heated for 19 h at 210°C. Chromatography (silica gel, hexane:EtOAc 95:5) afforded 30.1 mg of **2** (72%). TLC (hexane:EtOAc 3:1), $R_f$ = 0.48.

**5-[tert-Butyldimethylsilyl)oxy]-3-(4-methoxyphenyl)-6,7,9,10-tetrahydro-5,9-metha nobenzocycloocten-8(5H)-one 5.**[4] A solution of **4** (44 mg; 105 µmol) in **3** (2 mL) was heated under reflux for 5 h. The solvent was evaporated in vacuum and the residue purified by flash chromatography (petroleum ether:AcOMe 20:1) to afford 23.7 mg of **5** (51%) as a colorless oil. TLC (petroleum ether:AcOMe 4:1), $R_f$ = 0.54.

## B E R N T H S E N  Acridine Synthesis
Acridine synthesis from diphenylamine and carboxylic acids (see 1st edition).

| 1 | Bernthsen, A | *Liebigs Ann.* | **1878** | *192* | 1 |
| 2 | Popp, F.D. | *J. Org. Chem.* | **1962** | *27* | 2658 |
| 3 | Albert, F. | *J. Org. Chem.* | **1948** | | 1225 |
| 4 | Buu-Hoi, M.P. | *J. Chem. Soc.* | **1955** | | 1082 |

## B I G I N E L L I  Pyrimidone Synthesis
Pyrimidone synthesis from urea, an aldehyde and a β-keto ester.

| 1 | Biginelli, P. | *Chem. Ber.* | **1891** | *24* | 2962 |
| 2 | Folkers, K. | *J. Am. Chem. Soc.* | **1933** | *55* | 3361 |
| 3 | Swett, I. | *J. Am. Chem. Soc.* | **1973** | *95* | 8741 |
| 4 | Zaugg, H.E. | *Org. React.* | **1965** | *14* | 88 |
| 5 | Kappe, C.O. | *J. Org. Chem.* | **1997** | *62* | 7201 |
| 6 | Wipf, P. | *Tetrahedron Lett.* | **1995** | *36* | 7819 |

**Pyrimidone 4.**[5] Ethyl acetoacetate 1 (1.3 g; 10 mmol), PhCHO 2 (1.06 g; 10 mmol) and urea 3 (0.6 g; 10 mmol) in MeOH (5 mL containing 1-2 drops of conc. HCl) was stirred 2 h at 20°C. A precipitate appeared and stirring was continued for 3 h to afford 1.98 g of 4 (76%), mp 106-107°C.

## BESTMANN Cumulene Ylides

Phosphocumulenes ylides and phosphallene ylides in nucleophilic additions to C=C; C≡N and C≡C or cycloaddictions (2+2; 4+2; 1,3-dipolar)

| 1 | Bestmann, H.J. | *Angew.Chem.Int.Ed.* | **1974** | *13* | 875 |
|---|---|---|---|---|---|
| 2 | Bestmann, H.J. | *Liebigs Ann.* | **1977** | *16* | 349 |
| 3 | Bestmann, H.J. | *Angew.Chem.Int.Ed.* | **1976** | *15* | 115 |
| 4 | Bestmann, H.J. | *Angew.Chem.Int.Ed.* | **1965** | *4* | 585, 645, 830 |
| 5 | L'abbe, G. | *J.Org.Chem.* | **1974** | *39* | 3770 |

**Benzocoumarine (6).[3]** 1-Formyl-2-naphthol 5 (1.72 g, 10 mmol) is added slowly to a stirred solution of ylid 4 (3.02 g, 10 mmol) in PhH (30 mL). After 2-3 days stirring at 20°C or 24 h reflux, the solvent was removed in vacuum and the residue was crystallized from MeOH or i-PrOH. Recrystallization from i PrOH or PhH/MeOh afforded 1.43 g of 6 (73%), mp 117°C.

**Phosphorane (8).[3]** A solution of 1-dimethylaminomethyl-2-naphthol 7 (2.01 g, 10 mmol) and ylid 4 (3.02 g, 10 mmol) in anh. PhH (50 mL) was heated to reflux under stirring and $N_2$. After complete evolution of $Me_2NH$, the mixture was refluxed for 5 hours, then the solvent was removed in vacuum and the residue, after recrystallization from EtOAc or PhH/EtOAc, afforded 3.15 g of 8 (69%), mp 217°C.

## BIRCH-HÜCKEL-BENKESER Reduction

Reduction of aromatics, unsaturated ketones or conjugated dienes by alkali metals in liquid ammonia or amines.

| 1 | Hückel, W. | *Liebigs Ann.* | **1939** | *540* | 156 |
|---|---|---|---|---|---|
| 2 | Birch, A.I. | *J. Chem. Soc.* | **1944** | | 430 |
| 3 | Benkeser, R.A. | *J. Am. Chem. Soc.* | **1961** | *77* | 3230 |
| 4 | Benkeser, R.A. | *J. Org. Chem.* | **1964** | *29* | 955 |
| 5 | Moody, C.J. | *Tetrahedron Lett.* | **1986** | *27* | 5253 |
| 6 | Silverstein. R.M. | *Synthesis* | **1987** | | 922 |
| 7 | Robideau, P.W. | *Org. Reactions* | **1992** | *42* | 1 |
| 8 | Birch, A.I. | *Pure Appl. Chem.* | **1996** | *68* | 553 |

**5,8-Dihydro-1-naphtol 2.**[2] To 1-naphtol **1** (10.0 g; 69 mmol) was added powdered NaNH$_2$ (2.7 g; 69 mmol), liquid NH$_3$ (100 mL), *t*-BuOH (12.5 g) and then Na (3.2 g; 0.139 at) in small pieces. After evaporation of the NH$_3$, the residue was extracted with Et$_2$O. Acidification gave an oil which solidified. Recrystallization gave 89.5 g of **2** (85%), mp 71-74°C.

## BISCHLER-NAPIERALSKI Isoquinoline Synthesis

Isoquinoline synthesis from amides or phenethylamines (see 1st edition).

| 1 | Bischler, A.; Napieralski, B. | *Chem. Ber.* | **1893** | *26* | 1903 |
|---|---|---|---|---|---|
| 2 | Morrison, C.G | *J. Org. Chem.* | **1964** | *29* | 2771 |
| 3 | Ramesh, D. | *Synth. Commun.* | **1986** | *16* | 1523 |
| 4 | Thygarayan, B.S. | *Chem. Rev.* | **1954** | *54* | 1033 |
| 5 | Fodor, G. | *Angew. Chem. Int. Ed.* | **1972** | *11* | 919 |
| 6 | Govindachari, T.R. | *Org. React.* | **1951** | *6* | 74 |
| 7 | Ishikawa, T. | *Tetrahedron Lett.* | **1995** | *36* | 2795 |

## BLANC – QUELLET Chloroalkylation

Lewis acid catalyzed aromatic chloromethylation (Blanc), chloroalkylation (Quellet).

| 1 | Grassi, G., Masselli, C. | *Gazz. Chim. Ital.* | **1898** | 28 | 477 |
|---|---|---|---|---|---|
| 2 | Blanc, G. | *Bull. Soc. Chim. Fr.* | **1923** | 33 | 313 |
| 3 | Tashiro Masashi | *J. Org. Chem.* | **1978** | 43 | 1413 |
| 4 | Fuson, R. | *Org. React.* | **1942** | 1 | 63 |
| 5 | Quellet, R. | *C. R.* | **1932** | 195 | 155 |
| 6 | Quellet, R. | *Bull. Soc. Chim. Fr.* | **1940** | 7 | 196 |
| 7 | Neda, V. | *J. Soc. Chem. Ind. Jpn.* | **1944** | 47 | 565 |
| 8 | Mitchel, R. H. | *Synlett.* | **1989** | | 55 |

**2,2'-Di(chloromethyl) -4,4'-di(tert-butyl)diphenylmethane (3).[3]** To cooled (-5 °C) **1** (35 g, 125 mmol) and chloromethyl methyl ether **2** (80.5 g, 100 mmol) in $CS_2$ (150 mL) was added $TiCl_4$ (20 mL). The mixture was stirred for 1 h, poured into ice water (300 mL) and the organic layer extracted with PhH. Evaporation gave 36 g of **3** (76%), mp 90-91 °C(EtOH).

**2,4-Bis (bromomethyl)-mesitylene (9).[8]** Mesitylene **8** (120 g, 1 mol) was added to a mixture of 48% HBr (475 mL) and glacial acetic acid (125 mL), followed by 1,3,5-trioxane (60 g, 2 mol) and tetradecyltrimethylammomium bromide (5 g). The mixture was then well stirred such that only a single layer could by seen and then heated to a gentle reflux for 24 h. After cooling to 20 °C the white solid was filtered, washed (water) and extracted with hot hexane-$CH_2Cl_2$. Finally there were obtained 290 g of **9** (94%), mp 133-4 °C

## BLICKE-PACHTER Pteridines Synthesis

Condensation of aminopyrimidines with aldehydes and HCN followed by cyclization with NaOMe to pteridines.

| | | | | | |
|---|---|---|---|---|---|
| 1 | Blicke, F.F. | *J. Am. Chem. Soc.* | **1954** | 76 | 2798 |
| 2 | Pachter, I.J. | *J. Org. Chem.* | **1963** | 28 | 1191 |
| 3 | Pachter, I.J. | *J. Org. Chem.* | **1963** | 28 | 1203 |

**2,4,7-Triamino-6-phenylethynyl-pteridine 3.**[2] 2,4,5,6-Tetraaminopyrimidine **1** (2.5 g; 14 mmol) in MeOH (12 mL) and HOAc (12 mL) was treated with NaCN (1.5 g; 30 mmol) in water (6 mL) and phenylpropargylaldehyde **2** (2.5 g; 19 mmol) in MeOH (3 mL). After 10 min stirring and boiling, cooling deposited yellow crystals, washed (MeOH, water and MeOH), 1.9 g (28%) of **3** (acetate).

## BLOMQUIST Macrocycles Synthesis

Synthesis of large ring carbocycles by cyclization of bifunctional ketenes.

| | | | | | |
|---|---|---|---|---|---|
| 1 | Blomquist, A.T. | *J. Am. Chem. Soc.* | **1947** | 69 | 472 |
| 2 | Blomquist, A.T. | *J. Am. Chem. Soc.* | **1948** | 70 | 30 |

**1,8-Cyclotetradecanedione 2.**[2] Suberic acid **1** (3 g; 1.7 mmol) and SOCl$_2$ (0.4 g; 3.4 mmol) were heated at 55°C for 2 h and on a water bath until gas evolution ceased. Excess SOCl$_2$ was removed in vacuum and the acid chloride was diluted with Et$_2$O (200-300 mL). This was added to Et$_3$N (10-20 mL) in Et$_2$O (500-600 mL) over 26 h under gentle reflux. The decanted solution was washed with dil. HCl and water, dried (MgSO$_4$) and distilled. The yellow residue was treated with EtOH (5 mL) and KOH sol (1.8 g in 20 mL EtOH). After 10 h at 20°C and 2 h reflux, the mixture was diluted with water, extracted with Et$_2$O and the solvent evaporated to afford two crops of **2**, total yield 10%, mp 147.5-148°C.

## B L U M  Aziridine Synthesis

Synthesis of aziridines from epoxides via amino alcohols or azido alcohols and reaction with phosphines or phosphites (see 1st edition).

| 1 | Blum, J. | *J.Org.Chem.* | **1978** | *43* | 397, 4273 |
|---|----------|---------------|----------|------|-----------|
| 2 | Shudo, K | *Chem.Pharm.Bull.* | **1976** | *24* | 1013 |
| 3 | Hassner, A | *J.Am.Chem.Soc.* | **1970** | *92* | 3733 |
| 4 | Hassner, A | *J.Am.Chem.Soc.* | **1969** | *91* | 5046 |
| 5 | Blum, J. | *J.Heterocycl.Chem.* | **1994** | *31* | 837 |
| 6 | Chiappe, C. | *Tetrahedron Asymm.* | **1998** | *121* | 4079 |

**Threo-2-Azido-1,2-diphenylethanol (2).**[1] A mixture of cis-stilbene oxide **1** (3.92 g, 20 mmol) and NaN$_3$ (4.48 g, 70 mmol) in 50% aqueous acetone (60 mL) was refluxed for 3 h. The solvent was removed in vacuum and the residue extracted with CHCl$_3$. The organic solution was washed with water, dried (MgSO$_4$) and concentrated. Distillation of the residue afforded 3.70 g of **2** (77%) as a pale yellow oil, bp 122 °C/0.15 mm.

**Cis-2,3-Diphenylaziridine (3).** A solution of **2** (0.84 g, 3.5 mmol) and triphenylphosphine (0.92 g, 3.5 mmol) in dry Et$_2$O (25 mL) was refluxed for 1 h. Et$_2$O (50 mL) was added and the mixture was allowed to stand overnight at 5°C to allow complete precipitation of triphenyphosphine oxide. Column chromatography on silica gel yielded 0.53 g of **3** (77%).

## B O D R O U X - C H I C H I B A B I N  Aldehyde Synthesis

Aldehyde synthesis from Grignard reagents and trialkyl orthoformate; see also Bouveault (see 1st edition).

$$C_6H_5\text{-Br} \xrightarrow[Et_2O]{Mg} C_6H_5\text{-MgBr} \xrightarrow[\text{reflux, 6 h}]{\overset{HC(OEt)_3}{2}} C_6H_5\text{-CH=O}$$

**1**                                          **3** (60%)

| 1 | Chichibabin, A.E. | *J. Russ. Phys. Chem. Soc.* | **1903** | *35* | 1284 |
|---|---|---|---|---|---|
| 2 | Bodroux, F. | *C. R.* | **1904** | *138* | 92 |
| 3 | Smith, L.I. | *J. Org. Chem.* | **1941** | *6* | 437 |

## B O G E R - C A R B O N I - L I N D S E Y  Heterocycle Synthesis

Diels-Alder reactions of olefins, acetylenes, allenes with tetrazines or triazines to provide pyridazines or pyridines; reverse demand Diels-Alder reactions (see 1st edition).

**1**            **2**            **3**            **4** (60%)

**5**            **6**            **7** (71%)

| 1 | Carboni, R.A.; Lindsey, R.V. | *J. Am. Chem. Soc.* | **1959** | *81* | 4342 |
|---|---|---|---|---|---|
| 2 | Boger, D.L. | *J. Org. Chem.* | **1981** | *48* | 2179 |
| 3 | Boger, D.L. | *J. Org. Chem.* | **1982** | *47* | 3736 |
| 4 | Boger, D.L. | *J. Org. Chem.* | **1983** | *48* | 621 |
| 5 | Boger, D.L. | *J. Am. Chem. Soc.* | **1985** | *107* | 5745 |
| 6 | Boger, D.L. | *Chemtracts: Org. Chem.* | **1996** | *9* | 149 |

**3-Ethyl-4-n-propylpyridine 7.**[2] **5** (132 mg; 0.8 mmol) in CHCl$_3$ (0.5 mL) was added to a stirred solution of 1,2,4-triazine **6** (85 mg; 1.2 mmol) in CHCl$_3$ (0.5 mL) under N$_2$ at 25°C. The resulting dark orange solution was warmed at 45°C for 20 h. Chromatography (silica gel, 50% Et$_2$O in hexane) afforded 92 mg of pure **7** (71%).

**BOGER** Thermal Cycloadditions

Thermal cycloaddition of cyclopropenone ketal with olefinic acceptors to form cyclopentene derivatives.

(90)   3 (60%)   (10)

89%[7]

| 1 | Boger, D.L. | *J.Am.Chem.Soc.* | **1984** | *106* | 805 |
|---|---|---|---|---|---|
| 2 | Boger, D.L. | *Tetrahedron Lett.* | **1984** | *25* | 5611 |
| 3 | Boger, D.L. | *J.Org.Chem.* | **1985** | *50* | 3425 |
| 4 | Boger, D.L. | *Tetrahedron* | **1986** | *42* | 2777 |
| 5 | Boger, D.L. | *Tetrahedron Lett.* | **1984** | *25* | 5615 |
| 6 | Boger, D.L. | *J.Am.Chem.Soc.* | **1986** | *108* | 6695, 6713 |
| 7 | Boger, D.L. | *J.Org.Chem.* | **1988** | *53* | 3408 |

**cis-Benzyl methyl 2-phenyl-6,10-dioxaspiro[4,5]dec-3-ene 1,1-dicarboxylate(cis).[7]**
A solution of (Z)-benzyl methyl (phenyl methylene) malonate **2** (Z) (120 mg, 0.405 mmol) in MeCN-$d_3$ (0.4 mL) was treated with cyclopropenone 1,3-propanediyl ketal **1** (132 mg, 1.18 mmol, 2.9 equiv) under $N_2$. After 20 h heating at 80°C (shielded from light), the cooled mixture was concentrated in vacuum, and the residue filtered through a short column of $SiO_2$ ($CH_2Cl_2$). Evaporation of the solvent and chromatography ($SiO_2$ $CH_2Cl_2$) afforded: 8 mg of **2** (recovered), **1** (recovered) and a mixture of **3** (99 mg, 60%). Ratio cis:trans 90:10.

## B O R C H  Reduction

Reductive amination of aldehydes or ketones by cyanoborohydride (or triacetoxyborohydride)[6] anion. Selective reduction of carbonyls to alcohols, oximes to N-alkylhydroxylamines, enamines to amines (see 1st edition).

$$OHC\text{-}(CH_2)_3\text{-}CHO \ + \ CH_3\text{-}NH_2\cdot HCl \xrightarrow[\text{MeOH; 48 h}]{\text{NaBH}_3\text{CN; 25°C}}$$

| 1 | | 2 | | 3 (43%) |

$$+ \ H_2N\text{-}Ph \xrightarrow[\text{20 min; 20°C}]{\text{NaBH(OAc)}_3}$$

| 4 | 5 | 6 (95%) |

| 1 | Borch, R.F. | *J. Am. Chem. Soc.* | **1969** | *91* | 3996 |
|---|---|---|---|---|---|
| 2 | Borch, R.F. | *J. Am. Chem. Soc.* | **1971** | *93* | 2897 |
| 3 | Borch, R.F. | *J. Chem. Soc. Perkin I* | **1984** | | 717 |
| 4 | Lane, C.F. | *Synthesis* | **1975** | | 135 |
| 5 | Hutchins, R.O. | *Org. Prep. Proc. Int.* | **1979** | *11* | 20 |
| 6 | Abdel-Magid | *Tetrahedron Lett.* | **1990** | *31* | 5595 |

**Amine 6.** Aldehyde **4** (1.36 g; 10 mmol) and aniline **5** (1.023 g; 11 mmol) in dichloroethane (40 mL) was treated with sodium triacetoxyborohydride (3.18 g; 15 mmol) under $N_2$ at 20°C to afford 2.37 g of **6** hydrochloride (95%).

## B O U V E A U L T  Aldehyde Synthesis

Aldehyde synthesis from Grignard or Li derivatives with a formamide; see also Bodroux-Chichibabin (see 1st edition).

| 4 | 6 (65%) |

| 1 | Bouveault, L. | *C. R.* | **1903** | *137* | 987 |
|---|---|---|---|---|---|
| 2 | Bouveault, L. | *Bull. Soc. Chim. Fr.* | **1904** | *31* | 1306 (3) |
| 3 | Sice, J. | *J. Am. Chem. Soc.* | **1953** | *75* | 3697 |
| 4 | Einchorn, J. | *Tetrahderon Lett.* | **1986** | *27* | 1791 |

**5-Methoxy-2-thienaldehyde 6.**[3] 5-Methoxy-2-thienyllithium prepared from **4** (11.4 g; 0.1 mol) and Li in $Et_2O$ (125 mL) was added slowly to ice cooled DMF **5** (8.0 mL; 0.11 mol) in $Et_2O$ (75 mL) with efficient stirring and let stand at 20° overnight. The mixture was poured into ice, extracted with $Et_2O$ and distillation gave 9.27 g of **6** (65%), bp 79-81°C/0.9 mm; mp 24-26°C (petroleum ether).

## B O R S C H E – B E E C H Aromatic Aldehyde Synthesis

Synthesis of aromatic aldehydes and of alkyl aryl ketones from aldoximes or semicarbazones and aromatic diazonium salts (see 1st edition).

| | | | | | |
|---|---|---|---|---|---|
| 1 | Borsche, C. | *Chem. Ber.* | **1907** | *40* | 737 |
| 2 | Beech, W. F. | *J. Chem.Soc.* | **1954** | | 1297 |
| 3 | Woodward, R. B. | *Tetrahedron* | **1958** | *2* | 1 |

**Pyridine-3-aldehyde (3).**[2] 3-Aminopyridine **2** (23.5 g, 0.24 mol), 36% HCl (68 mL). NaNO$_2$ (17.5 g, 0.25 mol) and water (75 mL) was made neutral (NaOAc) and treated with formaldoxime **1**. The mixture was acidified (pH-3) and after FeCl$_3$ (150 g) was added, it was boiled for 1 h. Usual work up gave 3.6 g of **3** (14%), bp 95-100 °C/16 mm.

## B R E D E R E C K Imidazole Synthesis

Synthesis of imidazoles from formamide (acetamide) and α-diketones, α-ketols, α-aminoketones, α-oximinoketones (see 1st edition).

| | | | | | |
|---|---|---|---|---|---|
| 1 | Bredereck, H. | *Chem. Ber.* | **1953** | *86* | 88 |
| 2 | Grimmett, V. | *Adv. Heteroc. Chem.* | **1970** | *12* | 113 |
| 3 | Bredereck, H. | *Angew. Chem.* | **1959** | *71* | 753 |
| 4 | Schubert, H. | *Z. Chem.* | **1967** | *7* | 461 |
| 5 | Novelli, A. | *Tetrahedron Lett.* | **1967** | | 265 |

## B O U V E A U L T - B L A N C  Reduction

Reduction of esters to alcohols by means of sodium in alcohol (see 1st edition).

| 1 | | 2 (20%)[2] | | 3 |

| 1 | Bouveault, L.; Blanc, G. | C.R. | **1903** | *136* | 1676 |
|---|---|---|---|---|---|
| 2 | Paquette, L.A. | J. Org. Chem. | **1962** | 27 | 2274 |
| 3 | Ruhlmann, K. | Synthesis | **1972** | | 236 |
| 4 | Chaussar, J. | Tetrahedron Lett. | **1987** | 28 | 1173 |
| 5 | Rabideau, P.W. | Tetrahedron Lett. | **1980** | | 1401 |

## B O U V E A U L T - H A N S L E Y - P R E L O G - S T O L L  Acyloin Condensation

Condensation of two esters to an α-hydroxyketone by means of rapidly stirred (8000 rpm) Na suspension in boiling toluene or xylene (see 1st edition).

| **1** (0.2 mol) | | | **2** (75%) |

| 1 | Bouveault, L. | C. R. | **1905** | *140* | 1593 |
|---|---|---|---|---|---|
| 2 | Hansley, V.L. | U.S. Pat. 2.228.268; cf. Chem. Abstr., **1941**, *35*, 2354 | | | |
| 3 | Prelog, V. | Helv. Chim. Acta | **1947** | 30 | 1741 |
| 4 | Stoll, M. | Helv. Chim. Acta | **1947** | 30 | 1815 |
| 5 | Cramm, D.J. | J. Am. Chem. Soc. | **1954** | 76 | 2743 |
| 6 | Finley, K.T. | Chem. Rev. | **1964** | 64 | 573 |
| 7 | Ruhlmann, K.T. | Synthesis | **1971** | | 236 |

## B O Y L A N D - S I M S  o-Hydroxylaniline Synthesis

Oxidation of dialkylanilines or their N-oxides with persulfates to o-aminophenols (see 1st edition).

| 1 | Boyland, E.; Sims, P. | *J. Chem. Soc.* | **1953** | | 3623 |
|---|---|---|---|---|---|
| 2 | Boyland, E.; Sims, P. | *J. Chem. Soc.* | **1958** | | 4198 |
| 3 | Behrman, E.J. | *J. Am. Chem. Soc.* | **1967** | 89 | 2424 |
| 4 | Behrman, E.J. | *J. Org. Chem.* | **1992** | 57 | 2266 |
| 5 | Behrman, E.J. | *Org. React.* | **1988** | 35 | 432 |

## B R U Y L A N T S  Amination

Amination – alkylation of aldehydes via α–cyanoamines (see 1st edition).

| 1 | Bruylant, P. | *Bull. Soc. Chim. Belge* | **1924** | 33 | 467 |
|---|---|---|---|---|---|
| 2 | Bruylant, P. | *Bull. Soc. Chim. Belge* | **1926** | 35 | 139 |
| 3 | Bersch, H. W. | *Arch. Pharm.* | **1978** | 311 | 1029 |
| 4 | Ahlbrecht, H. | *Synthesis* | **1985** | | 743 |

**N-(2-Hexene-4-yl)-pyrrolidine (4).**[3] To **3** (10.57 g, 70 mmol) in THF (20 mL) under Ar, EtMgBr (1 molar, 22 mmol) in THF is added slowly at 0 °C. The mixture was stirred for 3 h at 20 °C, diluted with Et$_2$O (50 mL) and worked up to give 8.35 g of **4** (78%), bp 83 °C (19 mm).

## BRANDI-GUARNA Rearrangement

Synthesis of pyridine derivatives by rearrangement of isoxazolidone-5-spirocyclopropanes resulting from dipolar addition to methylenecyclopropanes.

| 1 | Brandi, A., Guarna, A. | *J.Chem.Soc.Chem.Commun.* | **1985** | | 1518 |
|---|---|---|---|---|---|
| 2 | Brandi, A., Guarna, A. | *J.Org.Chem.* | **1988** | *53* | 2426; 2430 |
| 3 | Brandi, A. | *J.Org.Chem.* | **1992** | *57* | 5666 |
| 4 | Brandi, A. | *Tetrahedron Lett.* | **1995** | *36* | 1343 |
| 5 | Brandi, A. deMeijere, A. | *J.Org.Chem.* | **1996** | *61* | 1665 |
| 6 | Brandi, A., Guarna, A. | *Synlett* | **1993** | | 1 |

**Spiro 4,5-dihydro-3-methylisoxazole-5,1'-2'-phenylcyclopropane (3).**[3] Nitroethane (1.3 g, 22 mmol) and Et₃N (262 mg, 2.6 mmol) in PhH (11 mL) was added over 1 h to a refluxing solution of 1-methylene 2-phenylcyclopropane **2** (1.88 g, 14.5 mmol) and methyl isocyanate **1** (1.24 g, 23 mmol) in PhH (10 mL) under stirring. After 18 h stirring at 20°C, the mixture was filtered and concentrated in vacuum. Unreacted **1** was recovered (45-65°C 0.5 torr) and the residue was chromatographed (CH₂Cl₂) to give 1 g of **3** (40%), mp 85°C.

**2-Methyl-6-phenyl-dihydropyrid-4-one (4).** Vapours of **3** (260 mg, 1.4 mmol) were passed at 0.04 Torr through a quartz tube heated at 400°C then led into a cold trap. Washing with petroleum ether afforded 216 mg of **4** (83%), mp 162°C (CHCl₃ - petroleum ether).

## von B R A U N Amine Degradation

Degradation of tertiary amines with cyanogen bromide (BrCN), or ethyl, benzyl or phenyl chloroformate (see 1st edition).

| 1 | V. Braun, J. | *Chem.Ber* | **1907** | *40* | 3914 |
|---|---|---|---|---|---|
| 2 | Elderfield, R.C. | *J.Am.Chem.Soc.* | **1950** | *72* | 1334 |
| 3 | Boekelheide, V. | *J.Am.Chem.Soc.* | **1955** | *77* | 4079 |
| 4 | Wright, W.B. | *J.Org.Chem.* | **1961** | *26* | 4057 |
| 5 | Calvert, B.J. | *J.Chem.Soc.* | **1965** | | 2723 |
| 6 | Rapoport, H. | *J.Am.Chem.Soc.* | **1967** | *89* | 1942 |
| 7 | Knabe, J. | *Arch. Pharm.* | **1964** | *259* | 135 |
| 8 | McCluskey, J.G. | *J.Chem.Soc. (C)* | **1967** | | 2015 |
| 9 | Hageman, H.A. | *Org.React.* | **1953** | *7* | 198 |

**4-Pipecoline (3).**[2] To a solution of BrCN (48 g, o.46 mol) in PhH (100 mL) was added 1-isopropyl-4-pipecoline **1** (58 g, 0.41 mol) in PhH (275 mL) over 1 h at 40°C. The mixture was heated for 45 min at 55-60°C and was maintained at 20°C for 36 h. The basic material was extracted with HCl (100 mL) and the solvent was distilled to give 44 g of residue. The neutral product **2** was refluxed with 48% HBr (300 mL) for 10 h. After distillation of HBr, the residue was leached in a mixture of EtOAc:EtOH (80:20). Filtration of insoluble NH₄Br and concentration gave **3**, mp 171-173°C.

**Phenyl 21-chlorodeoxydihydrochanoajmaline-N-carboxylate (5).**[8] 21-Deoxy ajmaline **4** (1.55 g, 5.06 mmol) in CH₂Cl₂ (50 mL) was treated with phenyl chloroformate (0.86 g, 5.5 mmol) at 20°C for 18 h. Usual work-up, and chromatography afforded 2.24 g of **5** (96%).

## B R O O K Silaketone Rearrangement

Rearrangement of silaketone to silyl ethers (with chirality transfer) (see 1st edition).

$$\text{Ph}_2\text{Si}\!-\!\underset{\underset{\text{O}}{\|}}{\text{C}}\!-\!\text{Ph} \quad\xrightarrow{\text{EtO}^-}\quad \text{Ph}_2\text{Si}\!-\!\underset{\underset{\text{OEt}}{|}}{\overset{\overset{\text{O}^-}{|}}{\text{C}}}\!-\!\text{Ph}_2 \quad\longrightarrow\quad \text{Ph}_2\text{Si}\!-\!\text{O}\!-\!\text{CHPh}_2$$

**1**                 OEt  **2**

| # | Author | Journal | Year | Vol | Page |
|---|--------|---------|------|-----|------|
| 1 | Brook, A.G. | *J.Org.Chem.* | **1962** | 27 | 2311 |
| 2 | Brook, A.G. | *Acc.Chem.Res.* | **1974** | 7 | 77 |
| 3 | Wilson, S.R.. | *J.Org.Chem.* | **1981** | 47 | 747 |
| 4 | Kuwajima, J. | *Tetrahedron Lett.* | **1980** | 21 | 623 |
| 5 | Mori, M. | *J.Org.Chem.* | **1996** | 61 | 1196 |
| 6 | West, R. | *J.Am.Chem.Soc.* | **1974** | 96 | 3214 |

**Benzhydryloxy ethoxy diphenyl silane 2.[2]** To a solution of benzoyltriphenylsilane **1** (2.5 g, 6.9 mmol) in PhH (25 mL) was added a solution of sodium ethoxide in EtOH (2 mL, 0.8 mmol). The solution was washed with water and the solvent removed in vacuum. The oily residue was dissolved in hot EtOH (15 mL) and cooled to give 2.1 g of **2** (74%), mp 67-75°C. Recrystallization from EtOH gave 1.8 g of **2** (64%), mp 77-78°C.

**Silyl amines 4 and 5.[5]** To a solution of **3** in THF was added BuLi at −78°C and the solution was stirred for 30 min at the same temperature. MeI was add4ed at −78°C and the mixture was stirred for another 30 min at the same temperature. After usual work-up are obtained 40% from **4** and 20% from **5**.

## **BROWN** Acetylene Zipper Reaction

Isomerization of internal acetylenes to the terminal position by means of potassium (or lithium) 3-aminopropylamide (KAPA).

$$HO-CH_2-(CH_2)_{11}-C\equiv C-(CH_2)_2-CH_3 \xrightarrow{KAPA} HO-CH_2-(CH_2)_{14}-C\equiv CH$$

$$\mathbf{1} \qquad\qquad\qquad \mathbf{2}\ (98\%)$$

$$HO-CH_2-(CH_2)_5-C\equiv C-(CH_2)_{15}-CH_3 \xrightarrow[NH_2(CH_2)_3NH_2]{Li} HO-CH_2-(CH_2)_{21}-C\equiv CH$$

$$\mathbf{3} \qquad\qquad\qquad\qquad \mathbf{4}\ (82\%)$$

| | | | | | |
|---|---|---|---|---|---|
| 1 | Brown, C.A. | *J.Am.Chem.Soc.* | **1975** | 97 | 891 |
| 2 | Brown, C.A. | *J.Chem.Soc.Chem.Commun.* | **1976** | | 959 |
| 3 | Macaulay, S.R. | *J.Org.Chem.* | **1980** | 45 | 734 |
| 4 | Becker, D. | *J.Org.Chem.* | **1984** | 49 | 2494 |
| 5 | Abrams, S.R. | *Can.J.Chem.* | **1984** | 62 | 1333 |

**16-Heptadecyn-1-ol (2).[4]** A mixture of potassium (190 mg, 4.8 mmol) in 1,3-propanediamine **1** (5 mL) with ferric nitrate (1 mg) was heated to 90°C in a ultrasound cleaning bath. After 10-15 min potassium disappears and a green-brown solution of KAPA was formed. This mixture was cooled to 0°C and 12-heptadecyn-1-ol **1** (190 mg, 0.75 mmol) in THF (1 mL) was added. After 30 min stirring at 0°C, the mixture was poured into water (125 mL) and extracted with hexane (3 x 100 mL). The extract was dried with MgSO₄ and after evaporation of the solvent, there was obtained 185 mg of 16-heptadecyn-1-ol **2** (98%), mp 41°C.

**23-Tetracosyn-1-ol (4).[5]** 1,3-Diaminopropane (10 mL) under N₂ was treated with Li (140 mg, 20 m at g) under heating (70°C) and stirring. After 2 h the mixture was cooled to 20°C, KO-t-Bu (1.3 g, 12 mmol) was added and stirring was continued for another 15 min when 7-tetracosyn-1-ol **3** (1.05 g, 3 mmol) was added. After 2 h stirring the mixture was quenched with water and normal work up gave after chromatography (silica gel, hexane : Et₂O 1:1) 860 mg of **4** (82%), mp 76-7°C.

## **BROWN** Hydroboration

Hydroboration-regioselective and stereoselective (syn) addition of $BH_3$ ($RBH_2$, $R_2BH$) to olefins. Synthesis of alcohols or amines including optically active ones from olefins. Also useful in synthesis of ketones by "stitching" of olefins with CO (see 1st edition).

| 1 | Brown, H.C. | *J.Am.Chem.Soc.* | **1956** | *78* | 2583 |
|---|---|---|---|---|---|
| 2 | Brown, H.C. | *J.Org.Chem.* | **1978** | *43* | 4395 |
| 3 | Masamune, S. | *J.Am.Chem.Soc.* | **1986** | *108* | 7401 |
| 4 | Hoffmann, R.W. | *Angew.Chem.Int.Ed.* | **1982** | *21* | 555 |
| 5 | Brown, H.C. | *J.Am.Chem.Soc.* | **1986** | *108* | 2049 |
| 6 | Srebnik, M. | *Aldrichimica Acta* | **1987** | *20* | 9 |
| 7 | Brown, H.C. | *J.Org.Chem.* | **1989** | *54* | 4504 |
| 8 | Brown, H.C. | *J.Org.Chem.* | **1995** | *60* | 41 |

**Isopinocampheol 6.**[2] To a hot solution of borane-methyl sulfide **1** (2 mL, 20 mmol) in $Et_2O$ (11.3 mL) was added (+)-α-pinene **2** (7.36 mL, 46 mmol), which led to quantitative formation of **3**. After addition of TMEDA (1.51 mL, 10 mmol), reflux was continued for 30 min. The adduct was filtered and washed with pentane to give 3.32 g of **4** (80%), mp 140-141°C ($Et_2O$). A solution of **4** (3.32 g, 8 mmol) in THF (16 mL) was treated with $BF_3.Et_2O$ (1.97 mL, 16 mmol). After 1 h, the solid $TMEDA.2BF_3$ was removed and the solution of **5** was oxidized with alkaline $H_2O_2$ to give **6** (100%).

**(-) 3-Hydroxytetrahydrofuran 8.**[5] To a suspension of (-) $Ipc_2BH$ (diisopinocamphenyl borane) **3** (7.1 g, 25 mmol) in THF, see above, at -25°C was added 2,3-dihydrofuran **7** (1.9 mL, 25 mmol). The reaction mixture was stirred at the same tempreature for 6 h. The solid **3** disappeared, and formation of trialkyl borane was complete. The mixture was brought to 0°C, acetaldehyde (5.6 mL, 100 mmol) was added dropwise and stirring was continued for another 6 h at 25°C. Excess acetaldehyde was removed in vacuum (25°C, 12 mm Hg), and 20 mL of THF was added. The boronate thus obtained was oxidized with 25 mL of 3N NaOH and 3.75 mL of 30% $H_2O_2$, and maintained for 5 h at 25°C. The aqueous layer was saturated with $K_2CO_3$, extracted with 3.25 mL $Et_2O$ and the organic layer dried ($MgSO_4$). The solvent was evaporated, the residue filtered through silica; pentane eluent removed - pinene, whereas the $Et_2O$ eluent afforded the alcohol **8** which on distillation yielded 1.87 g, bp 80°C/15 mm (92%), GC purity 99%, $α_D$ = -17.3°C (c 2.4 MeOH, 100% ee).

**(S)-(-)-(Trifluoromethyl)oxirane 12.**[8] B-chlorodiisopinocamphenylborane **10** (8.8 g, 27.5 mmol) in $Et_2O$ (25 mL) under $N_2$ was cooled to -25°C and **9** (4.7 g, 25 mmol) was added using a syringe. The reaction was followed by [11]B NMR ([11]B: 32 ppm) for 96 h, when the reaction was complete. At 0°C was added diethanolamine (5.3 mL, 55 mmol), then the mixture was heated to 20°C and stirred for 2 h, whereupon the borane precipitated as a complex which was filtered and washed with pentane. The solvent was removed, the residue added to 15 N NaOH (10 equiv.) and heated at 95-100°C to distill the epoxide. This afforded 1.536 g of **12** (64%, 96%ee).

**B R O W N**   Stereoselective Reduction

Stereoselective reduction of ketones to alcohols by means of borohydride reagents (Li s-Bu₃BH) or t-BuClBR* for formation of chiral alcohols.

| 1 | Brown, H.C. | *J. Am. Chem. Soc.* | **1970** | *92* | 709 |
| 2 | Brown, H.C. | *J. Am. Chem. Soc.* | **1972** | *94* | 1750 |
| 3 | Brown, H.C. | *Chem. Commun.* | **1972** | | 868 |
| 4 | Brown, H.C. | *J. Am. Chem. Soc.* | **1972** | *94* | 7159 |
| 5 | Brown, H.C. | *J. Org. Chem.* | **1989** | *54* | 4540 |
| 6 | Brown, H.C. | *J. Org. Chem.* | **1995** | *60* | 41 |

*Cis*-4-tert-butylcyclohexanol 2.[4] To 1M lithium trimethoxyaluminium hydride (LTMA) (5.0 mL) in THF under N₂, was added sec-butylborane (from 2-butene and diborane), 1.25 mL, 5 mmol. After 30 min the mixture was cooled to -78°C and 1 (390 mg; 2.5 mmol) was added. After 3 h, hydrolysis and oxidation (H₂O₂) gave 2 (96.5% *cis* and 3.5% *trans*).

(S)-Cyclohexylethanol 6.[5] To 5.5 mmol of 4 in THF (from Li-tBuBH₃, HCl followed by (-)-2-ethylapopinene 3, α_D= -42.78°) was added 5 (0.64 g; 5 mmol) under N₂. After 2 days the solvent was removed, the residue dissolved in Et₂O (20 mL), diethanolamine (2.2 equiv.) was added and stirred for 2 h. After filtration and washing with pentane, the filtrates were concentrated and chromatography gave 0.42 g of 6 (65%), 90% ee.

## BUCHNER-CURTIUS-SCHLOTTERBECK Homologation

Ring enlargement of benzene derivatives by carbenes generated from diazo compounds (better in the presence of a Rh catalyst). Conversion of aldehydes to ketones by diazo compounds (Schlotterbeck); see also Pfau-Platter (see 1st edition).

**1**                                                                    **2** (75%)

| | | | | | |
|---|---|---|---|---|---|
| 1 | Buchner, E.; Curtius, T. | *Chem. Ber.* | **1885** | *18* | 2371 |
| 2 | Buchner, E. | *Chem. Ber.* | **1896** | *29* | 106 |
| 3 | Sclotterbeck, F. | *Chem. Ber.* | **1907** | *40* | 479 |
| 4 | Ramonczay, J. | *J. Am. Chem. Soc.* | **1950** | *72* | 2737 |
| 5 | Doering, W.v. | *J. Am. Chem. Soc.* | **1957** | *79* | 352 |
| 6 | Anciaux, A.J. | *J. Org. Chem.* | **1981** | *46* | 873 |
| 7 | Manitto, P. | *J. Org. Chem.* | **1995** | *60* | 484 |

## BURTON Trifluoromethylation

Trifluoromethylation of aryl iodides or nitroarenes with Cd(Cu) reagents (see 1st edition).

| | | | | | |
|---|---|---|---|---|---|
| 1 | Burton, D.J. | *J. Am. Chem. Soc.* | **1985** | *107* | 5014 |
| 2 | Burton, D.J. | *J. Am. Chem. Soc.* | **1986** | *108* | 832 |
| 3 | Clark, J.H. | *J. Chem. Soc. Chem. Commun.* | **1988** | | 638 |
| 4 | Clark, J.H. | *Tetrahedron Lett.* | **1989** | *30* | 2133 |

**1-Trifluoromethyl-2,4-dinitrobenzene 2.**[4] A mixture of *m*-dinitrobenzene **1** (840 mg; 5 mmol), metallic Cu (1.905 g; 30 mat), dibromodifluoromethane (2.43 g; 11 mmol), charcoal (1 g) (dried at 280°C) in dimethylacetamide (7.5 mL) was heated to 100°C under $N_2$, to afford 1.026 g of **2** (87%).

## B U C H W A L D  Heterocyclization

Preparation of benzisothiazoles, butenolides or pyrroles using organo-zirconium reagents and acetylenes.

| 1 | Buchwald, S.L. | *J. Am. Chem. Soc.* | **1987** | *109* | 7137 |
| 2 | Buchwald, S.L. | *Tetrahedron Lett.* | **1988** | *29* | 3445 |
| 3 | Buchwald, S.L. | *J. Am. Chem. Soc.* | **1989** | *111* | 776 |
| 4 | Buchwald, S.L. | *J. Org. Chem.* | **1989** | *54* | 2793 |
| 5 | Buchwald, S.L. | *J. Am. Chem. Soc.* | **1991** | *113* | 4685 |
| 6 | Buchwald, S.L. | *Chem. Rev.* | **1988** | *88* | 1044 |
| 7 | Gribble, G.W. | *Contemp. Org. Synth.* | **1994** | *1* | 145 |

**Chiral butenolide 3.**[2] A mixture of **1** (995 mg; 2.79 mmol) and $Cp_2Zr(H)Cl$ **2** (791 mg; 3.07 mmol) in PhH (30 mL) were stirred at 20°C under Ar for 16 h. After degassing, the mixture was stirred under a $CO_2$ atm for 6 h. A solution of $I_2$ (708 mg; 2.79 mmol) in PhH (20 mL) was added and stirring was continued for 1 h. Usual work up and chromatography (radial), pentane:$Et_2O$ (9:1 to 7:3) gave 1.93 g of **3** (55%), 90% ee.

**7-Methoxy-2,3-dimethylbenzo[*b*]thiophene 7.**[4] To 2-bromoanisole **4** (385 mg; 2 mmol) in THF (10 mL) at -78°C was added BuLi (1.2 mL 1.68M; 2.2 mmol). After 15 min stirring, zirconocene(methyl)chloride **5** (570 mg; 2.1 mmol) in THF (10 mL) was added followed by 2-butyne **6** (130 mg; 2.4 mmol) and heated for 18 h at 80°C. Usual work up and recrystallization from pentane gave 274 mg of **7** (71%), mp 110-110.5°C.

## B U C H W A L D - H A R T W I G  Aryl Halide Amination

Amination of aryl halides in the presence of a base and $Pd_2(dba)_3$ + BINAP (Buchwald) or (DPPF)PdCl$_2$ (DPPF= 1,1'-bis(diphenylphosphino-ferrocene) (Hartwig).

(94%)

**1**          **2**          **3** (81%)

| 1 | Buchwald, S.L. | *J. Am. Chem. Soc.* | **1996** | *118* | 7215 |
|---|---|---|---|---|---|
| 2 | Hartwig, J.F. | *J. Am. Chem. Soc.* | **1996** | *118* | 7217 |
| 3 | Snieckus, V. | *Synlett* | **1998** | | 419 |

**Amide 3.**[3] **1** (505 mg; 1.97 mmol), **2** (0.21 mL; 2.30 mmol), NaO*t*Bu (266 mg; 2.77 mmol), $Pd_2(dba)_3$ (5 mg; 0.006 mmol), BINAP (11 mg; 0.017 mmol) and PhMe (5 mL) under $N_2$ were heated for 21 h at 90-100°C. Work up  and chromatography afforded 426 mg of **3** (81%), mp 74-76°C.

## B U R G E S S  Alcohol Dehydration

Thermolysis of tertiary and secondary alcohols with (carbomethoxysulfamoyl) triethylammonium inner salt **1** or polymer linked reagent[6] to give olefins; also conversion of amides to nitriles (see 1st edition).

**1** (56%)          (69%)

| 1 | Burgess, E.M. | *J. Org. Chem.* | **1973** | *38* | 26 |
|---|---|---|---|---|---|
| 2 | O'Grodnick, J.S. | *J. Org. Chem.* | **1974** | *39* | 2124 |
| 3 | Goldsmith, D.J. | *Tetrahedron Lett.* | **1980** | *21* | 3543 |
| 4 | Claremon, D.A. | *Tetrahedron Lett.* | **1988** | *29* | 2155 |
| 5 | Burgess, E.M. | *Org. Synth.* | **1977** | *56* | 40 |
| 6 | Wipf, P. | *Tetrahedron Lett.* | **1996** | *37* | 4659 |
| 7 | Wipf, P. | *Tetrahedron* | **1998** | *54* | 6987 |
| 8. | Wipf, P. | *Chem. Rev.* | **1995** | *95* | 2115 |

## CADOGAN – CAMERON WOOD Cyclization

Synthesis of indoles, pyrroles and others N-heterocycles by cyclization of nitro compounds with trialkyl phosphite.

| 1 Cadogan, J.I.G.; Cameron-Wood, M. | *Proc.Chem.Soc.* | **1962** | | 361 |
|---|---|---|---|---|
| 2 Taylor, E. G. | *J.Org.Chem.* | **1965** | *30* | 1013 |
| 3 Cadogan, J.I.G. | *Chem.Commun..* | **1966** | | 491 |
| 4 Buckl, P. | *Angew.Chem.Int.Ed.* | **1969** | *8* | 120 |
| 5 Amarnath, V. | *Synthesis* | **1974** | | 840 |

**1.3 – Dimethyl – 6 - (p-dimethylaminophenyl) - 5H - 2,4 (1H,3H) pyrrolo [3,2-d] pyrimidinedione (2).**[2] A mixture of 1,3-dimethyl-5-nitro-6-(p-dimethylamino) styryluracyl **1** (1.65 g, 5 mmol) and triethyl phosphite (5 mL, 4.85 g, 29 mmol) was refluxed under $N_2$ for 5.5 h. After 18 h at 20°C the volatile components were evaporated under vacuum and the residue recrystallized from DMF. Vacuum sublimation (240-250°C/0.05 mm) afforded 0.9 g of **2** (60 %), mp 310-318°C.

# C A N N I Z Z A R O   Oxidation - Reduction

A redox reaction between two aromatic aldehydes (or an aromatic aldehyde and formaldehyde) to a mixture of alcohol and acid (see 1st edition).

| 1 | Wöhler, F. | *Liebigs Ann.* | **1832** | *3* | 252 |
|---|---|---|---|---|---|
| 2 | Cannizzaro, S. | *Liebigs Ann.* | **1853** | *88* | 129 |
| 3 | Bruce, R:A: | *Org.Prep.Proced.Int.* | **1987** | | 19 |
| 4 | Geissmann, T.A. | *Org. React.* | **1944** | *2* | 92 |
| 5 | Moore, I.A. | *Org.Prep.Proced.Int.* | **1988** | *20* | 82 |

**o-Methoxybenzyl alcohol (3) and o-Methoxybenzoic acid (2).**[3]   To a solution of KOH (120 g, 2 mol) in water are added o-methoxybenzaldehyde **1** (136 g, 1 mmol) under efficient stirring and external cooling with water. Stirring was maintained until a stable emulsion was obtained. After 24 h at 30°C  the mixture was diluted with water and extracted with Et$_2$O. Evaporation of the solvent and vacuum distillation of the residue afforded 55 g of **3** (79%), bp 245-255°C. Acidification of the aqueous solution, extraction with Et$_2$O and evaporation of the solvent gave **2**, mp 98-99°C.

**Dicarboxylic acids (5) and (6).**[5]  1,6,1',6'-Tetraformylbiphenyl **4** (25.8 g, 96.9 mmol) was dissolved in 6N NaOH (400 mL) at 25°C; The mixture warmed by the heat of reaction. After 30 min, conc HCl was added dropwise to the stirred solution until the pH of the mixture reached pH=1. The creamy colored precipitate was collected and recrystallized from water, to afford 18.7 g of **5** and **6** (64%), mp 204-206°C, tlc (EtOH) R$_f$ (**5**)=0.56  R$_f$ (**6**)=0.54.

# **CARGILL** Rearrangement

Rearrangement of unsaturated ketones catalyzed by acids

| 1 | Cargill, R. L. | *Tetrahedron Lett.* | **1967** | | 169 |
|---|---|---|---|---|---|
| 2 | Cargill, R. L. | *J.Org.Chem.* | **1970** | *35* | 356 |
| 3 | Narasaka, K. | *Chem.Lett.* | **1993** | | 621 |
| 4 | Cargill, R. L. | *Acc.Chem.Res.* | **1974** | *7* | 106 |
| 5 | Fetizon, M. | *J.Chem.Soc.Chem.Comm.* | **1975** | | 282 |
| 6 | Kakiuchi, K. | *J.Am.Chem.Soc.* | **1980** | *111* | 3707 |

**Tricyclo(4.3.2.0$^{1,6}$)undec-10-en-2-one (4)$^2$.** A solution of bicyclo [4.3.0] non-1(6) – en – 2 – one **1** (2.6 g, 19.1 mmol) and a mixture of "E" and "Z" 1,2-dichloroethylene **2** (3 ml, 7.62 g, 78 mmol) in pentane (80 mL) was irradiated (Corex) for 30 min. The residue obtained after evaporation of volatiles, was dissolved in Et$_2$O (100 mL) and added to dry liquid NH$_3$ (2,000 mL). The solution was treated with Na until a blue color was obtained. After additional 10 min stirring, NH$_4$Cl was added and NH$_3$ was evaporated. Addition of water, extraction with Et$_2$O followed by distillation gave 2.38 g of **4** (77 %), bp 71-73°C/0.25 Torr.

**Tricyclo(3.3.3.0$^{1,5}$)undec-3-en-2-one (5).** A solution of **4** (1.92 g, 11.8 mmol) and p-TsOH.H$_2$O (0.8 g, 4.2 mmol) in PhH (50 mL) was refluxed for 10 min. After washing with NaHCO$_3$ solution and concentration, the residue after distillation afforded 1.32 g of **5** (68.7 %), bp 65°C/0.25 Torr.

## C A R R O L L   Rearrangement of Allyl Acetoacetic Esters

Thermal condensation of allyl alcohols with ethyl acetoacetate in the presence of a catalyst, with loss of $CO_2$; a one pot ester exchange-Claisen-Ireland rearrangement with loss of $CO_2$ (see 1st edition).

| 1 | Carroll, M.F. | *J.Chem.Soc.* | **1940** | | 704 |
|---|---|---|---|---|---|
| 2 | Cologne, J. | *Bull.Soc.Cim.Fr.* | **1955** | | 1312 |
| 3 | Kimel, W. | *J.Org.Chem.* | **1957** | 22 | 1611 |
| 4 | Kimel, W. | *J.Org.Chem.* | **1958** | 23 | 153 |
| 5 | Stephen, W. | *J.Org.Chem.* | **1984** | 49 | 722 |
| 6 | Podraza, K.F. | *J.Heterocycl.Chem..* | **1986** | 23 | 581 |
| 7 | Enders, D. | *Angew.Chem.Int.Ed.* | **1995** | 34 | 2278 |
| 8 | Sorgi, K.L. | *Tetrahedron Lett.* | **1995** | 36 | 3597 |

**Cinnamylacetone (3).**[1] A mixture of phenyl vinlyl carbinol **1** (26.8 g, 0.2 mmol) ethyl acetoacetate **2** (35.1 g, 0.27 mmol) and KOAc (0.3 g) was heated to 220°C for 3 h and maintained at this temperature for another 3 h. 15 mL of distillate (EtOH, 0.25 mol) was collected. Washing and distillation of the residue afforded EtOAc (10 g), an alcoholic fraction (2 g) and 26 g of **3** (75%), bp 125-130°C (4 mm Hg), $\alpha_D^{20}$ = 1.5475; oxime mp 87.5-89°C.

## CHAN   Reduction of Acetylenes

Stereospecific reduction of acetylenic alcohols to E- allylic alcohols by means of sodium bis(2-methoxyethoxy)aluminium hydride (SMEAH) (see 1st edition).

| 1 | Chan, Ka-Kong | *J.Org.Chem.* | **1976** | *41* | 62 |
|---|---|---|---|---|---|
| 2 | Chan, Ka-Kong | *J.Org.Chem.* | **1976** | *41* | 3497 |
| 3 | Chan, Ka-Kong | *J.Org.Chem.* | **1976** | *43* | 3435 |

## CHAPMAN   Rearrangement

O to N aryl migration in O-aryliminoethers (see 1st edition).

| 1 | Chapman, A.W. | *J.Chem.Soc.* | **1925** | *127* | 1992 |
|---|---|---|---|---|---|
| 2 | Dauben, W.G. | *J.Am.Chem.Soc.* | **1950** | *72* | 3479 |
| 3 | Crammer, F. | *Angew.Chem.* | **1956** | *68* | 649 |
| 4 | Roger, R. | *Chem.Rev.* | **1969** | *69* | 503 |
| 5 | Schulenberg, J.W. | *Org.React.* | **1965** | *14* | 1 |

## CHATGILIALOGLU Reducing agent

Tris(trimethylsilyl)silane (TTMSS) reducing agent for alkyl halides, ketones; an alternative to tributyltin hydride.

| 1 | Chatgilialoglu, C. | *J.Org.Chem.* | **1988** | 53 | 3641 |
|---|---|---|---|---|---|
| 2 | Giese, B. | *Tetrahedron Lett.* | **1989** | 30 | 681 |
| 3 | Chatgilialoglu, C. | *J.Org.Chem.* | **1991** | 56 | 678 |
| 4 | Chatgilialoglu, C. | *J.Org.Chem.* | **1989** | 54 | 2492 |
| 5 | Chatgilialoglu, C. | *Tetrahedron Lett.* | **1989** | 30 | 2733 |
| 6 | Giese, B. | *Tetrahedron Lett.* | **1990** | 31 | 6013 |
| 7 | Chatgilialoglu, C. | *Tetrahedron* | **1990** | 46 | 3963 |
| 8 | Arya, P. | *J.Org.Chem.* | **1990** | 55 | 6248 |

**Naphthalene (3).[5]** To a solution of 1-bromonaphthalene **1** (278 mg, 1 mmol) in monoglyme (3 mL) in a quartz tube with magnetic stirrer was added NaBH$_4$ (1.9 g) and under Ar were added TTMSS **2** (23.8 mg, 0.1 mmol) and p-methoxybenzoyl peroxide. The reaction mixture was photolyzed at 254 nm in a Rayonet reactor. GC analysis: yield 91%.

**1-Phenyl-3,4-dimethylcyclopentane (5 and 6).[6]** A solution of **4** (1.00 g, 5mmol) in PhMe (40 mL) was heated with stirring at 90°C under Ar. TTMSS and AIBN in PhMe (10 mL) was added slowly (over 3-4 h) via syringe pump. Evaporation of the solvent and chromatography (silica gel, pentane:Et$_2$O) afforded 78% of **5** and **6** in a ratio cis / trans 4.6 : 1.

## CHICHIBABIN Pyridine synthesis

Pyridine synthesis from aromatic acetaldehydes and ammonia (see 1st edition).

13%

| | | | | | |
|---|---|---|---|---|---|
| 1 | Chichibabin, A. | *J.Russ.Phys.Chem.Soc.* | **1906** | *37* | 1229 |
| 2 | Eliel, E.L. | *J.Am.Chem.Soc.* | **1953** | *75* | 4291 |
| 3 | Sprung, M.M. | *Chem.Rev.* | **1940** | *26* | 301 |
| 4 | Frank, R.L. | *Org.Synth.Coll.* | | *IV* | 451 |
| 5 | Mc Gill, C.K. | *Adv.Heterocycl.Chem.* | **1988** | *44* | 1 |

## CHICHIBABIN Amination

$\alpha$-Amination of pyridines, quinolines and other N-heterocycles in liq. $NH_3$ (see 1st edition).

| | | | | | |
|---|---|---|---|---|---|
| 1 | Chichibabin, A. | *J.Russ.Phys.Chem.Soc.* | **1914** | *46* | 1216 |
| 2 | van der Plas, H.C. | *J.Org.Chem.* | **1981** | *46* | 2134 |
| 3 | Bunnett, J.F. | *Chem.Rev.* | **1951** | *49* | 375 |
| 4 | Rykowscy, A. | *Synthesis* | **1985** | | 884 |
| 5 | Leffler, M.T. | *Org.React.* | **1942** | *1* | 19 |

## C I A M I C I A N  Photocoupling

Reductive photocoupling of ketones to diols (see 1st edition).

| 1 | Ciamician, G. | *Chem.Ber.* | **1900** | *33* | 2911 |
| 2 | De Mayo, P. | *Quart.Rev(London)* | **1961** | *15* | 415 |
| 3 | Goth, H. | *Helv.Chim.Acta* | **1965** | *48* | 1395 |

## C I A M I C I A N - D E N N S T E D T  Cyclopropanation

Cyclopropanation of alkenes with dichlorocarbene derived from $CHCl_3$ and sometimes subsequent ring enlargement of fused cyclopropanes (see 1st edition).

| 1 | Ciamician, G. Dennstedt, N. | *Chem.Ber.* | **1881** | *14* | 1153 |
| 2 | Parham, W.E. | *J.Am.Chem.Soc.* | **1955** | *77* | 1177 |
| 3 | Vogel, E. | *Angew.Chem.* | **1960** | *72* | 8 |
| 4 | Makosza. M. | *Angew.Chem.Int.* | **1974** | *13* | 665 |
| 5 | Skell, P.S. | *J.Am.Chem.Soc.* | **1958** | *80* | |
| 6 | Oddo, B. | *Gazz.Chim.Ttal.* | **1939** | *69* | 10 |

**1,1-Dichloro-2-phenylcyclopropane (2).**[4] To a solution of styrene **1** (10.4 g, 0.1 mol) in $CHCl_3$ (11.9 g, 0.1 mol) was added 50% NaOH followed under efficient stirring by dibenzo(18)-crown-**6** (0.36 g, 1 mmol). After a mild exotermic reaction, usual work-up gave 16.25 g of **2** (87%), bp 112°Cc/ 15 torr.

### C L A I S E N – G E U T E R – D I E C K M A N N Ester Condensation

Synthesis of open chain Claisen or cyclic Dieckmann β-ketoesters by aldol type condensation

$$2 \ Me_2N-\underset{O}{\overset{\parallel}{C}}-(CH_2)_8-CO_2Me \xrightarrow[\Delta \ 24 \ h]{NaOMe} Me_2N-\underset{O}{\overset{\parallel}{C}}-(CH_2)_8-\underset{\overset{\mid}{CO_2Me}}{\overset{\mid}{C}}H-(CH_2)_7-CONMe_2 \quad (75\%)$$

| # | Author | Journal | Year | Vol | Page |
|---|--------|---------|------|-----|------|
| 1 | Geuter, A. | *Arch.Pharm.* | **1863** | *106* | 97 |
| 2 | Claisen, L. | *Chem.Ber.* | **1887** | *20* | 651 |
| 3 | Dieckmann, W. | *Chem.Ber.* | **1894** | *27* | 965 |
| 4 | Cohen, H. | *J.Org.Chem.* | **1973** | *38* | 1425 |
| 5 | Bosch, J. | *Tetrahedron* | **1984** | *40* | 2505 |
| 6 | Thyagarajan, B.S. | *Chem.Rev.* | **1954** | *54* | 1029 |
| 7 | Schaefer, J.P. | *Org.React.* | **1967** | *15* | 1 |

**2-t-Butoxycarbonylcyclopentanone 2.**[4] To a stirred suspension of NaH (24 g, 1 mol) in PhH (400 mL) under $N_2$ was added **1** (5.0 g, 20 mmol) and t-BuOH (2.0 mL) in one portion and the mixture was boiled for 30 min. Another portion of **1** (120 g, 0.465 mol) in PhH (200 mL) was added dropwise for 45 min and reflux was continued 4.5 h. The mixture was neutralized (AcOH) and water (750 mL) was added followed by extraction with $Et_2O$ (2X500 mL). Evaporation of the solvent and distillation afforded 65.5 g of **2** (73%), bp 80-85°C/2 torr, Rf = 0.25 (silica gel, $Et_2O$:hexane 1:2).

**Ethyl 1-benzyl-3-oxo-4-piperidinecarboxylate 4.**[5] A solution of **3** (25 g, 78 mmol) in dioxane (100 mL) containing EtOH (6.8 mL) was added dropwise to a suspension of NaH (2.7 g, 117 mmol) in dioxane (100 mL). After 7 h refluxing, usual work up afforded 17.5 g of **4** (80%), mp 102-104°C ($Me_2CO$).

## C L A I S E N - I R E L A N D  Rearrangement

Rearrangement of allyl phenyl ethers to o-(or p)-allylphenols or of allyl vinyl ethers to γ,δ-unsaturated aldehydes or ketones (Claisen). Rearrangement of allyl esters as enolate anions or silyl enol ethers to γ,δ-unsaturated acids (Ireland). Also rearrangement of N-allylanilines (an aza-Cope rearrangement) (see 1st edition).

| 1 | Claisen, L. | Chem.Ber. | **1912** | 45 | 3517 |
|---|---|---|---|---|---|
| 2 | Rhoades, S.L. | J.Am.Chem.Soc. | **1955** | 73 | 5060 |
| 3 | Ireland, E. | J.Am.Chem.Soc. | **1972** | 94 | 5897 |
| 4 | Daub, D.W. | J.Org.Chem. | **1986** | 51 | 3404 |
| 5 | Anderson, W.K. | Synthesis | **1995** | | 1287 |
| 6 | Tarbell, D.S. | Org.React. | **1944** | 2 | 1 |

**2,6-Dimethyl-4-(α-methylallyl)phenol (2).[2]** The ether **1** (17.6 g, 0.1 mol) was heated in dimethylaniline for 3 h at reflux. After work-up are obtained 11.8 g of **2** (67%), bp 89-90°C /05 mm.

**7-Allylindoline (4).[5]** N-Allylindoline **3** (9.32 g, 58.54 mmol), sulfolane (20 mL) and BF₃•OEt₂ (3.6 mL, 29.27 mmol, 0.5 equiv) was heated at 200-210°C under Ar. After quenching with water, extraction and chromatography of the residue (EtOAc:hexane 1:10), there are obtained 890 mg of **3** (10%) and 4.38 g of **4** (47%), Rf = 0;47 (EtOAc:Hexane 1:5).

**4-Decenoic acid (6).[3]** N-Isopropylcyclohexylamine (1.7 g, 12.1 mmol) in THF (20 mL) at 0°C was treated with BuLi (5 mL, 11.1 mmol) in hexane. After 10 min **5** (1.64 g, 10 mmol) was added dropwise at -78°C. After 5 min stirring the mixture was warmed to 20°C poured into 5% NaOH (20 mL) and extracted with Et₂0. Acidification (HCl) and extraction with CH₂Cl₂ afforded 1.356 g of **6** (83%) 99.5% E.

## C L A U S O N - K A A S  Pyrrole synthesis

Preparation of N-substituted pyrroles from 2,5-dialkoxytetrahydrofurans and primary amines.

| 1 | Clauson-Kaas, N. | *Acta Chem.Scand.* | **1952** | 6 | 667 |
|---|------------------|--------------------|----------|----|------|
| 2 | Josey, A.D. | *J.Org.Chem.* | **1962** | 27 | 2466 |
| 3 | Patterson, J.M. | *Synthesis* | **1976** | | 281 |

**1-(2-Methoxycarbonyl)phenylpyrrole (3).**[2]  2,5-Diethoxytetrahydrofuran **1** (95.5 g, 0.59 mol) was added to a well stirred solution of methyl antranilate **2** (90 g, 0.59 mol) in AcOH (265 mL). During the exothermic reaction,the mixture became clear deep red. The mixture was heated to reflux for 1 h and the solvent was removed in vacuum. Fractional distillation in vacuum gave 95.8 g of **3** (80%), bp 90-95$^0$C.

## C L A Y - K I N N E A R - P E R R E N  Phosphonyl Chloride Synthesis

Synthesis of alkyl phosphonyl chlorides from alkyl chlorides or from ethers with $PCl_3$ – $AlCl_3$ (see 1st edition).

$$C_2H_5Cl + AlCl_3 + PCl_3 \xrightarrow[24\,h]{4^0C} C_2H_5ClAlCl_3PCl_3 \xrightarrow[0^0C]{32\%\ HCl} C_2H_5\text{-}PCl_2$$
$$\quad\ \ 3 \qquad\ \ 1 \qquad\ 2 \qquad\qquad\qquad 4 \qquad\qquad\qquad 5\ (43\%)$$

$$(C_2H_5)_2O\ +\ 1\ +\ 2 \xrightarrow{0^0C} \xrightarrow[7\,h]{100^0C} 5\quad (43\%)$$
$$\qquad 6$$

| 1 | Clay, J.P. | *J.Org.Chem.* | **1951** | 16 | 892 |
|---|-----------|---------------|----------|----|------|
| 2 | Kinnear, M.M.; Perren, E.A. | *J.Chem.Soc.* | **1952** | | 3434 |
| 3 | Hamilton, C.S. | *Org.Synth.Coll.vol* | IV | | 950 |

**Ethylphosphonyl dichloride (5).**[1]  **From diethyl ether:** $Et_2O$ **6** (18.5 g, 0.25 mol) was added to a mixture of **1** (66.5 g, 0.5 mol) and **2** (68.5 g, 0.5 mol) at 0$^0$C. The mixture was heated for 7 h at 100$^0$C (sealed tube). The crystalline product was dissolved in $CH_2Cl_2$ and hydrolyzed with water. After filtration and distillation 28 g of **5** (43%) was isolated.

## C L E M M E N S E N  Reduction

Reduction of ketones or aldehydes to hydrocabons by means of zinc amalgam and acid (see 1st edition).

| 1 | Clemmensen, E. | *Chem.Ber.* | **1913** | 46 | 1838 |
|---|---|---|---|---|---|
| 2 | Dauben, W.G. | *J.Am.Chem.Soc.* | **1954** | 76 | 3864 |
| 3 | Starschewsky, W. | *Angew.Chem.* | **1959** | 71 | 726 |
| 4 | Yamamura, S. | *Bull.Chem.Soc.Jpn.* | **1972** | 45 | 364 |
| 5 | Sanda, G. | *Tetrahedron Lett.* | **1983** | 24 | 4425 |
| 6 | Vedejs, E. | *Org.React.* | **1975** | 22 | 401 |

**Cis-9-Methyldecalin (2).**[2] cis-10-Methyl-2-decalone **1** (8.0 g, 48.2 mmol) was heated under reflux with amalgamated zinc (40 g, 0:61 at g) in AcOH (35 mL) and 32% HCl (17.5 mL). Reflux was maintained for 17 h and every 2 h there was added HCl (2 mL). Water (60 mL) was added and the mixture steam distilled. Neutralization of the distillate with $Na_2CO_3$, extraction with pentane, evaporation of the solvent, followed by distillation from potassium afforded 6.57 g of **2** (90%), bp 91.5-92.0°C / 20 mm.

**Cholestane (4).**[4] To a solution of cholestan-3-one **3** (500 mg, 1.3 mmol) in EtOH saturated with HCl gas (75 mL) at 0°C, was added active Zn powder (5.0 g) (in portions) under stirring. After being stirred for 1 h at 0°C, the reaction mixture was basified ($Na_2CO_3$) and extracted with $Et_2O$. The residue obtained after removal of the solvent, was chromatographed (silica gel, PhH) to give 431 mg of **4** (89%), mp 77.5-79°C.

## C L I V E – R E I C H – S H A R P L E S S Olefination

Organoselenium compounds in synthesis of terminal olefins, unsaturated ketones

| 1 | Clive, D.L.J. | *J.Chem.Soc.Chem.Commun.* | **1973** | | 695 |
|---|---|---|---|---|---|
| 2 | Reich, H.J. | *J.Am.Chem.Soc.* | **1973** | *95* | 5813 |
| 3 | Sharpless, K.B. | *J.Org.Chem.* | **1975** | *40* | 947 |
| 4 | Krief, A. | *Bull.Soc.Chim.Fr.* | **1997** | *134* | 869 |

**Cyclohex-2-en-1-one 4.**[1] Enol acetate of cyclohexanone **1** (1 equiv) in Et$_2$O at 0°C in the presence of AgOCOCF$_3$ (1.2 equiv) and **2** (1.1 equiv) afforded after hydrolysis **3** in 70% yield. Oxidation of **3** with NaIO$_4$ gave **4** (92%).

**Acrylophenone 7.**[2] To a solution of LDA under N$_2$ in THF was added 1,4-diphenyl-1-butanone. After 10 min stirring, **2** was added dropwise at –78°C. To the solution at 0°C, H$_2$O$_2$ was added and the reaction mixture was stirred for 30 min at 20-25°C. Usual work up and chromatography afforded **7** in 85% yield.

**1-Dodecene 11.**[3] To a solution of selenide **10** (0.2 mmol) in MeOH/THF/H$_2$O containing NaHCO$_3$ (3 equiv) at 20°C was added NaIO$_4$ (0.3 mmol). After 6 h the reaction mixture was evaporated in vacuum. Usual work up afforded the olefin in 72% yield.

**C L O K E - W I L S O N** Cyclopropylketone Rearrangement

Rearrangement of cyclopropyl ketones or imines to dihydrofurans or dihydropyrroles, thermally, photochemically, or by Lewis acids (see 1st edition).

| 1 | Cloke, J.B. | *J.Am.Chem.Soc.* | **1929** | *51* | 1174 |
|---|---|---|---|---|---|
| 2 | Wilson,C.L. | *J.Am.Chem.Soc.* | **1947** | *69* | 3002 |
| 3 | Alonso, M.E. | *J.Org.Chem.* | **1980** | *45* | 4532 |
| 4 | Hudlicky, T. | *Org.React.* | **1986** | *33* | 247 |

**C O M B E S** Quinoline Synthesis

Quinoline synthesis from anilines and β-diketones (see 1st edition).

| 1 | Combes, A. | *Bull.Soc.Chim.Fr.* | **1882** | *49* | 89(2) |
|---|---|---|---|---|---|
| 2 | Johnson, W.S. | *J.Am.Chem.Soc.* | **1944** | *66* | 210 |
| 3 | Born,J.L. | *J.Org.Chem.* | **1972** | *37* | 3952 |
| 4 | Bergstrom, F.W. | *Chem.Rev.* | **1944** | *35* | 156 |
| 5 | Seifert, W. | *Angew.Chem.Int.Ed.* | **1962** | *1* | 215 |

**2,4-Dimethylbenzo(g)quinoline (4).** A mixture of **3** (13.4 g, 0.059 mol) in HF (300 ml) was maintained for 24 h at 20°C. The residue obtained after removing the HF was neutralized with 10% $K_2CO_3$ solution, extracted with $Et_2O$ and the solvent was evaporated to yield 11.75 g of **4** (96%), mp 91-92.5°C.

## C O L L M A N Carbonylation Reagent

Dipotassium or disodium iron tetracarbonyl in the synthesis of aldehydes and ketones from alkyl halides (see 1 st edition).

$$Fe(CO)_5 \xrightarrow[\text{MeOH}]{\text{KOH}} K_2[Fe(CO)_9] \xleftarrow{Fe(CO)_5} K(sec\ Bu)_3BH$$

| 1 | | 3 (90%) | | 2 |

$$4 \xrightarrow{3} 5$$

$$\xrightarrow{3}$$

| 1 | Collman, J.P. | *Acc. Chem. Res* | **1986** | *1* | 136 |
| 2 | Collman, J.P. | *J. Am. Chem. Soc* | **1973** | 95 | 4089 |
| 3 | Collman, J.P. | *Acc. Chem. Res.* | **1975** | 8 | 342 |
| 4 | Collman, J.P. | *J. Am. Chem. Soc.* | **1977** | 99 | 2515 |
| 5 | Glaisy, J.A. | *J. Am. Chem. Soc.* | **1978** | 100 | 2545 |
| 6 | Glaisy, J.A. | *J. Org. Chem.* | **1978** | 43 | 2280 |
| 7 | Burnett, J.J. | *Syn. Commun.* | **1997** | 27 | 1473 |

**Dipotassium iron tetracarbonyl (catalyst) 3.**[7] $Fe(CO)_5$ **1** (1.5 mL, 11 mmol) was syringed into a degassed sol. of KOH (1.47 g, 26 mmol) in MeOH (15 mL). After 1 h stirring at 25°C the solvent was evaporated and the residue was stirred with THF (10 mL). The new solvent was evaporated and the operation repeated to remove MeOH. Finally, the residue was extracted with THF, filtered to remove $KHCO_3$ to obtain a pale pink filtrate (90-95% yield).

**Nonanal (5).** Octyl bromide **4** (89.44 mg, 0.46 mmol), **3** (94.5 mg, 0.0384 mmol) and $Et_3P$ (132.5 mg, 0.508 mmol) were stirred for 12 h. Glacial AcOH (200 mL) and tridecane (100 mL) (as reference standard) was added. GC analysis indicated 100% yield of **5**.

## C O L V I N  Alkyne Synthesis

Reaction of ketones with lithium trimethylsilyldiazomethane **2** (Peterson olefination) to give after rearrangement the homologous alkynes.

| 1 | Colvin, E.V. | *J. Chem. Soc. Chem. Commun.* | **1973** | 151 |
| 2 | Colvin, E.V. | *J. Chem. Soc. Perkin Trans. I* | **1977** | 869 |
| 3 | Colvin, E.V. | *J. Chem. Soc. Chem. Commun.* | **1992** | 721 |
| 4 | Aoyama, T.; Shioiri, T. | *Tetrahedron Lett.* | **1994** | 107 |

**p-Methoxyphenylpropyne  4.[4]**  To LDA in THF (8 mL) was added trimethylsilyldiazomethane **2** 1.9M in hexane (0.63 mL; 1.2 mmol) at -78°C under Ar. After 30 min **1** (150 mg; 1 mmol) in THF (2 mL) was added dropwise at -78°C. After 1 h the mixture was refluxed 3 h, quenched ($H_2O$) and extracted with $Et_2O$. Evaporation and chromatography provided 199.7 mg of **4** (82%), bp 85-88°C/0.9 mm.

## C O M I N S  Triflating Reagent

N-(5-Chloro-2-pyridyl)triflimide **3**, a reagent for introduction of the triflyl ($CF_3SO_2$) group.

| 1 | Comins, D.L. | *Tetrahedron Lett.* | **1992** | 33 | 6299 |
| 2 | O'Neil, I.A. | *Synlett* | **1995** | | 151 |

**Enol triflate 5.[2]** Under $N_2$ at -78°C γ-thio-butyrolactone **4** (0.17 mL; 2 mmol) in THF (5 mL) was treated with KHMDS (4.4 mL; sol. of 0.5M in PhMe). After 1 h stirring **3** (780 mg; 2 mmol) in THF (2 mL) was added. After 3 h at -78°C, quenching ($H_2O$), extraction ($Et_2O$), evaporation and chromatography ($Al_2O_3$ neutral) gave 342 mg of **5** (73%).

## C O N I A  Cyclization

Thermal cyclization of dienones, enals, ynones, diones, ketoesthers to monocyclic, spirocyclic bicyclic derivatives (ene reaction of unsaturated enol) (see 1st edition).

| 1 | Conia, J.M. | *Tetrahedron Lett.* | **1965** | | 3305; 3319 |
|---|---|---|---|---|---|
| 2 | Conia, J.M. | *Bull. Soc. Chim. Fr.* | **1966** | | 278; 281 |
| 3 | Krapcho, A.P. | *Synthesis* | **1974** | | 416 |
| 4 | Conia, J.M. | *Angew. Chem. Int. Ed.* | **1975** | *14* | 473 |

## C O R E Y - K I M  Oxidizing Reagent

Oxidation of alcohols to ketones by means of N-chlorosuccinimide (NCS) or NBS and Me$_2$S (see 1st edition).

| 1 | Corey, E.J.; Kim, C.U. | *J. Am. Chem. Soc.* | **1972** | *94* | 7586 |
|---|---|---|---|---|---|
| 2 | Corey, E.J. | *Tetrahedron Lett.* | **1973** | | 919 |
| 3 | Corey, E.J. | *J. Org. Chem.* | **1973** | *38* | 1223 |
| 4 | Dalgard, N.K. | *Acta Chim. Scand.* | **1984** | *38B* | 423 |
| 5 | Jamauki, M. | *Chem. Lett.* | **1989** | | 973 |

**Ketone 4.**[1] To a stirred NCS 1 (400 mg; 3 mmol) in PhMe (10 mL) was added 2 (0.3 mL; 4.1 mmol) at 0°C under Ar; a white precipitate appeared. At -25°C 3 (312 mg; 2 mmol) in PhMe (2 mL) was added dropwise, then Et$_2$O (20 mL). The organic layer was washed with 1% HCl (5 mL) and twice with water (15 mL). Evaporation left 310 mg of 4 (100%), mp 44-47°C.

## C O O P E R - F I N K B E I N E R  Hydromagnesiation

Ti catalyzed formation of Grignard reagents from olefins or acetylenes.

| 1 | Cooper, G.D; Finkbeiner, H.L | *J. Org. Chem.* | **1962** | 27 | 3395 |
|---|---|---|---|---|---|
| 2 | Sato, F. | *J. Chem. Soc. Chem. Commun.* | **1981** | | 718 |
| 3 | Sato, F. | *Tetrahedron Lett.* | **1983** | 24 | 1804 |
| 4 | Sato, F. | *J. Chem. Soc. Chem. Commun.* | **1983** | | 162 |
| 5 | Sato, F. | *Tetrahedron Lett.* | **1984** | 25 | 5063 |
| 6 | Adam, W. | *Synthesis* | **1994** | | 567 |

$\beta$-($\Delta^3$-Cyclohexenyl)ethanol 3.[1] To 1, from Mg 13.2 g and PrBr 61.3 g in Et$_2$O (150 mL) was added 2 (54 g; 0.5 mol) followed by TiCl$_4$ (1 mL). After 2 h reflux and heating with more TiCl$_4$ (0.5 mL), the mixture was oxidized with air and distilled to give 25 g of 3 (40%), bp 92-94°C.

3-Trimethylsilyl-2-ethylfuran 6.[5] Cp$_2$TiCl$_2$ (0.12 g; 0.48 mol) was added to iBuMgBr in Et$_2$O (43 mL; 0.4 M) under Ar at 0°C. 4 (0.18 g; 6.8 mmol) was added and the mixture was stirred 6 h at 25°C. EtCN (0.48 g; 8.8 mmol) was added and the mixture was stirred 2 h at 25°C. Usual work up and chromatography (silica gel) afforded 0.94 g of 6 (82%).

(E)-4-(Tributylstannyl)-3-penten-2-ol 8.[6] Cp$_2$TiCl$_2$ (1.74 g; 7 mmol) was added to iBuMgBr (2.1 equiv.) and stirred 10 min at 0°C. 7 (5.89 g; 70 mmol) was added and the mixture was stirred for 15 min at 20°C followed by reflux for 3 h. The solvent was evaporated and the residue dissolved in THF and treated with Bu$_3$SnCl (25.1 g; 77 mmol) at 0°C. Stirring for 1 h at 25°C and under reflux for 2 h gave after chromatography (silica gel, pentane:Et$_2$O) 11.6 g of 8 (44%).

# C O P E  Rearrangement

Thermal 3,3-sigmatropic rearrangement of 1,5-dienes (see 1st edition).

(34%)                    (66%)

*cis*

| 1 | Cope, A.C. | *J. Am. Chem. Soc.* | **1940** | *62* | 441 |
|---|---|---|---|---|---|
| 2 | McDowell, D.W. | *J. Org. Chem.* | **1986** | *51* | 183 |
| 3 | Baldwin, J.E. | *J. Org. Chem.* | **1987** | *52* | 676 |
| 4 | Vogel, E. | *Liebigs Ann.* | **1958** | *615* | 1 |
| 5 | Lutz, R.P. | *Chem. Rev.* | **1984** | *84* | 205 |
| 6 | Blechert, S. | *Synthesis* | **1989** | | 71 |

# C O P E - M A M L O C - W O L F E N S T E I N  Olefin Synthesis

Olefin formation by *syn*-elimination from tert. amine N-oxides (see 1st edition).

1                    2                    3 (90%)

| 1 | Mamloc, L.; Wolfenstein, R. | *Chem. Ber.* | **1900** | *33* | 159 |
|---|---|---|---|---|---|
| 2 | Cope, A.C. | *Tetrahedron Lett.* | **1949** | *71* | 3929 |
| 3 | Bluth, M. | *Tetrahedron Lett.* | **1984** | *25* | 2873 |
| 4 | De Puy, C.H. | *Chem. Rev.* | **1960** | *60* | 448 |
| 5 | Fujita, J. | *Synthesis* | **1978** | | 934 |
| 6 | Cope, A.C. | *Org. Synth. Coll.* | **1963** | *IV* | 612 |

## C O R E Y  Homologative Epoxidation

Reaction of ketones with S-ylides derived from $Me_3S^+I^-$ (from DMSO+MeI) or $Me_3SO^+I^-$ to give epoxides (see 1st edition).

| 1 | Corey, E.J. | *J. Am. Chem. Soc.* | **1962** | *84* | 866 |
|---|---|---|---|---|---|
| 2 | Kuhn, R. | *Angew. Chem.* | **1957** | *68* | 570 |
| 3 | Kuhn, R. | *Liebigs Ann.* | **1958** | *611* | 117 |
| 4 | Olah, G.A. | *Synthesis* | **1990** | | 887 |
| 5 | Nesmeyanov, A.N. | *Tetrahedron* | **1987** | *43* | 2600 |

**2-Methyleneadamantane epoxide 4.**[4] Ketone **4** (1.5 g; 10 mmol), **1** (2.20 g; 10 mmol) and *t*-BuOK (97% 1.15 g; 10 mmol) in DME (50 mL) was refluxed with good stirring under $N_2$ for 8 h. Quenching ($H_2O$), extraction ($Et_2O$) and evaporation gave 1.57 g of **4** (96%), mp 176°C.

## C O R E Y  Oxidizing Reagents

Pyridinium chlorochromate (PCC) **1** or $CrO_3$-dimethylpyrazole **4** reagents for oxidation of alcohols to ketones or aldehydes.

| 1 | Corey, E.J. | *Tetrahedron Lett.* | **1973** | | 2647 |
|---|---|---|---|---|---|
| 2 | Dauben, W.G. | *J. Org. Chem.* | **1977** | *42* | 682 |
| 3 | Corey, E.J. | *Tetrahedron Lett.* | **1979** | | 399 |
| 4 | Luzzio, F.A. | *Org. Prep. Proc. Int.* | **1988** | *20* | 559 |

**Isophorone 3.**[2] To a slurry of **1** (from 6M HCl, $CrO_3$ and pyridine at 0°C)[1] (4.30 g; 20 mmol) in $CH_2Cl_2$ (30 mL) was added in one portion **2** (1.40 g; 10 mmol) in $CH_2Cl_2$ (10 mL) at 20°C. After 3 h stirring, extraction ($Et_2O$), washing (5% NaOH, 5% HCl, $NaHCO_3$), evaporation and bulb to bulb distillation afforded 1.33 g of **3** (92%), bp 213-214°C.

## C O R E Y  Enantioselective Borane Reduction

Enantioselective reduction of ketones by borane or catecholborane catalyzed by oxazaborolidine **3** (see 1st edition).

| | | | 1 | | | 3 (94%)[2] | | | 5 (82%; 94% ee) | |
|---|---|---|---|---|---|---|---|---|---|---|

| 1 | Corey, E.J. | | | *J. Am. Chem. Soc.* | | | **1987** | *109* | 5551 |
|---|---|---|---|---|---|---|---|---|---|
| 2 | Corey, E.J. | | | *J. Org. Chem.* | | | **1988** | *53* | 2861 |
| 3 | Corey, E.J. | | | *Tetrahedron Lett.* | | | **1989** | *30* | 6275 |
| 4 | Corey, E.J. | | | *Tetrahedron Lett.* | | | **1990** | *31* | 611 |
| 5 | Todd, K.J. | | | *J. Org. Chem.* | | | **1991** | *56* | 763 |

*R*-(+)-3-**Chloro-1-phenyl-1-propanol 5.**[3] β-Chloropropiophenone **4** (0.162g; 1 mmol) in THF was added to 0.6 equiv. of $BH_3$ and 0.1 equiv. of **3** at 0°C in THF over 20 min. After 30 min, one adds MeOH and 1.2 equiv. of HCl in $Et_2O$, followed by removal of the volatiles. Addition of PhMe precipitated **1**. Concentration afforded 0.162 g of **5** (99%), 94%ee, recrystallyzed (hexane), mp 57-58°C, $[\alpha]_D^{25} = +24°$ (c=1, $CHCl_3$).

## C O R E Y - F U C H S  Alkynes Synthesis

Chain extension of aldehydes to 1,1-dibromoalkenes followed by elimination to alkynes by means of BuLi or RMgX.

$$n\text{-}C_6H_{13}\text{-CHO} \rightarrow n\text{-}C_6H_{13}\text{-CH=CBr}_2 \rightarrow n\text{-}C_6H_{13}\text{-C≡CH}$$

| 1 | Corey, E.J.; Fuchs, P.L. | | | *Tetrahedron Lett.* | | **1972** | | 3769 |
|---|---|---|---|---|---|---|---|---|
| 2 | Ma, P. | | | *Synth. Commun.* | | **1995** | *25* | 364 |

*D*-(+)-3,4-*O*-**isopropylidenebutyne-3,4-diol 3.**[2] To ice-cooled $Ph_3P$ (5.19 g; 19.8 mmol) in $CH_2Cl_2$ (11 mL) was added $CBr_4$ (3.29 g; 9.9 mmol) in $CH_2Cl_2$ (4 mL) below 15°C. At 0°C the aldehyde **1** (1 g; 7.63 mmol) and $Et_3N$ (1.06 mL; 7.63 mmol) in $CH_2Cl_2$ (1 mL) was added dropwise. After 30 min at 0°C hexane (10 mL) was added. Filtration, evaporation, dissolving the residue in hexane, filtration and concentration gave 1.96 g of **2** (95%), bp 70-72°C/0.5 mm. **2** (1.084 g; 4 mmol) in THF (2 mL) was treated with EtMgBr (1M in THF, 8 mL; 4 mmol) at 25-30°C. After 30 min quenching with solid $NH_4Cl$ (0.53 g) afforded after vacuum distillation 0.428 g of **3** (85%), bp 70°C/735 mm, $[\alpha]_D^{25} = 33.8°$ (c=1.01, $CHCl_3$).

## C O R E Y – N I C O L A O U – G E R L A C H Macrolactonization

2-Pyridinethiol a reagent in the synthesis of large ring lactones.

| 1 | Corey, E.J., Nicolaou, K.C. | J.Am.Chem.Soc. | **1974** | 96 | 5614 |
|---|---|---|---|---|---|
| 2 | Gerlach, H. | Helv.Chim.Acta | **1974** | 57 | 2306; 2661 |
| 3 | Corey, E.J. | Tetrahedron Lett. | **1976** | | 3409 |
| 4 | Green, A.E. | J.Am.Chem.Soc. | **1980** | 102 | 7583 |
| 5 | Nicolaou, K.C. | J.Am.Chem.Soc. | **1997** | 119 | 3421 |
| 6 | Nicolaou, K.C. | Angew.Chem.Int.Ed. | **1998** | 37 | 2714 |

**Lactone 3.**[1] The ω-hydroxy acid **1** (129 mg, 0.5 mmol), 2,2'-dipyridyl disulfide **2** (165 mg, 0.75 mmol) and triphenyl phosphine (197 mg, 0.75 mmol) were stirred for 5 h at 25°C in xylene under Ar. The reaction mixture was diluted with xylene (10 mL) and the resulting solution was added over 15 h to xylene (200 mL) under reflux and in an Ar atmosphere. After an additional 10 h reflux (GLC 10 ft, 10% silicone SE-30 column) the solvent was removed in vacuum and the residue was purified by preparative TLC (silica gel 10% Et₂O in pentane) to furnish 96 mg of **3** (80%) and 6 mg of dilactone **4** (5%).

## C O R E Y – S E E B A CH  Dithiane Reagents

Dithianes as acyl anion equivalents useful for synthesis of carbonyl compounds.

| 1 | Corey, E. J., Seebach, D. | *Angew. Chem. Int. Ed.* | **1965** | *4* | 1075;1077 |
|---|---|---|---|---|---|
| 2 | Corey, E. J., Seebach, D. | *J. Org. Chem.* | **1966** | *31* | 4097 |
| 3 | Corey, E. J., Seebach, D. | *J. Org. Chem.* | **1968** | *33* | 300 |
| 4 | Seebach, D. | *Synthesis* | **1969** | | 17 |
| 5 | Seebach, D. | *Synthesis* | **1977** | | 357 |
| 6 | Seebach, D. | *Angew. Chem. Int. Ed.* | **1979** | *18* | 239 |
| 7 | Seebach, D., Corey, E. J. | *Org. Synth.* | **1970** | *50* | 487 |

**1, 3-Dithiane 3.**[2] To a refluxing solution of $BF_3 \cdot Et_2O$ (10 mL) in AcOH (360 mL) and $CHCl_3$ (600 mL) under stirring, was added a solution of 1, 3-propandithiol **1** (150 mL, 1.5 mol) and methylal **2** (145 mL, 1.65 mol) in $CHCl_3$ (2.25 mL) at a constant rate over 8 h. Usual work up afforded after recrystallization from MeOH (300 mL), 130-140 g of **3** (70 %), mp 52-53 °C.

**2-(ω-Chloroalkyl)-1, 3-dithiane 5.** To a solution of **3** in THF at –40 °C are added n-BuLi (5.5 excess). Stirring was continued for 1-2 h at –25 °C. To this solution an equimolar amount of neat dihalide was added under $N_2$ at –50 °C. After 12 h at –20 °C, work up afforded **5** in 60-80 % yield.

**Chloro-aldehyde 6.** To $HgCl_2$ (2.18 g, 10.3 mmol) and $CaCO_3$ (1.68 g, 9.8 mmol) under $N_2$ was added **5** (4.92 mmol) in water (2.5 mL) and MeCN (47.5 mL). After 7.5 h stirring at 50 °C the mixture was concentrated to dryness. Extraction with $CHCl_3$, and evaporation of the solvent afforded **6** (80%).

## C O R E Y - W I N T E R - E A S T W O O D   Olefination of Diols

Alkene synthesis from glycols via cyclic 1,2-thionocarbonates (Corey-Winter) or 1,3-dioxolanes (Eastwood) (see 1st edition).

|     |     |     |
| --- | --- | --- |
| **1** | **2** | 3 (60%)[3] |

$$Ph-CH-CH-Ph \ + \ HC(OEt)_3 \ \xrightarrow[100\text{-}160°]{PhCOOH} \ Ph-CH=CH-Ph$$

**4**                                        5 (88%, mainly *E*)

| 1 | Corey, E.J. | *J. Am. Chem. Soc.* | **1963** | 85 | 2677 |
|---|---|---|---|---|---|
| 2 | Corey, E.J. | *J. Am. Chem. Soc.* | **1965** | 87 | 934 |
| 3 | Carr, R.I. | *Org. Prep. Proc. Int.* | **1990** | 22 | 245 |
| 4 | Eastwood, F.W. | *Austral. J. Chem.* | **1964** | 17 | 1392 |
| 5 | Eastwood, F.W. | *Austral. J. Chem.* | **1968** | 21 | 2013 |
| 6 | Eastwood, F.W. | *Tetrahedron Lett.* | **1970** |  | 5223 |

*E*-Stilbene **5**.[5] **4** (10 g; 46 mmol) and ethyl orthoformate (7.2 g; 48 mmol) was heated in the presence of PhCOOH (1 g; 8.2 mmol) for 2 h at 100-105°C. More PhCOOH was added and all was heated to 160-170°C then dissolved in $Et_2O$, washed (aq. $Na_2CO_3$) and evaporated. Extraction with hexane afforded 0.1 g of *Z*-**5** (1.2%). Evaporation gave 7.35 g of *E*-**5** (88%), bp 74-76°C/0.2 mm.

## C O R N F O R T H   Oxazole Rearrangement

Thermal rearrangement of 4-carbonyl substituted oxazoles via nitrilium ylide **2** (see 1st edition).

**1**                         **2**                         **3**

| 1 | Cornforth, J.W. | *The Chemistry of Penicillin* | **1949** | 689 | 705 |
|---|---|---|---|---|---|
| 2 | Dewar, M.J.S. | *J. Chem. Soc. Chem. Commun.* | **1973** |  | 925 |
| 3 | Dewar, M.J.S. | *J. Am. Chem. Soc.* | **1974** | 96 | 6148 |
| 4 | Dewar, M.J.S. | *J. Org. Chem.* | **1975** | 40 | 1521 |
| 5 | L'Abbé, G. | *J. Chem. Soc. Perkin I* | **1993** |  | 2259 |

## CRABBÉ Allene Synthesis

Synthesis of terminal allenes from propargylic acetates.

|   |              |                              |      |    |      |
|---|--------------|------------------------------|------|----|------|
| 1 | Crabbé, P.   | *J. Chem. Soc. Chem. Commun.* | **1976** |    | 183  |
| 2 | Nantz, M.H.  | *Synthesis*                  | **1993** |    | 577  |
| 3 | Niemstra, H. | *J. Org. Chem.*              | **1997** | 62 | 8862 |

**Allene 2.**[2] To a suspension of PhSCu (4.41 g; 25.5 mmol) in $Et_2O$ (100 mL) at -35°C was added 2.47M BuLi in hexane (9.91 mL; 2.45 mmol). After 20 min at -30°C **2** (2.26 g; 10.2 mmol) in $Et_2O$ (35 mL) was added dropwise at -78°C. After 1 h stirring the mixture was quenched with 2 mL of sat. $NH_4Cl$ at a rate of 0.16 mL/min. After 6 h at -78°C and warming to 20°C, the solids were filtered. The organic phase after washing, evaporation and chromatography afforded 1.51 g of **2** (91%).

## CRIEGEE Glycol Oxidation

Oxidation of 1,2-glycols to two carbonyl moieties by lead tetraacetate (LTA) (see 1st edition).

|   |                   |                    |      |     |      |
|---|-------------------|--------------------|------|-----|------|
| 1 | Criegee, R.       | *Liebigs Ann.*     | 1930 | 481 | 263  |
| 2 | Criegee, R.       | *Chem. Ber.*       | 1931 | 64  | 260  |
| 3 | Chi-yi, H.        | *J. Am. Chem. Soc.* | 1939 | 61  | 3589 |
| 4 | Criegee, R.       | *Angew. Chem.*     | 1958 | 70  | 173  |
| 5 | Michailovici, M.L. | *Synthesis*        | 1970 |     | 209  |
| 6 | Nakajima, N.      | *Chem. Ber.*       | 1956 | 89  | 2274 |

## CRIEGEE Rearrangement

Rearrangement of hydroperoxides to ester ketals or 1,3-diols.

| 1 | Criegee, R. | *Chem. Ber.* | **1944** | *77* | 722 |
|---|-------------|--------------|----------|------|-----|
| 2 | Criegee, R. | *Liebigs Ann.* | **1948** | *560* | 127 |
| 3 | Brückner, R. | *Synlett* | **1993** | | 901 |
| 4 | Kishi, Y. | *J. Org. Chem.* | **1994** | *59* | 5125 |

**Triol 3.**[3] A solution of **1** (330 mg; 0.84 mmol) in THF (5 mL) was treated at -78°C with MeLi (1.04 mL; 1.62 mmol) in Et$_2$O. After 15 min a sat. solution of NaHCO$_3$ was added, followed by extraction with tBuOMe. After evaporation of the solvent, the residue was treated with H$_2$O$_2$ (0.3 mL of 85%) and a catalytic amount of pyridinium *p*-toluene-sulfonate in THF (4 mL). After 20 min the mixture was extracted with petroleum ether and the crude hydroperoxide **2** was dissolved in THF (3 mL). Et$_3$N (0.34 mL), *p*-nitrobenzenesulfonylchloride (197 mg; 0.888 mmol) were added and after 29 min the mixture was diluted with tBuOMe and washed with NaHCO$_3$ sol. The solvent was exchanged with THF, Et$_3$N (0.78 mL; 5.7 mmol), Ac$_2$O (0.4 mL; 4 mmol) and a catalytic amount of DMAP were added. After 2 h, work up and chromatography (silica gel, tBuOMe:petroleum ether 1:10) afforded 303 mg of **3** (80%).

## C U R T I U S  Rearrangement

Degradation of acid hydrazides or acyl azides to amines or amine derivatives (see 1st edition).

| | | | | | |
|---|---|---|---|---|---|
| 1 | Curtius, T. | *Chem. Ber.* | **1890** | *23* | 3023 |
| 2 | Caldwell, W.T. | *J. Am. Chem. Soc.* | **1939** | *61* | 3584 |
| 3 | Newcastle, G.W. | *Synthesis* | **1985** | | 220 |
| 4 | Thornton, T.J. | *Synthesis* | **1990** | | 295 |
| 5 | Saunders, J.M. | *Chem. Rev.* | **1948** | *43* | 205 |
| 6 | Cohen, L.D. | *Angew. Chem.* | **1961** | *73* | 259 |
| 7 | Smith, P.A.S. | *Org. React.* | **1946** | *3* | 337 |
| 8 | Pfister, J.R. | *Synthesis* | **1983** | | 39 |

**3,5-Dimethoxyaniline 4.**[8] **1** (5.65 g; 28 mmol) in $CH_2Cl_2$ (50 mL) and TBAB (20 mg) were cooled and treated with $NaN_3$ (2.5 g; 38.5 mmol) in $H_2O$ (10 mL) with stirring over 2 h at 0°C. After extraction ($Et_2O$), the extract was added to TFA (2.5 mL; 43 mmol) and refluxed for 40 h to give 5.63 g of **3** (80%), mp 99°C. **3** (4.5 g; 18 mmol), $K_2CO_3$ (4.2 g; 30 mmol) and water (80 mL) were stirred under $N_2$ for 20 h at 20°C. Work up and distillation gave 2.6 g of **4** (94%), bp 85-110°C/0.2 torr, mp 48°C.

## D A N H E I S E R  Annulation

Regiocontrolled synthesis of five membered rings from silylallenes and Michael acceptors in the presence of $TiCl_4$ (see 1st edition).

| | | | | | |
|---|---|---|---|---|---|
| 1 | Danheiser, R.L. | *J. Am. Chem. Soc.* | **1981** | *103* | 1604 |
| 2 | Danheiser, R.L. | *Tetrahedron* | **1983** | *39* | 935 |
| 3 | Danheiser, R.L. | *Org. Synth.* | **1988** | *66* | 8 |

**Cyclopentene 3.**[1] $TiCl_4$ (0.283 g; 1.5 mmol) was added to **1** (0.126 g; 1 mmol) and **2** (0.07 g; 1 mmol) in $CH_2Cl_2$ at -78°C. The mixture was stirred for 1 h at -78°C. Work up and chromatography afforded 0.125-0.144 g of **3** (68-75%).

## D A K I N   Phenol Oxidation

Oxidation of aldo- or keto-phenols to polyphenols by $H_2O_2$ (a Bayer-Villiger oxidation) (see 1st edition).

| 1 | Dakin, H.D. | *Am. Chem. J.* | **1909** | 42 | 477 |
|---|---|---|---|---|---|
| 2 | Baker, J. | *J. Chem. Soc.* | **1953** | | 1615 |
| 3 | Criegee, R. | *Liebigs Ann.* | **1948** | 560 | 127 |
| 4 | Seshadri, T.R. | *J. Chem. Soc.* | **1959** | | 1660 |
| 5 | Rosenblat, D.H. | *J. Am. Chem. Soc.* | **1953** | 75 | 4607 |
| 6 | Jung, M.E. | *J. Org. Chem.* | **1997** | 62 | 1553 |
| 7 | Lee, J.B. | *Quart. Rev.* | **1969** | 21 | 454 |
| 8 | Varma, R.S. | *Org. Lett.* | **1999** | 1 | 189 |

**Phenol 2.**[6] To **1** (96 mg; 0.24 mmol) in $CH_2Cl_2$ (3 mL) were added (PhSe)$_2$ (3 mg; 0.01 mmol) and 30% $H_2O_2$ (0.062 mL; 0.614 mmol). After 18 h stirring at 20°C water and EtOAc were added and the organic layer was evaporated. The residue in 3 mL MeOH was treated with $NH_3$ to give 73 mg of **2** (78%).

## D A K I N - W E S T Ketone Synthesis

An acylative decarboxylation of α-amino or α-thio acids (see 1st edition).

| 1 | Dakin, H.; West, R. | *J. Biol. Chem.* | **1928** | 78 | 91 |
|---|---|---|---|---|---|
| 2 | Dyer, E. | *J. Org. Chem.* | **1968** | 33 | 880 |
| 3 | Buchanan, G.L. | *Chem. Soc. Rev.* | **1988** | 17 | 91 |
| 4 | Fischer, L.E. | *Org. Prep. Proc. Int.* | **1990** | 22 | 467 |
| 5 | Kawase, M. | *J. Chem. Soc. Chem. Commun.* | **1998** | | 641 |

**Purine 2.**[2] A suspension of acid **1** (1.0 g; 4.4 mmol) in $Ac_2O$ (30 mL) was refluxed for 5 h and stirred overnight at 20°C. The residue on evaporation was triturated with $Et_2O$, dried (KOH) and extracted (hexane, 9x40 mL) to afford 0.66 g of **2** (57%), mp 98-99°C.

## D A N I S H E F S K Y  Dienes

Silyloxydienes in regio- and stereo-controlled Diels-Alder and hetero Diels-Alder reactions (see 1st edition).

**2** (72%)[1]  R: H

(91%)[4]

| 1 | Danishefsky, S. | *J. Am. Chem. Soc.* | **1974** | *96* | 7807 |
|---|---|---|---|---|---|
| 2 | Danishefsky, S. | *J. Am. Chem. Soc.* | **1978** | *100* | 6536; 7098 |
| 3 | Danishefsky, S. | *J. Am. Chem. Soc.* | **1982** | *104* | 6457 |
| 4 | Vorndam, P.E. | *J. Org. Chem.* | **1990** | *55* | 3693 |
| 5 | Nakagawa, N.; Aino, T. | *J. Org. Chem.* | **1992** | *57* | 5741 |
| 6 | Cativiela, C. | *Synthesis* | **1995** | | 671 |
| 7 | Danishefsky, S. | *Acc. Chem. Res.* | **1981** | *14* | 400 |

**3-Phenyl-4-benzamidophenol 6.**[6] Danishefsky diene 1 (468 mg; 4 mmol) was added to oxazolone 3 (474 mg; 2 mmol) in PhH (25 mL) and the mixture was refluxed for 48 h with stirring. After evaporation the cycloadducts 4 and 5 were treated with 0.005N HCl in 20 mL THF (1:4) for 7 h at 20°C. Work up and chromatography (silica gel, hexane:EtOAc 1:1) gave 410 mg of 6 (71%).

## DARZENS Epoxide Synthesis

Synthesis of glycidic esters, amides or ketones from an aldehyde or ketone and an α-haloester, amide or ketone (see 1st edition).

$$\text{Ph-CHO} + \text{ClCH}_2\text{-CONEt}_2 \xrightarrow[10°-50°]{\text{t-BuOK; t-BuOH}}$$

| 1 | 2 | 3 (88%) |

**4** → **6 (88%; 43% ee)**

cat. **5**

| 1 | Darzens, G. | *C. R.* | **1904** | *139* | 1214 |
|---|---|---|---|---|---|
| 2 | Tung, T.T. | *J. Org. Chem.* | **1963** | *28* | 1514 |
| 3 | Gladiale, S. | *Synth. Commun.* | **1982** | *12* | 355 |
| 4 | Corey, E.J. | *Tetrahedron Lett.* | **1991** | *32* | 2857 |
| 5 | Pridgen, L.N. | *J. Org. Chem.* | **1993** | *58* | 5107 |
| 6 | Maillard, B. | *J. Org. Chem.* | **1994** | *59* | 4765 |
| 7 | Töke, L. | *Synlett* | **1997** | | 291 |
| 8 | Balester, M. | *Chem. Rev.* | **1955** | *55* | 283 |
| 9 | Newman, M.S. | *Org. React.* | **1949** | *5* | 414 |

**cis- and trans-Epoxide 3.**[2] tBuOK (K, 16 g; t-BuOH, 400 mL) was added to a mixture of **1** (42.4 g; 0.4 mol) and **2** (59.8 g; 0.4 mol) under $N_2$ at 10°C over 90 min. After stirring the solvent was removed at 50°C. Work up gave a viscous oil (87.1 g; 99%) which treated with $Et_2O$ (150 mL) and hexane (300 mL) gave 77 g of **3** (88.4%), mp 43-47°C.

**1-Benzoyl-2-phenylethene oxide 6.**[7] A toluene solution of phenacyl chloride **4** (0.2 g; 1.3 mmol) was treated with PhCHO **1** (0.2 g; 1.9 mmol) and catalyst **5** (0.1 mmol) in 30% NaOH (0.6 mL). The mixture was stirred for 4 h at 20°C under Ar. Usual work up followed by chromatography (preparative TLC, $CH_2Cl_2$) gave 262 mg of **6** (90%; 43% ee).

## D A V I E S   Asymmetric synthesis

Iron chiral auxiliary for asymmetric aldol reaction, Michael addition, β-amino acid and β-lactam synthesis.

| 1 | Davies, S.G. | *Chem. Commun.* | **1982** | | 1303 |
| 2 | Davies, S.G. | *Chem. Commun.* | **1985** | | 607 |
| 3 | Davies, S.G. | *J. Organometal. Chem.* | **1985** | 296 | C40 |
| 4 | Davies, S.G. | *Tetrahedron* | **1986** | 42 | 1759 |
| 5 | Davies, S.G. | *Tetrahedron* | **1986** | 42 | 5123 |
| 6 | Davies, S.G. | *Aldrichimica Acta* | **1990** | 23 | 31 |

For synthesis of **1** see ref. 3 and 4.

(*RR/SS*)-[(η$^5$-C$_5$H$_5$)Fe(CO)(PPh$_3$)COCH$_2$CH(Me)NHCH$_2$Ph] **2**.$^5$ n-BuLi (0.4 mL; 0.64 mmol) was added to PhCH$_2$NH$_2$ (70 mg; 0.66 mmol) in THF (20 mL) at -20°C to give a purple solution. After 1 h stirring at -20°C this was added to **1** (250 mg; 0.52 mmol) in THF (30 mL) at -78°C. MeOH (66.5 mg; 2.08 mmol) was added and the mixture further stirred 1 h at -78°C. After evaporation of the solvent, the residue dissolved in CH$_2$Cl$_2$ was filtered through Celite and chromatographed (Alumina I, CH$_2$Cl$_2$:EtOAc:MeOH 10:9:1) to afford 690 mg of **2** in 90% single diastereoisomer, [α]$_D^{21}$ = +143.0°.

(**4S**)-(-)-**4-Methyl-N-benzyl-β-lactam 3**. Oxidation of **2** with Br$_2$ in CH$_2$Cl$_2$ at -40°C followed by chromatography on silica gel (Merck 60 H), hexane:Et$_2$O 2:1 gave the iron complex. Elution with the same solvents 1:2 gave 106 mg of **3** (65%), [α]$_D^{21}$ = -38.5° (c 2.1, MeOH).

## D A V I S  Oxidizing Reagent

2-Sulfonyloxaziridines as aprotic neutral oxidizing reagents in oxidation of amines, sulfides, selenides and asymmetric oxidation (see 1st edition).

| | | | | | |
|---|---|---|---|---|---|
| 1 | Davis, F.A. | *J. Org. Chem.* | **1982** | *47* | 1174 |
| 2 | Davis, F.A. | *Tetrahedron Lett.* | **1983** | *24* | 1213 |
| 3 | Davis, F.A. | *J. Org. Chem.* | **1986** | *51* | 4083; 4240 |
| 4 | Zajak, W.W. | *J. Org. Chem.* | **1988** | *53* | 5856 |
| 5 | Davis, F.A. | *J. Org. Chem.* | **1990** | *55* | 3715 |
| 6 | Davis, F.A. | *J. Am. Chem. Soc.* | **1990** | *112* | 6679 |
| 7 | Chen, D.C. | *Org. Prep. Proc. Int.* | **1996** | *28* | 115 |
| 8 | Dimitrenco, G.I. | *J. Am. Chem. Soc.* | **1997** | *119* | 1159 |

***cis*-4-(Nitromethyl)cyclohexanecarboxylic acid 3.**[4] To a solution of 2-(phenylsulfonyl)-3-phenyloxaziridine **2** (0.523 g; 2.0 mmol) in CHCl₃ (10 mL) was added 3-azabicyclo[3.2.2]nonane **1** (0.125 g; 1 mmol). The reaction mixture was stirred for 15 min, then the solvent was removed by rotary evaporation and replaced by CH₂Cl₂. This solution was ozonized at -78°C. The CH₂Cl₂ solution was then extracted with saturated NaHCO₃ solution. The aqueous layer was neutralized with HCl and then extracted with CH₂Cl₂. The CH₂Cl₂ solution was rotary evaporated and the residue subjected to PLC. The major fraction that was isolated was recrystallized from EtOH to provide 0.123 g of **3** (66%), mp 83-85°C.

## DAVID-MUKAIYAMA-UENO  Selective Diol Oxidation

Regiospecific oxidation of diols to ketoalcohols by $Br_2$ via Sn derivatives.

$$Ph\text{-}CH\text{-}CH_2OH \xrightarrow[CH_2Cl_2;\ 20°]{(Bu_3Sn)_2O;\ Br_2} Ph\text{-}CH\text{-}CH_2OH$$

          OH                                    O
          **1**                              **2 (76%)**

          **3**                **4**                **5 (72%)**

| 1 | Mukaiyama, T. | *Chem. Lett.* | **1975** |  | 145 |
|---|---|---|---|---|---|
| 2 | Mukaiyama, T. | *Bull. Soc. Chim. Japan* | **1976** | *49* | 1656 |
| 3 | Ueno, Y. | *Tetrahedron Lett.* | **1976** |  | 4597 |
| 4 | David, S. | *Nouveau J. Chem.* | **1979** | *3* | 63 |
| 5 | David, S. | *C. R. Acad. Sci. Paris (C)* | **1974** | *278* | 1051 |
| 6 | David, S. | *J. Chem. Soc. Perkin I* | **1979** |  | 1568 |

**Hydroxyacetophenone 2.**[3] To **1** (570 mg; 4 mmol) and hexabutyl-distannoxane (2.7 mL; 5.2 mmol) in $CH_2Cl_2$ was added dropwise $Br_2$ (0.27 mL; 5.2 mmol) in $CH_2Cl_2$ (5 mL) under Ar. After 3 h stirring evaporation and crystallization gave 410 mg of **2** (76%), mp 84-86°C.

## DAVID-THIEFFRY  Monophenylation of Diols

Selective phenylation of one hydroxyl group of glycols by triphenylbismuth diacetate.

$$(\pm)\ Me\text{-}CH\text{-}CH\text{-}Me \xrightarrow[]{Ph_3Bi(OAc)_2 \quad 3} (\pm)\ Me\text{-}CH\text{-}CH\text{-}Me$$

          OH OH                                    OH OPh
          **1**                              **2 (86%)**

          **6**                **7 (60%)**

| 1 | David, S.; Thieffry, A. | *Tetrahedron Lett.* | **1981** | 22 | 2885 |
|---|---|---|---|---|---|
| 2 | David, S.; Thieffry, A. | *Tetrahedron Lett.* | **1981** | 22 | 5063 |
| 3 | David, S.; Thieffry, A. | *J. Org. Chem.* | **1983** | 48 | 441 |

**3-Phenoxybutan-2-ol 2.**[3] **1** (90 mg; 1 mmol), triphenylbismuth diacetate **3** (558 mg; 1 mmol) in $CH_2Cl_2$ (5 mL) were refluxed for 4-5 h (TLC). Evaporation and chromatography afforded 142 mg of **2** (86%).

## **D A V I D S O N** Oxazole Synthesis

Synthesis of triaryloxazoles from α-hydroxyketones (see 1st edition).

| | | | | | | |
|---|---|---|---|---|---|---|
| 1 | Davidson, D. | *J. Org. Chem.* | **1937** | 2 | | 328 |
| 2 | Cornforth, J.W. | *J. Chem. Soc.* | **1953** | | | 93 |
| 3 | Theilig, S. | *Chem. Ber.* | **1953** | 86 | | 96 |
| 4 | Budevich, M. | *Chem. Ber.* | **1954** | 87 | | 700 |
| 5 | Willey, R.H. | *Chem. Rev.* | **1945** | | | 93 |

## **D I M R O T H** Rearrangement

Migration of an alkyl or aryl group from a heterocyclic to an exocyclic N (first descovery by Rathke) (see 1st edition).

| | | | | | |
|---|---|---|---|---|---|
| 1 | Rathke, B. | *Chem. Ber.* | **1888** | 21 | 867 |
| 2 | Dimroth, O. | *Liebigs Ann.* | **1909** | 364 | 183 |
| 3 | Brown, D.J. | *J. Chem. Soc.* | **1963** | | 1276 |
| 4 | Brown, D.J. | *Nature* | **1961** | 189 | 828 |
| 5 | Korbonits, D. | *J. Chem. Soc.* | **1986** | | 2163 |
| 6 | Katritzky, A.R. | *J. Org. Chem.* | **1992** | 57 | 190 |
| 7 | Saito, T. | *Chem. Pharm. Bull.* | **1993** | 41 | 1850 |
| 8 | Loakes, D. | *J. Chem. Soc. Perkin I* | **1999** | 1 | 1333 |

**2-(Ethylamino)pyrimidine 3.**[3] **2** (0.25 g; 1 mmol) in 1N NaOH (10 mL) was heated for 15 min on a water bath. The pH was corrected to 5 and all was added to a picric acid solution to afford 0.23 g of picrate **3** (70%), mp 167°C.

## DE KIMPE Amidine Synthesis

Conversion of aldehydes to keteneimines (see **6**) and amidines (see **7**) via α-cyano-enamines.

| 1 | De Kimpe, N. | *Tetrahedron* | **1976** | *32* | 3063 |
| 2 | De Kimpe, N. | *Synthesis* | **1978** | | 895 |
| 3 | De Kimpe, N. | *J. Org. Chem.* | **1978** | *43* | 2670 |
| 4 | De Kimpe, N. | *Synth. Commun.* | **1979** | *9* | 901 |
| 5 | De Kimpe, N. | *Chem. Ber.* | **1983** | *116* | 3846 |
| 6 | De Kimpe, N. | *Can. J. Chem.* | **1984** | *62* | 1812 |

**2-Isopropylimino-3-methylbutanenitrile 4.**[2] NaHSO$_3$ (10.9 g; 105 mmol) in water (50 mL) was added with stirring to **1** (7.1 g; 100 mmol). After 2 h at 20°C, KCN (14.3 g; 220 mmol) in water (25 mL) was added and stirring was continued for 5 h. Extraction with Et$_2$O and vacuum distillation afforded 10 g of **2** (72%), bp 75-76°C/13 torr. To a solution of **2** (10 g; 70 mmol) in PhH (100 mL) at 0°C was added a solution of tBuOCl (8.7 g; 80 mmol) in PhH (15 mL). After 1 h stirring at 0°C Et$_3$N (8.4 g; 84 mmol) or the same amount of DABCO was added. Stirring was continued 1 h at 20°C and 18 h at 50°C. Usual work up afforded 5.9 g of **4** (61%), bp 47°C/12 torr.

**N$^1$-Phenyl-N$^2$-isopropyl-2-methylpropanamidine 7.**[3] A solution of **4** (6.9 g; 50 mmol) in Et$_2$O was treated with MeMgI (87.5 mmol) in Et$_2$O followed by quenching (NH$_4$Cl) and extraction to give keteneimine **6**. This with PhNH$_2$ (4.5 g; 50 mmol) afforded 6.15 g of amidine **7** (60%).

## DE MAYO Photocycloaddition

Photochemical 2+2 cycloaddition (see 1st edition).

| | | | | | |
|---|---|---|---|---|---|
| 1 | De Mayo, P. | *Proc. Chem. Soc. London* | **1962** | | 119 |
| 2 | De Mayo, P. | *Can. J. Chem.* | **1962** | *41* | 440 |
| 3 | De Mayo, P. | *J. Org. Chem.* | **1969** | *34* | 794 |
| 4 | De Mayo, P. | *Acc. Chem. Res.* | **1971** | *4* | 41 |
| 5 | Weedon, A.C. | *The Chemistry of Enols (Wiley)* | **1990** | | 591 |

## DESS-MARTIN Oxidizing Reagent

Oxidation of alcohols to aldehydes or ketones by means of periodinanes, e.g. 1 (see 1st edition).

| | | | | | |
|---|---|---|---|---|---|
| 1 | Dess, P.B.; Martin, J.C. | *J. Am. Chem. Soc.* | **1978** | *100* | 300 |
| 2 | Dess, P.B.; Martin, J.C. | *J. Am. Chem. Soc.* | **1979** | *101* | 5294 |
| 3 | Yagupolsky, L.M. | *Synthesis* | **1977** | | 574 |
| 4 | Dess, P.B.; Martin, J.C. | *J. Org. Chem.* | **1983** | *48* | 4155 |
| 5 | Robins, J.C. | *J. Org. Chem.* | **1990** | *55* | 5186 |
| 6 | Wipf, P. | *Synlett* | **1997** | | 1 |

**Formylaziridine 3.**[6] **2** (1.15 g; 4.76 mmol) in $CH_2Cl_2$ (24 mL) was added to a suspension of **1**[4] (2.35 g; 5.7 mmol) in $CH_2Cl_2$ (24 mL). After 1 h stirring at 20°C, usual work up and chromatography (silica gel, 28% EtOAc in hexane) afforded 0.91 g of **3** (80%).

## D E L E P I N E   Amine Synthesis

Synthesis of primary amines from alkyl halides with hexamethylenetetramines (see 1st edition).

$$PhCH_2Br \; + \; 3 \; \xrightarrow{\text{NaI}} \; PhCH_2(N_4C_6H_{12})^+ \; I^- \; \xrightarrow[\text{2) NaOH}]{\text{1) HCl(g)}} \; PhCH_2NH_2$$
$$(82\%)$$

| 1 | Delepine, M. | *Bull. Soc. Chim. Fr.* | **1885** | *13* | 356 |
|---|---|---|---|---|---|
| 2 | Galat, A. | *J. Am. Chem. Soc.* | **1939** | *61* | 3585 |
| 3 | Henry, A. | *J. Org. Chem.* | **1990** | *55* | 1796 |
| 4 | Angyal, S.T. | *Org. Synth.* | Coll. Vol. | *IV* | 121 |


## D E M J A N O V   Rearrangement

Deamination of primary amines to rearranged alcohols (via diazonium compounds) with ring contraction or enlargement for alicyclic amines (see 1st edition).

| 1 | Demjanov, N.J. | *J. Russ. Phys. Chem. Soc.* | **1903** | *35* | 26 |
|---|---|---|---|---|---|
| 2 | Kottany, R. | *J. Org. Chem.* | **1965** | *30* | 350 |
| 3 | Smith, P.A. | *Org. React.* | **1960** | *11* | 154 |

## DIELS-ALDER  Cyclohexene Synthesis

4+2 Thermal cycloaddition between a diene and an activated alkene or alkyne, sometimes catalyzed by Lewis acids (see 1st edition).

| 1 | Diels, O.; Alder, K. | *Liebigs Ann.* | **1928** | *460* | 98 |
|---|----------------------|-----------------------|----------|-------|------|
| 2 | House, H.O. | *J. Org. Chem.* | **1963** | *28* | 27 |
| 3 | Johnson, C.R. | *J. Org. Chem.* | **1987** | *52* | 1493 |
| 4 | Wenkert, E. | *Chem. Rev.* | **1990** | *22* | 131 |
| 5 | Waldmann, H. | *Tetrahedron Asymm.* | **1991** | *2* | 1231 |
| 6 | Jorgensen, K.A. | *J. Org. Chem.* | **1995** | *60* | 6851 |
| 7 | Fowler, F.W. | *J. Org. Chem.* | **1997** | *62* | 2093 |
| 8 | Oppolzer, W. | *Angew. Chem.* | **1984** | *96* | 840 |
| 9 | Boger, D.L. | *Chem. Rev.* | **1986** | *86* | 781 |
| 10 | Bieker, W. | *Tetrahedron Lett.* | **2001** | *42* | 419 |

**Indolizines 5 and 6.**[7] **4** (100 mg; 0.6 mmol) in PhH (4 mL) in a thick-walled glass tube, under Ar was heated (oil bath, 110°C) with stirring for 24 h. The residue obtained after evaporation was chromatography (silica gel, heptane:Et$_2$O 1:1) afforded **5** and **6** (4:1), 94 mg (94%).

## D I M R O T H  Triazole Synthesis

Synthesis of 1,2,3-triazoles from alkyl or aryl azides and active methylene compounds.

| | | | | | |
|---|---|---|---|---|---|
| 1 | Dimroth, O. | *Chem. Ber.* | **1902** | *36* | 1029; 4041 |
| 2 | Hoover, J.R.E. | *J. Am. Chem. Soc.* | **1956** | *78* | 5832 |
| 3 | L'abbé, G. | *Ind. Chim. Belge* | **1971** | *36* | 3 |
| 4 | Olsen, C.E. | *Tetrahedron Lett.* | **1968** | | 3805 |
| 5 | Tolman, R.L. | *J. Am. Chem. Soc.* | **1972** | *94* | 2530 |
| 6 | L'abbé, G. | *Angew. Chem. Int. Ed.* | **1975** | *14* | 779 |

**Triazole 3.**[2] To Na (4.6 g; 0.2 atg) in MeOH (500 mL) were added cyanoacetamide **1** (16.82 g; 0.2 mol) and benzyl azide **2** (26.6 g; 0.2 mol). After 1 h reflux, the mixture was cooled to afford 35 g of **3** (81%), mp 230-232°C.

## D J E R A S S I - R Y L A N D E R  Oxidation

RuO$_4$ in oxidative cleavage of phenols or alkenes, oxidation of aromatics to quinones, oxidation of alkyl amides to imides or of ethers to esters (see 1st edition).

| | | | | | |
|---|---|---|---|---|---|
| 1 | Djerassi, C.; Engle, R.R. | *J. Am. Chem. Soc.* | **1953** | *75* | 3838 |
| 2 | Pappo, R.; Becker, A. | *Bull. Res. Council Isr.* | **1956** | *A5* | 300 |
| 3 | Rylander, P.N. | *J. Am. Chem. Soc.* | **1958** | *80* | 6682 |
| 4 | Caputo, J.A. | *Tetrahedron Lett.* | **1962** | | 2729 |
| 5 | Caspi, E. | *J. Org. Chem.* | **1969** | *34* | 112; 116 |
| 6 | Tanaka, K. | *Chem. Pharm. Bull.* | **1987** | *35* | 364 |
| 7 | Tamura, O. | *Synlett* | **2000** | | 1553 |

**Imide 4.**[6] **3** (1.04 g; 6 mmol) in EtOAc (20 mL) was added to RuO$_4$·H$_2$O (100 mg) and 10% NaIO$_4$ (30 mL) under vigorous stirring at 20°C (TLC). Extraction with EtOAc, addition of iPrOH, filtration of RuO$_2$ gave 1.054 g of **4** (96%).

## D O E B N E R - M I L L E R Quinoline Synthesis

Quinoline synthesis from anilines and aldehydes (see 1st edition).

| | | | | | |
|---|---|---|---|---|---|
| 1 | Doebner, O.; Miller, W. | *Ber.* | **1883** | *16* | 2464 |
| 2 | Leir, C.M. | *J. Org. Chem.* | **1977** | *42* | 911 |
| 3 | Corey, J.E. | *J. Am. Chem. Soc.* | **1981** | *103* | 5599 |
| 4 | Bergstom, F.W. | *Chem. Rev.* | **1944** | *35* | 153 |
| 5 | Johnson, W.S. | *J. Am. Chem. Soc.* | **1944** | *66* | 210 |

## D O E R I N G - L A F L A M M E Allene Synthesis

Allene synthesis from olefins via gem-dihalocyclopropanes (see 1st edition).

| | | | | | |
|---|---|---|---|---|---|
| 1 | Doering, v.W. | *J. Am. Chem. Soc.* | **1954** | *76* | 6162 |
| 2 | La Flamme, P.M. | *Tetrahedron* | **1958** | *2* | 75 |
| 3 | Moore, W.R. | *J. Org. Chem.* | **1962** | *27* | 4182 |
| 4 | Chinoporos, E. | *Chem. Rev.* | **1963** | *63* | 235 |

**1,1,3-Trimethyl-2,2-dibromo-cyclopropane 2.**[1,2] To a solution of 2-methyl-2-butene **1** (14.0 g; 0.2 mol) in a solution of KOtBu (22.4 g; 0.2 mol) in tBuOH was added under stirring and cooling CHBr$_3$ (50.6 g; 0.2 mol). The mixture was poured into water, extracted with pentane and distilled to give 24.4 g of **2** (50%), bp 63-65°C/15 mm.

**2-Methyl-2,3-pentadiene 3.**[1,2] **2** (24.4 g; 0.1 mol) in THF (50 mL) was added to Mg turnings (4.86 g; 0.2 atg) in THF. Hydrolysis with water and fractionation afforded 2.75 g of **3** (34%), bp 72.5°C.

## D O N D O N I Homologation

Homologation of aldehydes, ketones, acyl chlorides via 2-(trimethylsilyl) thiazole addition, also two carbon homologation (see 1st edition).

| 1 | Dondoni, A. | *Angew. Chem. Int. Ed.* | **1986** | 25 | 835 |
|---|---|---|---|---|---|
| 2 | Dondoni, A. | *J. Org. Chem.* | **1989** | 54 | 693 |
| 2 | Dondoni, A. | *J. Org. Chem.* | **1997** | 62 | 6261 |
| 3 | Dondoni, A. | *Synthesis* | **1998** | 1681 | |
| 4 | Dondoni, A. | *J. Chem. Soc. Chem. Commun.* | **1999** | 2133 | |
| 5 | Vasella, A. | *Helv. Chim. Acta.* | **1998** | 81 | 889 |
| 6 | Nicolaou, A. | *Angew. Chem. Int. Ed.* | **1999** | 38 | 3345 |

**1,3,4,6-Tetra-*O*-acetyl-2-*O*-benzyl-L-gulopyranose (5).**[3] To a cooled (-20 °C), stirred solution of crude *aldehydo*-L-xylose diacetonide **3** (3.53 g, ca. 15.3 mmol) in anhydrous CH$_2$Cl$_2$ (60 mL) was added 2-(trimethylsilyl) thiazole **2** (3.2 mL,19.9 mmol) during 15 min. The solution was stirred at 0 °C for an additional hour and concentrated. A solution of the residue in anhydrous THF (60 mL) was treated with *n*–Bu$_4$NF.3H$_2$O (4.48 g, 15.3 mmol) at room temperature for 30 min and then concentrated. The residue was dissolved in CH$_2$Cl$_2$ (300 mL), washed with H$_2$O (3×50 mL), dried (Na$_2$SO$_4$), and concentrated to give the *anti* adduct **4** (4.50 g, 80% from **3**) containing 5% of the *syn* isomer. Crystallization of the crude product from AcOEt-cyclohexane afforded pure **4** (3.42 g, 61% from **3**). The transformation of **4** to **5** was carried out by the following reaction sequence: a) benzylation (BnBr, NaH, DMF); b) aldehyde liberation by cleavage of the thiazole ring (*N*-methylation, reduction, hydrolysis); c) deacetonization (AcOH, H$_2$O); d) exhaustive acetylation (Ac$_2$O).

## D Ö T Z  Hydroquinone Synthesis

Hydroquinone synthesis (regiospecific) from alkynes and carbonyl carbene chromium complexes (see 1st edition).

| 1 | Dötz, K.H. | *Angew. Chem. Int. Ed.* | **1975** | *14* | 644 |
|---|---|---|---|---|---|
| 2 | Dötz, K.H. | *Chem. Ber.* | **1988** | *121* | 665 |
| 3 | Hofmann, P. | *Angew. Chem. Int. Ed.* | **1989** | *28* | 908 |
| 4 | Dötz, K.H. | *New J. Chem.* | **1990** | *14* | 433 |
| 5 | Dötz, K.H. | *Synlett* | **1991** | | 381 |
| 6 | Schmaltz, H.G. | *Angew. Chem. Int. Ed.* | **1994** | *33* | 303 |

**DOWD** Ring Expansion

Ring expansion of cyclic ketones mediated by free radicals.

| 1 | Dowd, P. | *J.Am.Chem.Soc.* | 1987 | *109* | 3493 |
|---|----------|------------------|------|-------|------|
| 2 | Dowd, P. | *Tetrahedron* | 1989 | *45* | 77 |
| 3 | Dowd, P. | *J.Org.Chem.* | 1992 | *52* | 7163 |
| 4 | Dowd, P. | *Chem.Rev.* | 1993 | *93* | 2091 |

**Methyl 2-Bromomethylcyclopentanone-2-carboxylate 3.**[2] A solution of 2-carbomethoxycyclopentanone **1** (0.43 g, 3 mmol) in THF (2 mL) was added to a suspension of NaH (127 mg, 3.6 mmol) in THF (5mL) containing HMPA (645 mg, 3.6 mmol) at 20°C. After 1 h stirring, was added $CH_2Br_2$ **2** (2.6 g, 15 mmol). After 10 h reflux, water was added followed by usual work up. Column chromatography (silica gel 8 g, hexane:EtOAc 4:1) gave 435 mg of **3** (67%).

**3-Carboxymethoxycyclohexanone 4**. To **3** (100 mg, 0.43 mmol) in PhH (80 mL) was added tri-n-butyltin hydride (116 mg, 0.4 mmol) and AIBN (7 mg, 0.04 mmol). Under stirring the mixture was heated to reflux for 24 h. Evaporation of the solvent, extraction with $CH_2Cl_2$ (30 mL), washing with 10% KF (1 x 10 mL) and column chromatography (silica gel 2 g; hexane:EtOAc 2:1) afforded 49.4 mg of **4** (75%), $R_f$=0.31 (hexane:EtOAc 2:1).

## D U F F Aldehyde Synthesis

Formylation of phenols and anilines with hexamethylenetetramine **2** (see 1st edition).

|   |               |                      |          |     |      |
|---|---------------|----------------------|----------|-----|------|
| 1 | Duff, J.C.    | *J. Chem. Soc.*      | **1932** |     | 1987 |
| 2 | Duff, J.C.    | *J. Chem. Soc.*      | **1934** |     | 1305 |
| 3 | Ogata, Y.     | *Tetrahedron*        | **1968** | 24  | 5001 |
| 4 | Wada, F.      | *Bull. Soc. Chim. Jpn.* | **1980** | 53  | 1473 |
| 5 | Jacobsen, E.N.| *J. Org. Chem.*      | **1994** | 59  | 1939 |
| 6 | Ferguson, L.N.| *Chem. Rev.*         | **1946** | 38  | 230  |

**Aldehyde 3.**[5] **1** (125 g; 0.61 mol) and **2** (170 g; 1.21 mol) in HOAc (300 mL) were heated to 130°C with stirring and kept at 130°C (± 5°C) for 2 h.. At 75°C, 33% $H_2SO_4$ (300 mL) was added and the mixture heated to 105-110°C for 1 h. Work up afforded 56-71 g of **3** (40-50%), mp 53-56°C.

**D U T H A L E R - H A F N E R**  Enantioselective Allylation

Cyclopentadienyldialkoxyallyltitanium complex **1**[4] in enantioselective allylation of aldehydes.

(S)-**3** (93%; 95% ee)

(R,R)-**1**

| 1 | Duthaler, R.O. | Helv. Chim. Acta | **1990** | 73 | 353 |
|---|---|---|---|---|---|
| 2 | Duthaler, R.O; Hafner, A | Pure Appl. Chem. | **1990** | 62 | 631 |
| 3 | Hafner, A; Duthaler, R.O. | Eur. Pat. Appl. Ep. 387,196; C.A., **1991**, 114, 122718h | | | |
| 4 | Hafner, A. | J. Am. Chem. Soc. | **1992** | 114 | 2321 |
| 5 | Duthaler, R.O; Hafner, A. | Chem. Rev. | **1992** | 92 | 827 |
| 6 | Duthaler, R.O; Hafner, A. | Inorg. Chem. Acta | **1994** | 222 | 95 |

**(1S)-1-Phenyl-3-buten-1-ol 3.**[4] **2** in THF (5.3 mL; 0.8 M 4.25 mmol) was added slowly (10 min) at 0°C under Ar to a solution of (R,R)-**1** (3.06 g; 5 mmol) in Et$_2$O (60 mL). After 1.5 h stirring at 0°C, the mixture was cooled to -78°C and benzaldehyde (403 mg; 3.8 mmol) in Et$_2$O (5 mL) was added over 5 min. After 3 h stirring at -74°C the mixture was quenched with 45% NH$_4$F (20 mL) and after separation of 1.68 g of ligand, chromatography on silica gel (CH$_2$Cl$_2$:hexane:Et$_2$O 4:4:1) afforded 521 mg of (S)-**3** (93%, 95% ee).

## E C K E R T  Hydrogenation Catalysts

Metal phthalocyanines MPc (M=V, Mn, Fe, Co, and especially Pd) as very stable and selective hydrogenation or hydrogenolysis catalysts with adjustable chemospecificity, sometimes pH dependent.

$$Ph-CH=CH-CN \xrightarrow[20°, 1 h]{NaBH_4, CoPc} Ph-CH_2-CH_2-CN$$

1               2 (82%)

5 (88%)[2]           3          4 (96%)[1]

6        7          8 (60%) 4

| 1 | Eckert, H. | *Angew. Chem. Int. Ed.* | **1981** | 20 | 473 |
|---|---|---|---|---|---|
| 2 | Eckert, H. | *Angew. Chem. Int. Ed.* | **1983** | 22 | 881 |
| | | *Angew. Chem. Suppl.* | **1983** | | 1291 |
| 3 | Eckert, H. | *Angew. Chem. Int. Ed.* | **1986** | 25 | 159 |
| 4 | Eckert, H. | *Z. Naturforsch.* | **1991** | 46b | 339 |

**p-Choroaniline 4.**[1] To a well stirred mixture of NaBH$_4$ (2.7 g, 70 mmol) and Co-phthalocyanine, Co Pc catalyst, (0.5 g, 0.9 mmol) in ethanol (50 ml) 3 (1.58 g, 10 mmol) was added and stirred for 2 h at r.t.. Under ice cooling 5 N HCl was added until a pH=6-7. The catalyst was removed by filtration over a layer of sodium sulfate, the solvent evaporated and the residue partitioned with 1 N NaOH and ether. Drying and concentration of the organic layer afforded 1.22 g of 4 (96 %). For re-use the catalyst is washed with water and dried.

## E H R L I C H - S A C H S  Aldehyde Synthesis

Formation of *o*-nitrobenzaldehydes from *o*-nitrotoluenes and nitrosodimethylaniline (see 1st edition).

| 1 | Ehrlich, P.; Sachs, F. | *Chem. Ber.* | **1899** | *32* | 2341 |
|---|---|---|---|---|---|
| 2 | Sachs, F. | *Chem. Ber.* | **1900** | *33* | 959 |
| 3 | Ruggli, P. | *Helv. Chim. Acta* | **1937** | *20* | 271 |
| 4 | Adams, R | *Org. Synth. Coll.* | | *II* | 214 |
| 5 | Millich, F. | *Org. Proc. Prep. Int.* | **1996** | *28* | 366 |

## E L B S  Oxidation

Oxidation of monophenols to polyphenols or oxidation of aromatic methyl groups by persulfates (see 1st edition).

| 1 | Elbs, K. | *J. Prakt. Chem.* | **1893** | *48* | 179 |
|---|---|---|---|---|---|
| 2 | Bergmann, E.J. | *J. Am. Chem. Soc.* | **1958** | *80* | 3717 |
| 3 | Neumann, M.S. | *J. Org. Chem.* | **1980** | *45* | 4275 |
| 4 | Sethna, S.M. | *Chem. Rev.* | **1951** | *49* | 91 |
| 5 | Wallace, T.W. | *Synthesis* | **1983** | | 1003 |

**2,5-Dihydroxypyridine 2.**[2] To **1** (38.0 g; 0.4 mol) and NaOH (80.0 g; 2 mol) in water (1500 mL) at 0°C was added FeSO$_4$ (2.0 g) in water (20 mL) and potassium peroxydisulfate (135.0 g; 0.5 mol). After 20 h at 20°C and filtration, conc. H$_2$SO$_4$ was added (cooling) to pH=0.75 and the mixture was heated to 100°C under N$_2$ for 30 min. The cooled solution was neutralized by 10N NaOH to pH=6.5. Extraction (Soxhlet) with iPrOH and evaporation afforded 19 g of **2** (42%).

## E N D E R S Chiral Reagent

Asymmetric electrophilic substitution of aldehydes and ketones via (S) or (R) 1-amino-2-methoxymethylpyrrolidine (SAMP or RAMP) hydrazone or by N-N bond cleavage via Raney nickel promoted hydrogenolysis to alkylamines (see 1st edition).

| 1 | Enders, D. | *Angew. Chem. Int. Ed.* | **1976** | *15* | 549 |
|---|---|---|---|---|---|
| 2 | Enders, D. | *J. Am. Chem. Soc.* | **1979** | *101* | 5654 |
| 3 | Enders, D. | *Angew. Chem. Int. Ed.* | **1979** | *18* | 397 |
| 4 | Enders, D. | *Tetrahedron* | **1984** | *40* | 1345 |
| 5 | Enders, D. | *Helv. Chim. Acta* | **1995** | *78* | 970 |
| 6 | Enders, D. | *Synlett* | **1996** | | 126 |
| 7 | Enders, D. | *Synlett* | **1998** | | 1182 |
| 8 | Enders, D. | *O. P. P. I.* | **1985** | *17* | 1 |
| 9 | Nicolaou, K.C. | *J. Am. Chem. Soc.* | **1981** | *103* | 6967; 6999 |
| 10 | Enders, D. | *Org. Synth.* | **1987** | *65* | 173; 183 |

**Ferrocenecarboxaldehyde SAMP hydrazone (S)-2.**[6] A mixture of **1** (15 g; 70 mmol), molecular sieves (4Å) (15 g) and SAMP (10 g; 77 mmol) in $Et_2O$ (70 mL) was stirred at 0°C for 24 h and then diluted with $Et_2O$ (130 mL). Usual work up and chromatography ($SiO_2$, $Et_2O$:petroleum ether 2:1) gave 22.6 g of (S)-2 (99%).

**Hydrazine (S,R)-3.** A solution of 2 in $Et_2O$ was treated with organolithium reagent at -100°C under Ar. Upon warming up to 20°C overnight the solution was quenched at 0°C with water, dried and concentrated in vacuum. The air sensitive (S,R)-3 was used without further purification.

**(R)-1-Ferrocenylalkylamines (R)-4.** A solution of (S,R)-3 in MeOH was hydrogenated (Raney nickel, $H_2$, 10 bar, 45°-60°C). Usual work up and chromatography ($SiO_2$, MeOH) under Ar gave 4, R=n-hexyl, 55%, 91% ee (R).

## E S C H E N M O S E R  Methylenation Reagent

An isolable imminium salt **3** for α-methylenation of carbonyl compounds, analogous to the Mannich reaction (see 1st edition).

$$Me_3N \xrightarrow[EtOH]{CH_2I_2} Me_3\overset{+}{N}-CH_2I \xrightarrow{160°} Me_2\overset{+}{N}=CH_2 \ I^- \qquad Et_2NH \xrightarrow[2)\ MeSiCl_3]{1)\ CH_2O} Et_2\overset{+}{N}=CH_2 \ Cl^-$$

$$\textbf{1} \qquad\qquad \textbf{2 (89\%)} \qquad\qquad \textbf{3 (81\%)} \qquad\qquad \textbf{4} \qquad\quad ^5 \qquad\qquad \textbf{3a (97\%)}$$

| | | | |
|---|---|---|---|
| 1 | Eschenmoser, A. | *Angew. Chem. Int. Ed.* | **1971** | 10 | 330 |
| 2 | Roberts, J.L. | *Tetrahedron Lett.* | **1977** | | 1621 |
| 3 | Hiyama, T. | *Bull. Chem. Soc. Jpn.* | **1983** | 56 | 3093 |
| 4 | Duboudin, F. | *Synthesis* | **1986** | | 228 |

**Dimethyl(methylene)ammonium iodide 3.**[1] Me₃N **1** (20 g; 0.36 mol), CH₂I₂ (120 g; 0.73 mol) and EtOH were kept closed in the dark for 100 h at 20°C. Filtration, washing and drying for 1 h at 70°C in high vacuum afforded 98 g of **2** (89%), mp 190°C. **2** (40 g; 0.122 mol) in sulfolane (120 mL) was heated under N₂ to 160°C and MeI was distilled. Filtration, washing (CCl₄) and drying to 50°C in vacuum gave 18.4 g of **3** (81%), mp 240°C.

**3a.**[4] Et₂NH **4** (36.5 g; 0.5 mol) in EtOH (51 g; 1 mol) and K₂CO₃ (82.8 g; 0.6 mol) were stirred at 0°C for 5 min. CH₂O (0.4 mol) was added and the mixture was stirred for 24 h. Distillation afforded a gem-aminoether. The aminoether (25 mmol) was added to MeSiCl₃ **5** (25 mmol) in MeCN (10 mL) under cooling (ice bath). Evaporation in vacuum and washing with Et₂O afforded **3a** in 97% yield, mp 124°C.

**α-Methylenebutyrolactone 7.**[2] iPr₂NH (2.02 g; 20 mmol) in THF (20 mL) and BuLi (2.55M; 20 mmol) were stirred at -78°C for 15 min. Lactone **6** (1.6 g; 19 mmol) and **3** (7.4 g; 40 mmol) were added. Evaporation of the solvent and treatment of the residue in MeOH with MeI gave after 24 h stirring and chromatography 1.21 g of **7** (61%).

## E S C H E N M O S E R  Sulfide Contraction

Synthesis of enamino ketones from thioamides or of β-dicarbonyl derivatives from thioesters.

| 1 | Eschenmoser, A. | *Angew. Chem. Int. Ed.* | **1969** | 8 | 343 |
| 2 | Eschenmoser, A. | *Pure Appl. Chem.* | **1969** | 20 | 1 |
| 3 | Eschenmoser, A. | *Quart. Rev.* | **1970** | 24 | 366 |
| 4 | Eschenmoser, A. | *Helv. Chim. Acta* | **1971** | 54 | 710 |
| 5 | Horikawa, H. | *Tetrahedron Lett.* | **1994** | 35 | 2187 |
| 6 | Shiosaki. K. | *Compreh. Org. Synth.* | **1991** | 2 | 865-892 |

**Thioester 2.**[4] To a solution of thiobutyric acid **1** (3.16 g; 30 mmol) and Et$_3$N (4.20 mL) in Et$_2$O was added 1-bromobutan-2-one (3.06 mL; 30 mmol). After 2 h reflux, the mixture was filtered through Celite, the solvent evaporated and the residue distilled (Kugelrohr, 110°C/0.3 torr) to afford 4.925 g of **2** (95%).

**3,5-Octandione 3**. To a solution of **2** (442 mg; 2.54 mmol) and anh. LiBr (259 mg; 2.83 mmol) in MeCN was added bis(3-dimethylaminopropyl) phenylphosphine (2.2 mL). The reaction mixture was heated under N$_2$ for 17 h at 70°C. The cooled mixture (0°C) was quenched with ice water (10 mL) and conc. HCl (1.3 mL). Extraction with Et$_2$O:CH$_2$Cl$_2$ (5:1), evaporation of the solvent and distillation (Kugelrohr, 90°-105°C/10 torr) afforded 336 mg of **3** (93%).

## ESCHENMOSER-MEERWEIN  Allylic Acetamidation

Reaction of allyl and benzyl alcohols with 1-dimethylamino-1-methoxy-ethene **2** leading to acetamidation-rearrangement, proceeding via ether exchange followed by Claisen rearrangement (enamine SN$_2$' displacement).

| # | Author | Journal | Year | Vol | Page |
|---|--------|---------|------|-----|------|
| 1 | Eschenmoser, A. | *Helv. Chim. Acta* | **1964** | *47* | 2425 |
| 2 | Meerwein, H. | *Liebigs Ann.* | **1961** | *641* | 1 |
| 3 | Eschenmoser, A. | *Helv. Chim. Acta* | **1969** | *52* | 1030 |
| 4 | Stevenson, P.J. | *Tetrahedron Lett.* | **1991** | *32* | 4199 |
| 5 | Coudert, G. | *Synth. Commun.* | **1994** | *24* | 1781 |

**2-Methyl-1-naphtylacetic acid N,N-dimethylamide 3.**[3] A mixture of 2-naphtylcarbinol **1** (1.0 g; 6.33 mmol) and 1-dimethylamino-1-methoxy-ethene **2** (1.278 g; 12.66 mmol) in DMF (10 mL) was heated for 24 h at 160°C with stirring. The mixture was extracted with Et$_2$O/CH$_2$Cl$_2$ and the extract was washed with phosphate buffer (pH=5), brine and dried over Na$_2$SO$_4$. Evaporation of the solvent gave 1.613 g of crude **3**. Chromatography (Kieselgel, Ph:Et$_2$O 1:1) afforded, after recrystallization from MeOAc:petroleum ether, 1.27 g of **3** (90%), mp 114-115°C.

### E S C H W E I L E R - C L A R K  Amine Methylation

Reductive methylation of amines by a mixture of formaldehyde and formic acid (see 1st edition).

| 1 | Eschweiler, W. | *Chem. Ber.* | **1905** | *38* | 880 |
|---|---|---|---|---|---|
| 2 | Clark, H.T. | *J. Am. Chem. Soc.* | **1933** | *55* | 4571 |
| 3 | Cope, A.C. | *J. Org. Chem.* | **1965** | *30* | 2163 |
| 4 | Borch, R.F. | *J. Org. Chem.* | **1972** | *37* | 1673 |
| 5 | Moore, M.L. | *Org. React.* | **1949** | *5* | 301 |
| 6 | Casanova, I. | *Synth. Commun.* | **1993** | | 245 |

**N,N-Dimethyl-5-amino-1-hexene 2.**[3] **1** (8.5 g; 85.5 mmol) in 91% formic acid (24 g) and 37% formaldehyde was heated on a steam bath for 6 h, cooled and poured onto ice. The mixture was made strongly basic with 20% NaOH and extracted with $Et_2O$. Evaporation and distillation gave 6.3 g of **2** (60%), bp 135-136°C.

### F E I S T - B E N A R Y  Furan Synthesis

Synthesis of furans by base catalyzed condensation of an α-halocarbonyl compound with an enol, derived from a 1,3-dicarbonyl compound (see 1st edition).

| 1 | Feist, F. | *Chem. Ber.* | **1902** | *35* | 1539 |
|---|---|---|---|---|---|
| 2 | Benary, E. | *Chem. Ber.* | **1911** | *44* | 489 |
| 3 | Reichstein, T. | *Helv. Chim. Acta* | **1931** | *14* | 1270 |
| 4 | Reichstein, T. | *Helv. Chim. Acta* | **1933** | *15* | 268; 1105; 1112 |
| 5 | Cambie, R.C. | *Synth. Commun.* | **1990** | *20* | 1923 |

## **E V A N S**  Chiral Auxiliary

Enantioselective aldol condensation by means of an oxazolidone chiral auxiliary and boron enolate (see 1st edition).

88%; 99% ee

**1**          **2**          **3a** (*S*)   95 : 5   **3b** (*R*)

| 1 | Evans, D.A. | *J. Am. Chem. Soc.* | **1979** | *101* | 6120 |
|---|---|---|---|---|---|
| 2 | Evans, D.A. | *J. Am. Chem. Soc.* | **1981** | *103* | 2127; 2876 |
| 3 | Newmann, M.S. | *J. Am. Chem. Soc.* | **1951** | *71* | 4199 |
| 4 | Evans, D.A. | *J. Am. Chem. Soc.* | **1986** | *108* | 6757 |
| 5 | Evans, D.A. | *Tetrahedron Lett.* | **1987** | *28* | 39 |
| 6 | Evans, D.A. | *J. Am. Chem. Soc.* | **1990** | *112* | 4011 |
| 7 | Evans, D.A. | *Aldrichimica Acta* | **1982** | *15* | 23 |
| 8 | Evans, D.A. | *Org. Synth.* | **1988** | *68* | 89 |

**[3-(2S,4S)]-3-(2-Bromo-3-phenyl-1-oxopropyl)-4-(phenylmethyl)-2-oxazolidinone (3).**[6] The boronic enolate formed from acyloxazolidinone **1** (1.5 g; 4.85 mmol), dibutylboryl triflate (1.4 g; 5.09 mmol) and diisopropylethylamine (752 mg; 5.82 mmol) in $CH_2Cl_2$ (10 mL) was added to NBS (1.04 g; 5.82 mmol) in $CH_2Cl_2$ (10 mL). After 1.25 h stirring at -78°C the reaction mixture was quenched ($NaHSO_4$ aq), extracted (EtOAc) and flash chromatographed. The product **3a** + **3b** was stable for several months at -16°C, *(S):(R)* ratio = 95.4:4.6.

## FAVORSKI-WALLACH Rearrangement

Rearrangement of α-haloketones or α,α-dihaloketones to carboxylic acids or acrylic acids via cyclopropanones (see 1st edition).

| 1 | Favorski, A.E. | *J. Prakt. Chem.* | **1895** | *51* | 553 |
|---|---|---|---|---|---|
| 2 | Wallach, O. | *Liebigs Ann.* | **1918** | *414* | 296 |
| 3 | Wagner, R.B. | *J. Am. Chem. Soc.* | **1950** | *72* | 972 |
| 4 | Nace, H.R. | *J. Org. Chem.* | **1967** | *32* | 3438 |
| 5 | De Kimpe, M.D. | *J. Org. Chem.* | **1986** | *51* | 3938 |
| 6 | Sosnowsky, C. | *J. Org. Chem.* | **1995** | *60* | 3414 |
| 7 | Bekington, M. | *Synth. Commun.* | **1996** | *26* | 1097 |
| 8 | Kende, A.S. | *Org. React.* | **1960** | *11* | 261 |

**Ethyl cyclopentanecarboxylate 2.**[7] To a suspension of $Pb(OAc)_4$ (9.0 g; 20 mmol) in $(EtO)_3CH$ (50 mL) prepared at 5°C was added sequentially cyclohexanone **1** (2.0 g; 20 mmol) in $(EtO)_3CH$ (20 mL) and 70% $HClO_4$ (2.0 mL). The reaction mixture was stirred for 28 h at 20°C. After evaporation of the solvent in vacuum, the residue was dissolved in $CHCl_3$, the insoluble matter was removed by filtration and the filtrate washed (water), dried ($MgSO_4$) and the solvent was removed in vacuum. Chromatography (silica gel, hexane) afforded 1.65 g of **2** (70%).

## FELDMAN  Vinylcyclopentene Synthesis

Vinylcyclopentane synthesis via phenylthio radical catalyzed alkenylation or alkynylation of vinylcyclopropanes (see also Felkin).

| 1 | Feldman, K.S. | *J. Am. Chem. Soc.* | **1986** | *108* | 1328 |
|---|---|---|---|---|---|
| 2 | Feldman, K.S. | *J. Am. Chem. Soc.* | **1988** | *110* | 3300 |
| 3 | Feldman, K.S. | *J. Am. Chem. Soc.* | **1989** | *111* | 4878 |
| 4 | Feldman, K.S. | *Tetrahedron Lett.* | **1989** | *30* | 5845 |
| 5 | Feldman, K.S. | *Tetrahedron* | **1989** | *45* | 2969 |
| 6 | Feldman, K.S. | *J. Org. Chem.* | **1992** | *57* | 100 |
| 7 | Singleton, D.A. | *Synlett* | **1994** | | 272 |

**Vinylcyclopentenes 3a and 3b.**[4] To a solution of vinylcyclopropane 1 (150 mg; 0.68 mmol) and methyl butynoate 2 (1.26 g; 12.9 mmol) in PhH (6 mL) under Ar was added dropwise a deoxygenated solution of $Ph_2S_2$ (190 mg; 0.86 mmol) and AIBN (27 mg; 0,17 mmol) in PhH (40 mL) in cca 30 h under sunlamp irradiation. After addition, the mixture was maintained at 20°C till 1 was consumed (TLC). Evaporation of the solvent in vacuum and flash chromatography (silica gel, 5% $Et_2O$ in hexane) afforded 96 mg of 3a and 3b in a 1.5:1 ratio as a colorless oil (41%). By additional chromatography in the same system individual stereoisomers could be isolated.

## FELKIN Cyclization

Nickel and Grignard catalyzed stereoselective synthesis of *cis* and *trans* 2-alkyl-vinylcyclopentanes from telemerization of butadiene. Cyclization (ene reaction) of unsaturated allyl Grignard reagents, see also Feldman (see 1st edition).

3 *cis* (67%)

| | | | | | |
|---|---|---|---|---|---|
| 1 | Felkin, H. | *Tetrahedron Lett.* | **1972** | | 1433 |
| 2 | Felkin, H. | *Tetrahedron Lett.* | **1972** | | 2285 |
| 3 | Felkin, H. | *J. Chem. Soc. Chem. Commun.* | **1975** | | 243 |
| 4 | Oppolzer, W. | *Angew. Chem. Int. Ed.* | **1989** | 28 | 32 |

*cis*-3.[3] A mixture of $(PPh_3)_2NiCl_2$ (32.6 g; 5 mmol), butadiene 1 (12.42 g; 0.23 mol) and a solution of Pr-MgBr 1.9M (0.25 mol) was refluxed (25°, solid $CO_2$ condenser) for 24 h. After deuterolysis one obtains 16.9 g of *cis*-3 (67%). By heating the Grignard mixture, for 24 h in a sealed tube, the thermally more stable *trans* isomer of 3 was obtained. The Ni catalyzed ene cyclization also can be performed starting with octadienyl halides and conversion to 2.[2,4]

## FORSTER-DECKER Amine Synthesis

Selective monoalkylation of primary amines via imines. An alternative method is the reaction of 1 and 2 in the presence of $NaCNBH_4$ or triacetoxyborohydride (Borch reduction).[4]

$$Ph\text{-}CH_2\text{-}CH_2\text{-}NH_2 \ + \ OHC\text{-}Ph \xrightarrow[- H_2O]{PhMe;\ 110°} Ph\text{-}CH_2\text{-}CH_2\text{-}N{=}CH\text{-}Ph \xrightarrow[2)\ NaOH\ pH\ 7]{1)\ Me_2SO_4\ refl.} Ph\text{-}CH_2\text{-}CH_2\text{-}NH\text{-}Me$$

| | | 2 | 1 | | 3 | | 4 (80%)[3] |

| | | | | | |
|---|---|---|---|---|---|
| 1 | Forster, M.O. | *J. Chem. Soc.* | **1899** | 75 | 934 |
| 2 | Decker, H. | *Liebigs Ann.* | **1913** | 395 | 362 |
| 3 | Morrison, A.L. | *J. Chem. Soc.* | **1950** | | 1478 |
| 4 | Borch, R.F. | *J. Am. Chem. Soc.* | **1971** | 93 | 2897 |

## FERRARIO-AKERMANN Thiocyclization

Synthesis of phenoxathiines, phenothiazines by S insertion (see 1st edition).

$$\text{1} \xrightarrow[100°]{S_8 / AlCl_3} \text{2 (69\%)}$$

| | | | | | |
|---|---|---|---|---|---|
| 1 | Ferrario, E. | *Bull. Soc. Chim. Fr.* | **1911** | *9* | 536 |
| 2 | Akermann, F. | Ger. Pat. 234,743 | **1910** | | |
| 3 | Sutter, C.N. | *J. Am. Chem. Soc.* | **1936** | *58* | 717 |
| 4 | Lasco, E. | *J. Chem. Soc.* | **1956** | | 2408 |
| 5 | Coic, J.M. | *J. Heterocyclic Chem.* | **1974** | *11* | 287 |
| 6 | Deasy, C.L. | *Chem. Rev.* | **1943** | *32* | 174 |

**2,3,7,8-Tetramethylphenoxathiine 2.**[5] Diphenyl ether 1 (6.8 g; 30 mmol), sulfur (0.74 g) and $AlCl_3$ (1.54 g) were heated on a water bath. The cooled mixture was extracted with $Et_2O$, the extract washed and the solvent evaporated to give 5.4 g of 2 (69%), mp 172°C (PhH).

## FERRIER Chiral Cyclohexanone Synthesis

Transformation of unsaturated glycosides (cf. 1) into chiral cyclohexanone derivatives (or further into phenols) by heating in aqueous acetone with mercury (II) salts (see 1st edition).

| | | | | |
|---|---|---|---|---|
| 1 | Ferrier, R.J. | *J. Chem. Soc. Perkin Trans I* | **1979** | 1455 |
| 2 | Ferrier, R.J. | *J. Chem. Soc. Perkin Trans I* | **1985** | 2413 |
| 3 | Sakairi, N. | *J. Chem. Soc. Perkin Trans I* | **1990** | 1301 |

## FERRIER  Carbohydrate Rearrangement

Allylic rearrangement of unsaturated carbohydrates (glucals) (see 1st edition).

| | | | | | |
|---|---|---|---|---|---|
| 1 | Ferrier, R.J. | *Adv. Carbohydrate Chem.* | **1965** | 20 | 67 |
| 2 | Ferrier, R.J. | *J. Chem. Soc. (C)* | **1968** | | 974 |
| 3 | Ferrier, R.J. | *J. Chem. Soc. (C)* | **1969** | | 570 |
| 4 | Fraser-Reid, B. | *J. Org. Chem.* | **1995** | 60 | 3851 |
| 5 | Fraser-Reid, B. | *J. Chem. Soc. Perkin I* | **1998** | | 631 |
| 6 | Balasubramanian, K.K. | *Tetrahedron Lett.* | **2000** | | 1271 |

**Ethyl 4,6-di-O-acetyl-2,3-dideoxy-α-D-erythro-hex-2-enopyranoside 2.**[2] A solution of tri-O-acetyl-*D*-glucal **1** (5.0 g; 18 mmol) in PhH (20 mL) and EtOH (1.8 mL; 31 mmol) was treated with BF$_3$·Et$_2$O (1 mL). After 25 min the optical rotation changed from -35° to +20.25°. Neutralization of the catalyst, filtration of the solids and removal of the solvent left a syrup which on trituration with EtOH gave 2.8 g of **2** and a second crop of 0.5 g (70% yield). The pure product melted at 78-79°C and had [α] = +104° (PhH).

**Phenylthiopyranoside 3.**[4] To a stirred and cooled (-20°C) solution of tri-O-acetyl-*D*-glucal **1** (12.4 g; 45.6 mmol) in CH$_2$Cl$_2$ (45 mL) were added thiophenol (4.68 mL; 45.6 mmol) and a catalytic amount of BF$_3$·Et$_2$O (0.1 mL). The reaction was allowed to warm up to 20°C, was stirred for 2 h and then neutralized by addition of Na$_2$CO$_3$. After the solution was stirred for 30 min, the solid was filtered, the filtrated evaporated in vacuum and the residue chromatographed (petroleum ether:EtOAc 8:2). This afforded a mixture of 11.5 g of **4a** and **4b** (78%) in a ratio of 8:1. Recrystallization from hexane:Et$_2$O gave pure **4a**.

## F I N E G A N Tetrazole Synthesis

Tetrazole synthesis from azides by dipolar cycloaddition with activated nitriles or intramolecularly with nitriles in the presence of acids (see 1st edition).

| | | | | | |
|---|---|---|---|---|---|
| 1 | Finegan, W. G. | *J. Am. Chem. Soc.* | **1956** | *80* | 3908 |
| 2 | Carpenter, W. R. | *J. Org. Chem.* | **1962** | *27* | 2085 |
| 3 | Kereszty, von K. | *Germ. Pat. 611.692, C. A.* | **1935** | *29* | 5994 |

## F I S C H E R – B O R S C H E – D R E C H S E L Indole Synthesis

Indole synthesis from phenylhydrazones of ketones (Fischer); tetrahydrocarbazoles from cyclohexanone (Borsche-Drechsel) phenylhydrazones (see 1st edition).

| | | | | | |
|---|---|---|---|---|---|
| 1 | Fischer, E. | *Chem. Ber.* | **1883** | *16* | 2241 |
| 2 | Iyosuke Simizu | *Chem. Pharm. Bull.* | **1971** | *19* | 2561 |
| 3 | Sarmicole, F. | *Tetrahedron Lett.* | **1984** | *25* | 3101 |
| 4 | Robinson, B. | *Chem. Rev.* | **1969** | *69* | 227 |
| 5 | Welch, W. M. | *Synthesis* | **1977** | | 845 |
| 6 | Drechsel, E. | *J. Prakt. Chem.* | **1888** | *38* | 69(2) |
| 7 | Borsche, W. | *Chem. Ber.* | **1904** | *20* | 378 |
| 8 | Campbell, N. N. | *Chem. Rev.* | **1947** | *40* | 361 |

## **F I S C H E R** Carbene Complexes

Cyclopropanation of alkenes with phenylmethoxy carbene complexes (e. g. **5**) of Cr, Mn, W.

$$Cr(CO)_5 \xrightarrow{PhLi} \quad \xrightarrow{Me_4NBr} (Me_4N)Cr(CO)_5COPh \xrightarrow{H_2SO_4} \xrightarrow{CH_2N_2} (CO)_5Cr=C-Ph$$

|   |   |   |   |   |
|---|---|---|---|---|
| **1** | **2** | **3** | **4** (81%) | **5** (55%) |

6     + **5**   $\xrightarrow[100°, 4 h]{THF}$   7   (2.9:1)   +   8

$$\text{(CO}_2\text{Me)} + \textbf{5} \xrightarrow[5.5 \text{ h, } 80\ ^\circ\text{C}]{C_6H_{12}}$$

71%      4%

| 1 | Fischer, E. O. | *Chem. Ber.* | **1967** | *100* | 2445 |
|---|---|---|---|---|---|
| 2 | Fischer, E. O. | *Chem. Ber.* | **1972** | *105* | 1356,3966 |
| 3 | Fischer, E. O. | *J. Organomet. Chem.* | **1974** | *81* | C20-C22 |
| 4 | Reissig, H. U. | *Tetrahedron Lett.* | **1988** | *29* | 2351 |
| 5 | Harvey, D. F. | *J. Am. Chem. Soc.* | **1991** | *113* | 8916 |
| 6 | Chan Kin Shing | *J. Org. Chem.* | **1994** | *59* | 3585 |
| 7 | Pulley, S. R. | *Org. Lett.* | **1999** | *1* | 1721 |

**Pentacarbonyl (methoxyphenyl) chrom (O) (5).**[1] $Cr(CO)_5$ **1** (2.2 g, 10 mmol) in $Et_2O$ (200 mL) was refluxed with PhLi **2** (10 mmol) in $Et_2O$. Insoluble $Cr(CO)_6$ was removed, the $Et_2O$ evaporated and the residue in water treated with $Me_4NBr$ (2.3 g, 15 mmol) to give 3.3 g of **4** (89%). A suspension of **4** (1.86 g, 5 mmol) in $Et_2O$ (200 mL) was treated with water and N $H_2SO_4$ (20 mL). After extraction with $Et_2O$ and drying ($MgSO_4$), the ether solution was treated with $CH_2N_2$. After evaporation of the solvent, the residue was extracted with hexane. Chromatography (silica gel, hexane), evaporation of the principal fraction and sublimation (55 °C/vacuum) afforded 850 mg of carbene complex **5** (55%), mp 46 °C.
**cis and trans 2-[(E)-1-Butenyl]-1-methoxy-1-phenylcyclopropane 7 and 8**[5] (E)-1,3-Hexadiene **6** (42.5 mg, 0.465 mmol) and carbene complex **5** (202 mg, 0.647 mmol) in THF (19 mL) were heated for 4 h at 100 °C in a sealed glass vial. Usual work up afforded 66.3 mg of **7** and **8** (71%) in a ratio of 2.9:1.

## FLEMING-MAH Anthracene Synthesis

Synthesis of anthracenes from bromobenzenes and ketene acetal (via $\alpha$-benzyne and benzocyclobutanol).

1      2      3 (44%) $\xrightarrow{\text{NaBH}_4}$ 4      5

3   X; Y = O; 4   X = OH, Y = H

| | | | | | |
|---|---|---|---|---|---|
| 1 | Fleming I.; Mah, T. | *J. Chem. Soc. Perkin I* | **1975** | | 964 |
| 2 | Olofson, R.A. | *J. Am. Chem. Soc.* | **1973** | 95 | 581 |
| 3 | Bubb, W.A. | *Austr. J. Chem.* | **1976** | 29 | 1807 |
| 4 | Liebeskind, L.S. | *J. Org. Chem.* | **1989** | 54 | 1435 |
| 5 | Stevens, R.V. | *J. Org. Chem.* | **1982** | 47 | 2393 |
| 6 | Olofson, R.A. | *Synth. Commun.* | **1992** | 22 | 1907 |
| 7 | Olofson, R.A. | *J. Org. Chem.* | **1992** | 57 | 7122 |

**1,2-Dihydrocyclobuta[1]phenanthren-1-ol 4.**[7] A mixture of 9-bromo-phenan-threne **1** (3.3 g; 12.9 mmol), NaNH$_2$ (1 g; 25.9 mmol) and ketene diethyl acetal **2** (3 g; 25.9 mmol) in THF (6 mL) was refluxed for 7 h. Hydrolysis (10% HCl, 12 h at 20°C) and chromatography afforded 1.2 g of **3** (44%), mp 165-167°C.
A solution of **3** (349 mg; 1.6 mmol) in THF (3 mL) was added slowly to NaBH$_4$ (295 mg; 7.8 mmol) in EtOH (10 mL) at 0°C. After 2 h stirring, work up and chromatography gave 317 mg of **4** (90%), mp 129-130°C.

**Dibenz[a,c]anthracene 5**. To a mixture of **4** (141 mg; 0.64 mmol) and chlorobenzene (72 mg; 0.64 mmol) in tetrahydropyran (THP) (1 mL) was added lithium tetramethylpiperidide (LTMP) (2.6 mmol; 6 mL THP) over 5 min under reflux and heating was continued for another 30 min. The cooled mixture was quenched with 10% HCl (50 mL), extracted (CH$_2$Cl$_2$) and chromatographed (hexane:EtOAc 8:2). Recrystallization from EtOH provided 97 mg of **5** (55%), mp 202-205°C.

## F R A N K E L - S H I B A S A K I Rearrangement

Stereocontrol in allylamine to enamine isomerisation in hydrogenation and 1,5-hydrogen shift in conjugated dienes, catalyzed by metal derivatives.

**Methyl N-*(E)*-5-(methoxycarbonylpentenyl)-N-*(1E,3Z)*-pentadienyl carbamate (2).**[5]

| 1 | Frankel, E.N. | *J. Am. Chem. Soc.* | **1968** | 90 | 2446 |
|---|---|---|---|---|---|
| 2 | Frankel, E.N. | *Tetrahedron Lett.* | **1968** | | 1919 |
| 3 | Shibasaki, M. | *Chem. Lett.* | **1984** | | 570 |
| 4 | Shibasaki, M. | *J. Am. Chem. Soc.* | **1990** | 112 | 4906 |
| 5 | Shibasaki, M. | *J. Org. Chem.* | **1991** | 56 | 4569 |
| 6 | Shibasaki, M. | *Synthesis* | **1993** | | 643 |
| 7 | Noyori, R. | *J. Am. Chem. Soc.* | **1990** | 112 | 4897 |
| 8 | Noyori, R. | *Acc. Chem. Res.* | **1990** | 23 | 345 |

**Methyl N-*(E)*-5-(methoxycarbonylpentenyl)-N-*(1E,3Z)*-pentadienyl carbamate (2).**[5]
A mixture of diene **1** (26.9 mg; 0.1 mmol), $Cr(CO)_3$ (2.72 mg; 0.02 mmol), naphtalene and $Me_2CO$ (1 mL) was degased through four freeze-pump-thaw cycles, stirred for 4h at 20°C and concentrated. Silica gel chromatography afforded 23.1 mg of **2** (86%).

**F O R S T E R** Diazo Synthesis

Formation of diazo derivatives from oximes (see 1st edition).

| 1 | Forster, M. C. J. | *J. Chem. Soc.* | **1915** | *107* | 260 |
|---|---|---|---|---|---|
| 2 | Meinwald, J. | *J. Am. Chem. Soc.* | **1959** | *81* | 4751 |
| 3 | Hassner, A. | *Tetrahedron Lett.* | **1962** | | 795 |
| 4 | Kirmse, M. | *Angew. Chem.* | **1957** | *69* | 106 |
| 5 | Rundel, W. | *Angew. Chem.* | **1962** | *74* | 469 |

**F R I T S C H – B U T T E N B E R G – W I E C H E L L** Acetylene Synthesis

Alpha elimination from haloethylenes leading via carbene rearrangement to acetylenes. (see 1st edition).

| 1 | Fritsch, P. | *Liebigs Ann.* | **1894** | *279* | 319 |
|---|---|---|---|---|---|
| 2 | Buttenberg, W. P. | *Liebigs Ann.* | **1894** | *279* | 327 |
| 3 | Wiechell, H. | *Liebigs Ann.* | **1894** | *279* | 337 |
| 4 | Curtin, D. Y. | *J. Am. Chem. Soc.* | **1958** | *80* | 4599 |
| 5 | Kobrich, G. | *Chem. Ber.* | **1972** | *105* | 1674 |
| 6 | Kobrich, G. | *Angew. Chem. Int. Ed.* | **1965** | *4* | 49 |

## F R E E M A N  Lithium Reagent

Lithium 4,4-di-t-butylbiphenylide **2** (LiDBB), a reagent more efficient than Li metal or other Li radical anions in halogen metal exchange or in cleavage of C-O; C-S; C-Se; C-C bonds.

**1**  **2**

**3**  **4**  **5 (56%)**

1) Me$_2$S•CuBr
2) Cyclopentenone
TMSCl
3) H$_2$O

| | | | | | |
|---|---|---|---|---|---|
| 1 | Freeman, P.K. | *Tetrahedron Lett.* | **1976** | | 1849 |
| 2 | Freeman, P.K. | *J. Org. Chem.* | **1980** | 45 | 1924 |
| 3 | Yus, M. | *Tetrahedron Lett.* | **1993** | 34 | 2011; 3487 |
| 4 | Freeman, P.K. | *J. Org. Chem.* | **1991** | 56 | 3646 |
| 5 | Krief, A. | *Across Organica Acta* | **1995** | 1 | 37 |
| 6 | Freeman, P.K. | *J. Org. Chem.* | **1983** | 48 | 4705 |
| 7 | Freeman, P.K. | *Tetrahedron* | **1996** | 52 | 8397 |

**LiDBB 2.**[2] Under Ar, a solution of 4,4-di-t-butylbiphenyl **1** (6.65 g; mmol) in THF (82 mL) was treated with small pieces of Li (146 mg; 21.1 mmol). The reaction mixture was stirred until all Li was consumed (cca 3 h at 0°C) to provide a solution of 0.25 mol/l of **2**.

**Diol 5**. A mixture of 2,3-dichloropropene **3** (111 mg; 1 mmol) and cyclohexanone **4** (49 mg; 0.5 mol) in THF cooled to 0°C was treated with **1** (13.3 mg; 5 mol%) and Li (59.5 mg). Usual work up afforded 70 mg of **5** (56%), mp 101-102°C.

## FREUNDERBERG-SCHÖNBERG  Thiophenol Synthesis

Conversion of phenols to thiophenols via rearrangement of thiocarbonates or thiocarbamates (see 1st edition).

$$ArO^- + \underset{\underset{S}{\parallel}}{Cl\text{-}C}\text{-OPh} \longrightarrow Ar\underset{S}{\overset{O}{\diagdown}}C\text{-OPh} \longrightarrow Ar\underset{S}{\overset{O}{\diagup}}C\text{-OPh} \longrightarrow Ar\text{-SH}$$

$$ArO^- + \underset{\underset{S}{\parallel}}{Cl\text{-}C}\text{-SR} \longrightarrow Ar\underset{S}{\overset{O}{\diagdown}}C\text{-S-R} \longrightarrow Ar\underset{S}{\overset{O}{\diagup}}C\text{-S-R} \longrightarrow Ar\text{-SH}$$

| 1 | Freundenberg, K. | *Chem. Ber.* | **1927** | *60* | 232 |
|---|---|---|---|---|---|
| 2 | Schönberg, A. | *Chem. Ber.* | **1930** | *63* | 178 |
| 3 | Wiersum, U.E. | *J. Org. Chem.* | **1989** | *54* | 5811 |
| 4 | Kwart, R. | *J. Org. Chem.* | **1966** | *31* | 410 |
| 5 | Newman, M.S. | *J. Org. Chem.* | **1966** | *31* | 3980 |
| 6 | Newman, M.S. | *J. Am. Chem. Soc.* | **1967** | *89* | 3412 |
| 7 | Reeles, H.M. | *J. Org. Chem.* | **1968** | *33* | 2249 |
| 8 | Kawata, | *Chem. Pharm. Bull.* | **1973** | *21* | 614 |
| 9 | Schulenberg, J.W. | *Org. React.* | **1965** | *14* | 1 |

**O-p-t-Butylphenyl dimethylthiocarbamate 3.**[6] To a solution of dimethylthiocarbamoyl chloride 2 (21 g; 0.17 mol) in DMF (140 mL) in an ice bath (14°C) was added, all at once dry sodium p-t-butylphenolate 1 (17.6 g; 0.1 mol) (exothermic, temp. 26°C). The mixture was stirred for 1.5 h at 30-34°C, added to water (300 mL) and extracted with PhH/Skellysolve B (4:1). Usual work up, evaporation of the solvent and recrystallization from MeOH (100 mL) afforded 21.4 g of 3 (90.5%), mp 97-99°C.

**Pyrolysis.** Heating 3 neat at 270°C until by TLC the starting material is absent, afforded 4 in a 95% yield.

**p-t-Butylphenylthiol 5.** A solution of 4 in MeOH was heated under $N_2$ with excess NaOH to give 5 (85%), bp 102-105°C/7-8 mm.

## F R I E D E L - C R A F T S  Alkylation Acylation

Alkylation or acylation of aromatic compounds by means of alkyl halides, alcohols, alkenes, acyl halides, acids in the presence of Lewis acids (see 1st edition).

| 1 | Friedel, C.; Crafts, J.N. | C.R. | **1877** | 84 | 1450 |
|---|---|---|---|---|---|
| 2 | Groggins, P.T. | *Ind. Eng. Chem.* | **1951** | 43 | 1970 |
| 3 | Kulka, M. | *J. Org. Chem.* | **1986** | 51 | 2128 |
| 4 | Olah, G.A. | *J. Org. Chem.* | **1991** | 56 | 3955 |
| 5 | Gore, P. | *Chem. Rev.* | **1955** | 55 | 229 |
| 6 | Pearson, D.E. | *Synthesis* | **1972** | | 533 |
| 7 | Price, C.C. | *Org. React.* | **1946** | 3 | 1 |
| 8 | Poliacoff, M. | *J. Chem. Soc. Chem. Commun.* | **1988** | | 359 |

**Ketone 5.**[4] To **4** (2.12 g; 10 mmol) in *p*-xylenes (15 mL) was added Nafion-H (640 mg; 30 wt%). After 12 h reflux the resin was filtered, the solvent evaporated and the residue recrystallized from hexane to give 1.87 g of **5** (90%), mp 32-35°C.

### FRIEDLÄNDER Quinoline Synthesis

Quinoline synthesis by base promoted condensation of *o*-aminoaryl aldehydes (ketones) with α-methylene aldehydes (ketones) (see 1st edition).

| 1 | Friedländer, P. | *Chem. Ber.* | **1882** | *15* | 2572 |
|---|---|---|---|---|---|
| 2 | Markgraf, J.H. | *J. Org. Chem.* | **1969** | *34* | 4131 |
| 3 | Coffen, D.L. | *J. Org. Chem.* | **1974** | *39* | 1765 |
| 4 | Bergstrom, F.W. | *Chem. Rev.* | **1944** | *35* | 151 |
| 5 | Eckert, K. | *Angew. Chem. Int. Ed.* | **1981** | *20* | 208 |
| 6 | Avendano, C. | *Synlett* | **1997** | | 285 |

**2-Ethyl-5-methoxy-3-methylquinoline 5.**[6] To a solution of N-BOC-3-methoxyaniline **4** (506 mg; 2.48 mmol) in dry THF (10 mL) at 0°C was added sec-BuLi (4.75 mL; 6.2 mmol). After 2 h stirring, DMF (0.29 mL; 3.71 mmol) was added and the reaction mixture was stirred for one more hour at 0°C and allowed to warm up to 22°C for 12 h. 3-Pentanone (0.05 mL; 2.5 mmol) and a 15% toluene solution of KHMDS (6.6 mL; 4.95 mmol) was added at 0°C and stirred at the same temperature for 10 min and for 2 h at 20°C. Quenching (saturated aq. sol. of NH₄Cl) and usual work up followed by flash chromatography (silica gel, petroleum ether:CH₂Cl₂ 1:1) gave 409 mg of **5** (82%).

**F R I E S** Phenol Ester Rearrangement

Rearrangement of phenol esters to *o*- or *p*-ketophenols, Lewis acid catalyzed or photochemical (ref. 5) (see 1st edition).

| 1 | Fries, K. | *Chem. Ber.* | **1908** | *41* | 4271 |
|---|---|---|---|---|---|
| 2 | Cremer, S.E. | *J. Org. Chem.* | **1961** | *26* | 3653 |
| 3 | Burdera, K. | *Synthesis* | **1982** | | 941 |
| 4 | Martin, A.R. | *Tetrahedron Lett.* | **1986** | *27* | 1959 |
| 5 | Alvaro, M. | *Tetrahedron* | **1987** | *43* | 143 |
| 6 | Weiss, R.G. | *J. Org. Chem.* | **1996** | *61* | 1962 |
| 7 | Blatt, A.H. | *Chem. Rev.* | **1940** | *27* | 429 |
| 8 | Effenberg, | *Angew. Chem. Int. Ed.* | **1973** | *12* | 776 |
| 9 | Blatt, A.H. | *Org. React.* | **1942** | *1* | 342 |

**4-Methyl-2-propanoyl phenol 2.**[3] 4-Methyl-1-propanoyloxybenzene **1** (231 g; 1.41 mol) was heated with anh. AlCl$_3$ (330.9 g; 2.48 mol) for 2 h at 70-80°C followed by heating to 120°C for 40 min. The cooled mixture was quenched with conc. HCl (450 mL) and ice (400 g). Extraction (CHCl$_3$), washing and evaporation of the solvent gave 215 g of crude **2** (93%). Vacuum distillation afforded 203 g of **2** (88%), bp 115-117°C/10 torr.

**Aryl alkyl ketone 4.**[5] A solution of ester **3** (500 mg; 2 mmol) in hexane (450 mL) was irradiated for 6 h at 20°C with a 125W medium pressure lamp. The solvent was removed in vacuum and the residue chromatographed to give 360 mg of **4** (72%).

## FUJIMOTO-BELLEAU  Cyclohexenone Synthesis

Synthesis of fused cyclohexenones from cyclic enol lactones with Grignard reagents (an alternative to the Robinson annulation) (see 1st edition).

| 1 | Fujimoto, G.I. | *J. Am. Chem. Soc.* | **1951** | *73* | 1856 |
|---|---|---|---|---|---|
| 2 | Belleau, B. | *J. Am. Chem. Soc.* | **1951** | *73* | 5441 |
| 3 | Weyl Raynal, J. | *Synthesis* | **1969** | | 49 |

## FUJIWARA  Arylation, Carboxylation

A mild Pd catalyzed arylation or carboxylation of a Pd activated double bond (see 1st edition).

| 1 | Moritari, I.; Fujiwara, Y. | *Tetrahedron Lett.* | **1967** | | 1119 |
|---|---|---|---|---|---|
| 2 | Yamamure, K. | *J. Org. Chem.* | **1978** | *43* | 724 |
| 3 | Fujiwara, Y. | *J. Organomet. Chem.* | **1984** | *266* | C44 |
| 4 | Fujiwara, Y. | *Chem. Lett.* | **1989** | | 1687 |

## F U J I W A R A   Lanthanide (Yb) reaction

Use of ytterbium or other lanthanoids in substitution, reduction and 1,2 addition (see 1st edition).

2-Oxo-1,3-triphenylpropan-1-ol 6.[3] Yb powder (173 mg; 1mmol) under $N_2$ was treated with a drop of MeI and was heated to activate the Yb. THF (2 mL) was added, followed by HMPA (1 mL). Under stirring benzophenone 1 (182 mg; 1 mmol) in THF (2 mL) was added, followed by phenylacetonitrile 5 (117 mg; 1 mmol). After 4 h stirring at 20°C the mixture was quenched with 2N HCl, extracted with $Et_2O$ and the product separated by medium pressure LC to afford 187 mg of 6 (65%) and 50 mg of 4 (35%).

| | | | | | |
|---|---|---|---|---|---|
| 1 | Fujiwara, Y. | *Chem. Lett.* | **1981** | | 1771 |
| 2 | Fujiwara, Y. | *J. Org. Chem.* | **1984** | 49 | 3237 |
| 3 | Fujiwara, Y. | *J. Org. Chem.* | **1988** | 53 | 6077 |
| 4 | Fujiwara, Y. | *J. Org. Chem.* | **1987** | 52 | 3524 |

## G A B R I E L  Amine Synthesis

Synthesis of primary amines from alkyl halides via imides (see 1st edition).

| 1 | Gabriel, S. | *Chem.Ber.* | **1887** | *20* | 2224 |
|---|---|---|---|---|---|
| 2 | Bradsher, C.H. | *J.Org.Chem.* | **1981** | *46* | 327 |
| 3 | Gibson, J.S. | *Angew.Chem.Int.Ed.* | **1968** | *7* | 919 |
| 4 | Ragnarsson, A. | *Acc.Chem.Res.* | **1991** | *24* | 285 |
| 5 | Allenstein, E. | *Chem.Ber.* | **1967** | *100* | 3551 |
| 6 | Han Yinglin | *Synthesis* | **1990** | | 122 |

**Sodium diformylamide 2.**[5] A mixture of formamide **1** (90 g, 2 mol) and NaOMe in MeOH (23.5 g Na in MeOH 200mL) was stirred at 20°C for 1 h, then was slowly evaporated on a Rotavap for 2 h at 80-90°C. The crystalline product after drying under vacuum for 3h afforded 95 g of **2** (100%) pure enough for the next step.

**p-Bromobenzyl amine 4.**[6] A mixture of bromobenzyl chloride **3** (20.55 g, 0.1 mol) and **2** (11.4 g, 0.12 mol) in EtOH (50 mL) was heated in an autoclave for 3 h at 80°C with stirring. The mixture was treated with conc HCl (10 mL) and refluxed with stirring for 2 h. After evaporation, the residue was treated under cooling with 50% NaOH and extracted with Et$_2$O. Evaporation of the solvent and distillation from KOH gave 14.88 g of **4** (80%), bp 247-250°C/760 Torr.

## GABRIEL–HEINE Aziridine Isomerization

Isomerization of N-acyl, N-double bond aziridines by acids, nucleophilic reagents or pyrolysis to oxazolines, imidazolines, thiazolines, triazolines

| 1 | Gabriel, S. | *Chem.Ber.* | **1895** | *28* | 2929 |
|---|---|---|---|---|---|
| 2 | Heine, H.W. | *J.Org.Chem.* | **1958** | *23* | 1554 |
| 3 | Heine, H.W. | *J.Org.Chem.* | **1960** | *25* | 461 |
| 4 | Heine, H.W. | *Angew.Chem.Int.Ed.* | **1962** | *1* | 528 |

**1-(N-p-Nitrophenylbenzimidoyl)aziridine 3.**[3] To a stirred mixture of aziridine **2** (1.1 g, 25.5 mmol), Et₃N (5.05 g, 50 mmol) in PhH (70 mL) was added in 1 h a solution of N-p-nitrophenylbenzimidoyl chloride **1** (6.52 g, 11.6 mmol). After 1h stirring at 20°C, the Et₃N·HCl was removed by filtration and the solvent evaporated to afford 6.6 g of crude **3**, mp 116-120°C. Recrystallization from i-PrOH gave **3**, mp 132-134°C.

**1-p-Nitrophenyl-2-phenyl-2-imidazole 4.** A mixture of **3** (100 mg, 0.37 mmol) in Me₂CO (50 mL) and KSCN (1 g) was refluxed for 47 h. After evaporation of the solvent, the residue was washed with water and filtered to afford 94 mg of **4** (94%), mp 169-174°C.

## G A R I G I P A T I Amidine Synthesis

Conversion of nitriles to amidine with $Me_3Al/NH_4Cl$ (methylchloroaluminium amide).

$$Me_3Al + NH_4Cl \xrightarrow[\text{5-25}^\circ]{\text{PhMe}} MeAl(Cl)NH_2$$

**1**          **2**                    **3**

**4**                                        **5** (64%)

| 1 | Garigipati, R.S. | *Tetrahedron Lett.* | **1990** | *31* | 1969 |
|---|---|---|---|---|---|
| 2 | Weinreb, S.M. | *Synth. Commun.* | **1982** | *12* | 989 |
| 3 | Moss, R.A. | *Tetrahedron Lett.* | **1995** | *36* | 8761 |

**Adamantane amidine hydrochloride 5.**[3] A cooled solution of $Me_3Al$ **1** (25 mL, 50 mmol) in PhMe under stirring, was added slowly to a suspension of $NH_4Cl$ **2** (2.9 g, 54 mmol) in dry PhMe (20 mL) at 5°C under $N_2$. After the addition, the mixture was warmed to 25°C and stirred for 2 h until gas evolution ($CH_4$) ceased. Adamantane carbonitrile **4** (4.83 g, 30 mmol) was added in PhMe (10 mL) and the mixture was heated to 80°C for 18 h under Ar, when TLC indicated the absence of **4**. The reaction mixture was poured into a slurry of $SiO_2$ (15 g) and $CHCl_3$ (50 mL) and stirred for 5 min. The $SiO_2$ was filtered off, washed with MeOH and the combined solvents were concentrated to a volume of 15 mL. The insoluble $NH_4Cl$ was removed by filtration and the filtrate was treated with MeOH/HCl (10 mL conc 2 g, 54 mmol) followed by $Et_2O$ (400 mL). After 10 h stirring the precipitate was filtered (5.8 g of crude **5**) and recrystallized from 4:1 iPrOH:$Me_2CO$ (150 mL). After 12 h stirring at 25°C the insoluble $NH_4Cl$ was removed by filtration, the filtrate was concentrated to15 mL and the product was precipitated with $Et_2O$ (300 mL), to afford 4.1 g of **5** (64%), mp 257-259°C.

**G A S S M A N** Oxindole Synthesis

Synthesis of oxindoles from anilines (see 1st edition)

| 1 | Gassman, P.G. | *J.Am.Chem.Soc.* | **1973** | 95 | 2718 |
|---|---|---|---|---|---|
| 2 | Gassman, P.G. | *J.Am.Chem.Soc.* | **1974** | 96 | 5506 |
| 3 | Johnson, P.D. | *J.Org.Chem.* | **1990** | 55 | 1374 |
| 4 | Wright, S.M. | *Tetrahedron Lett.* | **1996** | 37 | 4631 |

**Oxindole 7.**[2] To a stirred, cooled (-65°C) solution of aniline **1** (4.09g, 44 mmol) in $CH_2Cl_2$ (150 mL) was added dropwise t-butyl hypochlorite **2** (4.77 g, 44 mmol) in $CH_2Cl_2$ (20 mL). After 10 min, ethyl methylthioacetate **3** (5.89 g, 44 mmol) in $CH_2Cl_2$ (20 mL) was added (exothermic) and stirring was continued for 1 h. TEA **4** (4.44 g, 44 mmol) in $CH_2Cl_2$ (20 mL) was added. The mixture was allowed to warm to room temperature, water (50 mL) was added and the organic layer was evaporated. The residue was redissolved in $Et_2O$ (150 mL) and was stirred with 2N HCl (20 mL) for 24 h. Fitration afforded 6.61 g of **6** (84%). A solution of **6** (2.00 g, 11 mmol) in anh. EtOH (50 mL) was stirred and refluxed with W-2 Raney nickel (12 g) for 2 h. The supernatant and the washing solution were evaporated to dryness, The residue was dissolved in $CH_2Cl_2$ (20 mL), the solution dried ($MgSO_4$), filtered and evaporated to give 1.13 g of **7** (76%), mp 116-117°C.

**G A S T A L D I** Pyrazine Synthesis

Pyrazine synthesis from α-oximinoketones via α-aminoketones (see 1st edition)

| 1 | Gastaldi, G. | *Gazz.Chem.Soc.* | **1921** | 51 | 233 |
|---|---|---|---|---|---|
| 2 | Sharp, W. | *J.Chem.Soc* | **1948** | | 1862 |
| 3 | Krems, I., Spoeerri, P. | *Chem.Rev.* | **1947** | 40 | 301 |

## G ATTERMANN–KOCH Carbonylation

Synthesis of aromatic aldehydes or ketones using cyanide salts or CO-HCl and Lewis acids (see 1st edition).

| # | Author | Journal | Year | Vol | Page |
|---|--------|---------|------|-----|------|
| 1 | Gatterman, L., Koch, J. | *Chem. Ber.* | **1897** | 38 | 1622 |
| 2 | Gatterman, L. | *Chem. Ber.* | **1898** | 31 | 1194 |
| 3 | Adams, R. | *J. Am. Chem. Soc.* | **1923** | 45 | 2373 |
| 4 | Brunson, H.R. | *J. Org. Chem.* | **1967** | 32 | 3359 |
| 5 | Kreutzberg, A. | *Arh. Pharm.* | **1969** | 302 | 828 |
| 6 | Tanaka, M. | *J. Org. Chem.* | **1992** | 57 | 2677 |
| 7 | Tanaka, M. | *J. Org. Chem.* | **1995** | 60 | 2106 |
| 8 | Tanaka, M. | *J. Chem. Soc. Chem. Commun.* | **1996** | | 159 |
| 9 | Gore, P.M. | *Chem. Rev.* | **1955** | 55 | 235 |
| 10 | Truce, W.E. | *Org. React.* | **1957** | 9 | 37 |

**Resorcinol aldehyde 2.**[3] HCl gas was bubbled for 2 h into **1** (20 g, 0.18 mol) and Zn(CN)$_2$ (37 g, 0.27 mol) in Et$_2$O (150 mL). After decantation the residue was crystallized from water (100 mL) to give 12.5 g of **2** (50 %), mp 135-137°C.

**2-Methyl-2-phenylindanone 5.**[4] To an efficiently stirred suspension of AlCl$_3$ (42 g, 0.3 mol) in PhH **3** (140 g, 1.8 mol), was added 1,2,2-trichloropropane **4** (44.5 g, 0.3 mol) over 3 h at 24-27°C while CO was rapidly bubbled in. Usual workup, followed by vacuum distillation and crystallization from EtOH afforded 39 g of **5** (58 %), mp 111°C.

## G E W A L D  2-Aminoheterocycles Synthesis

Formation of 2-aminothiophenes by condensation of α-mercaptoaldehydes or ketones with an activated nitrile or by condensation of carbonyl derivatives with activated nitriles and sulfur. Also formation of 2-aminofurans or 2-aminopyrroles from α-hydroxy- or α-aminoketones (see 1st edition).

| 1 | Gewald, K. | *Angew. Chem.* | **1961** | 73 | 114 |
|---|---|---|---|---|---|
| 2 | Gewald, K. | *Chem. Ber.* | **1965** | 98 | 3571 |
| 3 | Gewald, K. | *Z. Chem.* | **1962** | 2 | 305 |
| 4 | Gewald, K. | *J. Prakt. Chem.* | **1973** | 315 | 39 |
| 5 | Peet, P.N. | *J. Heterocyclic Chem.* | **1968** | 23 | 129 |
| 6 | Sabnis, R.W. | *Sulfur Reports* | **1994** | 16 | 1 |

## G I E S E  Free Radical Synthesis

Carbon-carbon bond formation via free radicals formed from organotin or organomercury compounds.

| 1 | Giese, B. | *Chem.Ber.* | **1979** | *112* | 3766 |
|---|-----------|-------------|----------|-------|------|
| 2 | Baldwin, J.E. | *J.Chem.Soc.Chem.Commun.* | **1983** | | 944 |
| 3 | Danishefsky, S. | *J.Org.Chem.* | **1982** | *47* | 2232 |
| 4 | Neumann, W.P. | *J.Org.Chem.* | **1991** | *56* | 5771 |
| 5 | Neumann, W.P. | *J.Chem.Soc.Perkin 1* | **1992** | | 3165 |
| 6 | Curran, D.P. | *J.Chem.Soc.Perkin 1* | **1995** | | 3061 |
| 7 | Giese, B. | *Angew.Chem.Int.Ed.* | **1985** | *24* | 553 |
| 8 | Giese, B. | *Org.React.* | **1996** | *48* | 301 |
| 9 | Barluenga, J. | *Chem.Rev.* | **1988** | *88* | 487 |

**C-Glucoside 3.**[2] A mixture of selenoglucoside **1** (584 mg, 1.2 mmol) and methyl acrylate **2** (552 mg, 12 mmol) in PhMe (2 mL) at reflux was treated with $Ph_3SnH$ (1.26 g, 3.6 mmol) in PhMe added over a period of 13 h. Chromatography (silica gel) afforded 180 mg of **3** (40%).

**1-(Chloromercurymethyl)-2-methoxycyclohexane 5.**[1] To a solution of $Hg(OAc)_2$ (49.8 g, 156 mmol) in MeOH (700 mL) was added norcaran **4** (15 g, 156 mmol) at 20°C. After 6 days, the solvent was evaporated, the oily residue (61 g) extracted with $CH_2Cl_2$. After filtration and evaporation, the new residue was dissolved in MeOH and treated with NaCl, to afford finally 51 g of **5** (90%).

**2-Methoxy-1-cyclohexanebutanenitrile 6.** A solution of **5** (5.2 g, 1.5 mmol) and acrylonitrile (1.59 g, 30 mmol) in $CH_2Cl_2$ (10 mL) was treated with $NaBH_4$ (400 mg, 10 mmol) in water (1.5 mL) at 20°C. A second portion of $NaBH_4$ (100 mg, 2.5 mmol) was added with stirring for 1 h. Evaporation of the solvent afforded 217 mg of **6** (80%), bp 80°C/0.06 mm.

## GILMAN – LIPSHUTZ - POSNER Organocrupate Reagents

Improved organocuprate reagents, obtained from CuCN or CuSCN and organolithium (magnesium) compounds, used in addition, substitution, selective ligand transfer, epoxide opening.

$1 : 9 (92\%) (SN^{2'})$

| 1 | Gilman, H. | *J.Org.Chem.* | **1952** | 17 | 1630 |
|---|---|---|---|---|---|
| 2 | Posner, G.H. | *J.Am.Chem.Soc.* | **1972** | 94 | 5106 |
| 3 | Lipshutz, B.H. | *J.Org.Chem.* | **1983** | 48 | 546 |
| 4 | Lipshutz, B.H. | *Tetrahedron* | **1986** | 42 | 3361 |
| 5 | Dieter, R.K. | *Symlett* | **1997** | | 801 |
| 6 | Posner, G.H. | *Org. React.* | **1977** | 19 | 1093 |
| 7 | Lipshutz, B.H. | *Org. React.* | **1992** | 41 | |

**Pelargonitrile (2).[3]** To a slurry of CuCN (89.6 mg, 1 mmol) at –78°C in THF (1 mL) were added n-BuLi (0.8 mL, 2 mmol). 5-Bromovaleronitrile 1 (89 $\mu$ L, 0.77 mmol) was added at –50°C and after 2.5 h stirring at –50°C work up and chromatography (silica gel, 10% Et$_2$O in pentane) afforded 99 mg of 2 (92%).

**Ketone (5).[4]** CuCN (102 mg, 1.14 mmol) in THF (1mL) under Ar was cooled at –78°C. 2-Thienyllithium (from thiophene, 91 $\mu$ L, 1.14 mmol) in THF (1 mL) at –30°C and 1.14 mmol t-BuLi (0.47 mL, 2.44 mmol in hexane) was stirred at 0°C for 30 min. All was added to CuCN at –78°C over 30 min. Grignard reagent 4 (80 $\mu$ L, 1.42 M in THF, 1.14 mmol) cooled to –78°C, was added dropwise and the mixture was warmed to 0°C for 2 min and cooled back to –78°C. Cyclohexenone 3 (100 L, 1.03 mmol) was added for 2.25 h at –78°C and quenched with 5 mL NH$_4$Cl/NH$_4$OH. Usual work up and chromatography (Et$_2$O : Skellysolve) gave 186 mg of 5 (85 %).

## GILMAN-VAN ESS Ketone Synthesis

Synthesis of ketones directly from carboxylic acids and alkyl or aryl lithium via addition to lithium carboxylates.

$$F_3C-CO_2H \xrightarrow{\text{n-BuLi}} F_3C-COOLi \xrightarrow{\text{n-BuLi}} F_3C-\underset{\underset{OLi}{|}}{\overset{\overset{OLi}{|}}{C}}-C_4H_9\text{-}n \longrightarrow F_3C-\underset{\underset{O}{||}}{C}-C_4H_9\text{-}n$$

1        2        3        4

$$Ph-COOLi + PhLi \longrightarrow Ph-\underset{\underset{OLi}{|}}{\overset{\overset{OLi}{|}}{C}}-Ph \longrightarrow Ph-\underset{\underset{O}{||}}{C}-Ph$$

$$Ph-\underset{\underset{Me}{|}}{CH}-\underset{\underset{O}{||}}{C}-OLi \xrightarrow[\text{2) H}_2\text{O}]{\text{1) PhLi}} Ph-\underset{\underset{Me}{|}}{CH}-\underset{\underset{O}{||}}{C}-Ph$$

| | | | | | |
|---|---|---|---|---|---|
| 1 | Gilman, H., Van Ess, P.R. | *J.Am.Chem.Soc.* | **1933** | *55* | 1258 |
| 2 | Gilman, H. | *J.Am.Chem.Soc.* | **1949** | *71* | 1499 |
| 3 | Tegner, C. | *Acta Chem.Scand.* | **1952** | *6* | 782 |
| 4 | Zook, H.D. | *J.Am.Chem.Soc.* | **1955** | *77* | 4406 |
| 5 | Schöllkopf, U. | *Liebigs Ann.* | **1961** | *642* | 1 |

## GINGRAS Reagent

Tetrabutylammonium difluorotriphenylstannate, a fluorine source for nucleophilic displacement reactions and a phenyl transfer agent in coupling reactions.

$$Ph_3SnF \xrightarrow[CH_2Cl_2,\ 20°]{nBu_4NF.3H_2O} nBu_4N^+(Ph_3SnF_2)^-$$

**1**

85%

$$CH_3(CH_2)_6CHO + (Tf_2O) \xrightarrow{PTS} CH_3(CH_2)_6CH(OTf)_2 \xrightarrow{\ 1\ } CH_3(CH_2)_6\text{-}CHF_2$$

**2**                 **3**          **4 (77%)**[3]

| | | | | | |
|---|---|---|---|---|---|
| 1 | Gingras, M. | *Tetrahedron Lett.* | **1991** | *32* | 7381 |
| 2 | Garcia-Martinez, A. | *Synlett* | **1993** | | 587 |
| 3 | Garcia-Martinez, A. | *Tetrahedron Lett.* | **1992** | *33* | 7787 |
| 4 | Garcia-Martinez, A. | *Synlett* | **1994** | | 1047 |

**1,1-Difluorooctane (4).**[3] To a solution of gem-bistriflate **3** (676 mg, 2 mmol) in $CH_2Cl_2$ was added **1** (3.7 g, 6 mmol). After 2 h stirring at 20 °C, pentane (50 mL) was added slowly. The inorganics were separated and the solvent distilled (Vigreux 20 cm). Chromatography afforded 233 mg of **4** (77%).

## GRAHAM Diazirine Synthesis

Oxidation of amidines with sodium hypohalides to give alkyl, aryl or alkoxy-3-halodiazirines.

$$Me\text{-}C\begin{array}{c}NH\cdot HCl\\ \\NH_2\end{array} \xrightarrow[DMSO]{NaOCl} \quad \longrightarrow \quad \longrightarrow$$

**1**                                             **2 (60%)**

| | | | | | |
|---|---|---|---|---|---|
| 1 | Graham, W.H. | *J. Am. Chem. Soc.* | **1965** | 87 | 4396 |
| 2 | Moss, R.A. | *Tetrahedron Lett.* | **1995** | 36 | 8761 |

**Methylchlorodiazirine 2.**[1] To a solution of acetamidinium HCl **1** (2.36 g; 25 mmol) in DMSO (150 mL) containing LiCl (10 g) was added rapidly a solution of NaOCl (300 mL; 0.78M) containing NaCl (60 g). The volatile product was condensed in a series of U tubes cooled to -35°C; -80°C; -126°C and -196°C. Methylchloroazirine **2** was collected in tube III (-126°C), 1.36 g (60%).

## GLASER – SONDHEIMER - CHODKIEWCZ Acetylene Coupling

Coupling of acetylenes with other acetylenes or with unsaturated halides or triflates catalyzed by Cu(I) or Cu-Pd (see 1st edition).

$$2\ Me_2C-C\equiv CH \xrightarrow[\text{Pyr , MeOH}]{Cu_2Cl_2,\ O_2} Me_2C-C\equiv C-C\equiv C-CMe_2$$

with OH groups; **1** ; product **2** (90%)[3] with two OH groups.

$$Me-(CH_2)_3-C\equiv CMgBr\ +\ BrCH_2-C\equiv C-(CH_2)_3-Me \xrightarrow[\text{Et}_2O,\ N_2]{Cu_2Cl_2} (Me-(CH_2)_3-C\equiv C)_2\ CH_2$$

**3**         **4**         **5** (43%)

$$HC\equiv C-(CH_2)_4-C\equiv CH \xrightarrow[\text{NH}_4Cl,\ 55°]{Cu_2Cl_2,\ O_2}$$

**6**

ring structure **7** (7%)

$$p\text{-}MeO_2C-C_6H_4-OCH_2-C\equiv CH \xrightarrow[\text{MeCN, Ar}]{Cu(OAc)_2} (p\text{-}MeO_2C-C_6H_4-OCH_2-C\equiv C)_2$$

**8**         **9** (83%)[15]

$$PhI\ +\ HC\equiv CH \xrightarrow[\text{Et}_2NH]{CuI\ /\ PdCl_2\ (PPh_3)_2} Ph-C\equiv C-Ph$$

**10**      **11**         **12** (83%)[19]

$$\textbf{13} + HC\equiv C-(CH_2)_3-CH_3 \xrightarrow[\text{CuI / Et}_3N]{PdCl_2(PPh_3)_2} \textbf{15}\ (87\%)$$

**14**

| | | | | | |
|---|---|---|---|---|---|
| 1 | Glaser, C. | *Chem.Ber.* | **1869** | 2 | 422 |
| 2 | Chodkiewcz, W. | *Ann.Chim.Paris* | **1957** | 2 | 819 (13) |
| 3 | Stansbury, H.A. | *J.Org.Chem.* | **1962** | 27 | 320 |
| 4 | Walton, D.R.M. | *Synthesis* | **1974** | | 890 |
| 5 | Straus, F. | *Liebigs Ann.* | **1905** | 342 | 190 |
| 6 | Weedon, B.C.L. | *J.Chem.Soc.* | **1954** | | 1704 |
| 7 | Weedon, B.C.L. | *J.Chem.Soc.* | **1957** | | 3868 |
| 8 | Weedon, B.C.L. | *Proc.Chem.Soc.* | **1958** | | 303 |
| 9 | Sondheimer, F. | *J.Am.Chem.Soc.* | **1956** | 78 | 4178 |
| 10 | Sondheimer, F. | *J.Am.Chem.Soc.* | **1957** | 79 | 5817 |
| 11 | Sondheimer, F. | *Acc.Chem.Res.* | **1982** | 15 | 96 |
| 12 | Eglington, G. | *Adv.Org.Chem.* | **1963** | 4 | 225 |

| 13 Eglington, G. | *Proc.Chem.Soc.* | **1958** | | 350 |
|---|---|---|---|---|
| 14 Akiyama,S. | *Bull.Chem.Soc.Jpn.* | **1960** | *33* | 1293 |
| 15 Vogtle,F. | *Synthesis* | **1992** | | 58 |
| 16 Stephens,R.D., Castro C. | *J.Org.Chem.* | **1963** | *28* | 3313 |
| 17 Campbell, I.D. | *J.Chem.Soc.Chem.Commun.* | **1966** | | 87 |
| 18 Staab, H.E. | *Synthesis* | **1974** | | 424 |
| 19 Schintzer, D. | *Synthesis* | **1995** | | 299 |
| 20 Sonogashiro, K. | *Tetrahedron Lett.* | **1975** | | 4470 |
| 21 Quing, F.L. | *Tetrahedron Lett.* | **1997** | *38* | 6729 |
| 22 Hagihara, N. | *Synthesis* | **1980** | | 627 |
| 23 Rychnovsky, S.D. | *Tetrahedron Lett.* | **1996** | *37* | 7910 |

**Trideca-5,8-diyne (5).**[7] Hex-1-yne (5.14 g, 62.6 mmol) in $Et_2O$ (20 mL) was added to EtMgBr (from Mg 1.4 g, EtBr 6.2 g in $Et_2O$ 50 mL) under $N_2$. After 3h stirring and reflux, $Cu_2Cl_2$ (250 mg, 2.5 mmol) was added followed after 15 min by 1-bromohept-2-yne **4** (10 g, 57 mmol) in $Et_2O$. Stirring for 3 h at 20°C and 16 h reflux followed by usual work up, gave after distillation 4.8 g of **5** (43.5%), bp 60-62°C/$10^{-4}$ mm.

**Cyclohexadeca-1,3,9,11-tetrayne (7).**[10] A solution of octa-1,7-diyne **6** (25 g, 235 mmol) in EtOH was added to a mixture of $Cu_2Cl_2$ (50 g) and $NH_4Cl$ (80 g) in water (215 mL) containing 32% HCl (0.5 mL). The mixture was heated to 55°C and oxygen was bubbled through the mixture under efficient stirring (the condenser maintained at −40°C). After 6 h the product was extracted with PhH, the solvent evaporated and the residue chromatographed ($Al_2O_3$ petroleum ether : PhH). After recrystallization from petroleum ether, there was obtained 1.62 g of **7** (6.7 %), mp 160-162°C.

**1,6-Bis(4-methoxycarbonylphenoxy)hexa-2,4-diyne 9 .**[15] Methyl 4-(2-propynoxy) benzoate **8** (3.8 g, 20 mmol) and $Cu(OAc)_2.H_2O$ (20 g, 100 mmol) was dissolved in MeCN (500 mL) under Ar (750 mL) and stirred for 1 h. The cooled mixture was diluted with water. The precipitate was filtered and washed with water and dried. Chromatography (silica gel cyclohexane : $Et_2O$ 1 : 3) afforded 3.13 g of **9** (83%), mp 119°C.

**Diacetylene 15.** A mixture of triflate **13** (199 mg, 0.447 mmol), $nBu_4NI$ (495 mg, 1.34 mmol), $PdCl_2$ $(PPh_3)_2$ (31 mg), CuI (26 mg) and $Et_3N$/DMF (1:5) (2.3 mL0 was degassed and **14** (0.211 mL) was added. After 3 h stirring at 70°C usual work up and chromatography (silica gel, 30% $CH_2Cl_2$ in hexane) gave 120 mg of **15** (87%).

## GOLD Reagent

Reagent **3** for dialkylaminomethynilation of activated methylenes or $NH_2$ groups (see 1st edition).

$$\text{1} + 6\ Me_2NCHO \xrightarrow[\text{(- CO}_2\text{)}]{65°C\ \ dioxane} \underset{\text{3 (86\%)}}{Me_2\overset{+}{N}=CH-N=CH-NMe_2\ \ \overset{Cl^-}{}}$$

$$\underset{\text{4}}{p\text{-Br-}C_6H_4-CO-CH_3} \cdot +\ \text{3} \xrightarrow[\text{refl. 24 h}]{NaOMe\ /\ MeOH} \underset{\text{5 (74\%)}^{5}}{p\text{-Br-}C_6H_4-CO-CH=CH-NMe_2}$$

$$Ph-NH_2 \xrightarrow{\text{3}} Ph-N=CH-NMe_2 \qquad Ph-NH-NH_2 \xrightarrow{\text{3}} \underset{Ph-N}{}$$

| | | | | | |
|---|---|---|---|---|---|
| 1 | Gold, H. | *Angew.Chem.* | **1960** | *72* | 959 |
| 2 | Eschenmoser, A. | *Angew.Chem.Int.Ed.* | **1971** | *10* | 330 |
| 3 | Kunst, G. | *Angew.Chem.Int.Ed.* | **1977** | *15* | 239 |
| 4 | Bryson, T.A. | *J.Org.Chem.* | **1980** | *45* | 524 |
| 5 | Gupton, J.T. | *J.Org.Chem.* | **1980** | *45* | 4522 |

## GOMBERG – BACHMANN – GRAEBE - ULMANN Arylation

Aryl-aryl bond formation via diazonium salts. Carbazole synthesis by intramolecular aryl-aryl bond formation (see 1st edition).

| | | | | | |
|---|---|---|---|---|---|
| 1 | Gomberg, M., Bachmann, W.E. | *J.Am.Chem.Soc.* | **1924** | *42* | 2339 |
| 2 | Smith, P.A.S. | *J.Am.Chem.Soc.* | **1951** | *73* | 2452,2626 |
| 3 | Dermer, O.C. | *Chem.Rev.* | **1957** | *57* | 77 |
| 4 | Graebe, C., Ullman, F. | *Liebigs Ann.* | **1896** | *291* | 16 |
| 5 | Ashton, B.W. | *J.Chem.Soc.* | **1957** | | 4559 |
| 6 | Campbell, N | *Chem.Rev.* | **1948** | *40* | 360 |
| 7 | Alvarez Builla, J.. | *Tetrahedron Lett* | **1993** | *34* | 2673 |

## GRÄNACHER Homologation

Homologation of aromatic aldehydes to arylpropanoic acid derivatives, including arylalanines, via condensation with thiazolidone **2** (rhodanine **3**)

| 1 | Gränacher, Ch. | *Helv.Chim.Acta* | **1922** | 5 | 610 |
|---|----------------|------------------|----------|-----|----------|
| 2 | Gränacher, Ch. | *Helv.Chim.Acta* | **1923** | 5 | 458, 467 |
| 3 | Hibbert, H. | *J.Am.Chem.Soc.* | **1947** | 69 | 1208 |
| 4 | Heilbron, J. | *J.Chem.Soc.* | **1949** | | 2099 |

**Vanillalrhodanine 3.**[3] Vanillin **1** (100 g, 0.657 mol), 2-thioxo-4-thiazolidone **2** (87.5 g, 0.657 mol) and anh. NaOAc (150 g) were refluxed in AcOH (400 mL) for 1h, decanted in water (3000 mL) and stirred for 3 h. Filtration and drying afforded 169.5 g of **3** (97%), mp 227-8°C.

**α-Thioketo-β-4-hydroxy-3-methoxyphenyl pyruvic acid 4.** **3** (40 g, 0.15 mol) was heated in 15% NaOH sol. (260 mL) for 45 min at 100°C. The cooled (-15 °C) mixture, acidified with 10% HCl (278 mL) afforded after filtration 34 g of **4** (100%), mp 153-155°C or mp 157-158°C (MeOH).

**Oxime 5.** H₂NOH·HCl (48 g, 0.69 mol) basified with NaOMe, was added to **4** (50 g, 0.22 mol). The mixture was refluxed for 1 h, the solvent removed in vacuum and the residue dissolved in 5% NaOH (380 mL) and acidified with 10% HCl (360 mL) to give 49.5 g of **5** (100%), mp 138-139°C (water).

**Acetylhomovanillinonitrile 6.** **5** (51.5 g, 0.228 mol) heated in Ac₂O (220 mL) gave 39.7 g of **6** (84.5%), mp 51-52°C, bp 200°C/15 mm.

## GRIECO Organoselenides

Displacement of OH by an ArSe group. Reaction of aryl selenocyanates with alcohols, aldehydes or carboxylic acids to give alkyl aryl selenides, homologation of aldehydes or esters of arylselenols.

| 1 | Grieco, P.A. | *J.Org.Chem.* | **1976** | *41* | 1485 |
|---|---|---|---|---|---|
| 2 | Grieco, P.A. | *J.Am.Chem.Soc.* | **1977** | *99* | 5210 |
| 3 | Grieco, P.A. | *J.Org.Chem.* | **1978** | *43* | 1283 |
| 4 | Krief, A. | *Bull.Soc.Chim.Fr.* | **1997** | *134* | 869 |

**Selenide 4.**[1] A solution of alcohol **1** (781 mg, 0.62 mmol) in pyridine containing o-nitrophenyl selenocyanate **2** (168 mg, 0.78 mmol) under N$_2$ was treated with tri-n-butylphosphine **3** (150 mg, 0.74 mmol) at 20°C. After 30 min stirring the solvent was removed in vaccuum and the residue chromatographed (hexane – Et$_2$O 3:1) to afford 170 mg of **4** (98%).

**Acrylonitrile 6.**[2] A solution of aldehyde **5** in THF containing **2** (1.5 equiv.) was treated with tri-*n*-butylphosphine **3** (1.5 equiv.) in THF. Stirring for 2.5 h, evaporation of the solvent and filtration through silica gel, gave **6** in 96% yield.

**Benzeneselenol ester 9.**[3] To a solution of **3** (1.11 g, 5.5 mmol) and carboxilic acid **7** (5 mmol) in CH$_2$Cl$_2$ (20 mL) was added phenyl seleno cyanate (2 equiv.). Usual work up afforded **9** in 88% yield.

## GRIECO Reagent

Pyridinium p-toluenesulfonate (PPTS) as a catalyst for protection of alcohols as the tetrahydropyranyl ethers, as well as for cleavage of ethers in warm EtOH (see 1st edition).

| 1 | Grieco, P.A. | J.Org.Chem. | **1977** | 42 | 3772 |
|---|---|---|---|---|---|
| 2 | Pinnick, H.W. | Tetrahedron Lett. | **1978** | 44 | 4261 |
| 3 | Mori, K. | J.Chem.Soc.Perkin 1 | **1993** | | 169 |

**Dialcohol (5).[3]** A solution of compound **4** (1.71 g, 2.41 mmol) and PPTS (20 mg) in MeOH (40mL) was stirred at 25°C for 2 days, then diluted with EtOAc, neutralized with NaHCO$_3$ and filtered through Florisil. Evaporation in vacuo gave 1.01 g of **5** (100%).

## GRIESS Deamination

Deamination of aromatic amines via diazonium salts, by means of alcohols (Griess), hypophosphorous acid, PO$_2$H$_3$ or Sn(OH)$_2$ (see 1st edition).

| 1 | Griess, P. | Phil.Trans | **1864** | 154 | 683 |
|---|---|---|---|---|---|
| 2 | Griess, P. | Chem.Ber. | **1897** | 21 | 547 |
| 3 | Howe, R. | J.Chem.Soc. (C) | **1966** | | 478 |
| 4 | Fletcher, T.L. | Synthesis | **1973** | | 610 |
| 5 | Cowdry, W.A. | Quart.Rev. | **1952** | 26 | 358 |
| 6 | Kornblum, N. | Org.React. | **1944** | 2 | 262 |

## GRIGNARD Reagents

Organomagnesium reagents capable of reacting with active "H" compounds or in additions to C=X bonds; also nickel catalyzed coupling (see also Riecke) (see 1st edition).

$$\text{1} \xrightarrow{\text{CH}_3\text{MgI}} \text{2 (89\%)}$$

$$\text{Ph}\diagup\diagdown^{\text{Br}} + \text{t - BuMgX} \xrightarrow{\text{Ni (II)}} \text{Ph}\diagup\diagdown^{\text{tBu}}$$

$$\text{3} + \text{4} \xrightarrow[\text{0°C}]{\text{Mg / THF}} \text{5}$$

| | | | | | |
|---|---|---|---|---|---|
| 1 | Barbier, P. | *C.R.* | **1899** | *128* | 110 |
| 2 | Grignard, V. | *C.R.* | **1900** | *130* | 1322 |
| 3 | Kirmse, W. | *Synthesis* | **1983** | | 994 |
| 4 | Vanderzaude, D.J.M. | *J.Org.Chem.* | **1997** | *62* | 1473 |
| 5 | Sonntag, N.O.V. | *Chem.Rev.* | **1953** | *53* | 372 |
| 6 | Bogdanovichi, B. | *Angew.Chem.* | **1983** | *95* | 749 |
| 7 | Walborsky, H.M. | *Acc.Chem.Res.* | **1990** | *23* | 286 |
| 8 | Walling, C. | *Acc.Chem.Res.* | **1991** | *24* | 255 |

**exo-2-Methylbicyclo[3.2.0]heptan-endo-2-ol (2).[3]** To MeMgI prepared from MeI (2.3 g, 16 mmol), Mg turnings (0.4 g, 17 mmol) in Et$_2$O (60 mL) was added bicyclo [3.2.0] heptan-2-one **1** (1.7 g, 15 mmol) in Et$_2$O (10 mL). After 1 h reflux the mixture was hydrolyzed (25 mL water) and extracted with Et$_2$O (2 x 25 mL). Evaporation gave 1.7 g of **2** (89%), purity 98% by GLC, purified by preparative GLC (Carbowax + KOH, 110°C), mp~25°C.

**1,2-Dithienoylbenzene 5.[4]** 2-Bromothiophene **4** (4.5 mL, 46 mmol) in THF (50 mL) was added to Mg (1.2 g, 40 mmol) in THF (50 mL). After 3.5 h stirring, this solution was added to 1,2-dipyridinyl benzene dithioate **3** (7.95 g, 23 mmol) in THF (150 mL) at 0°C. After 30 min stirring followed by usual working crude **5** was obtained in 95% yield. Recrystallization (CHCl$_3$/n-hexane) gave white crystals, mp 148-9°C.

## G R O B – E S C H E N M O S E R Fragmentation

An elimination reaction leading to fragmentation. An organic molecule containing a leaving group and a heteroatom undergoing acid, base or heat catalyzed fragmentation.

| 1 | Eschenmoser, A., Frey, A. | *Helv.Chim.Acta* | **1952** | *35* | 1660 |
| 2 | Grob, C.A. | *Angew.Chem.Int.Ed.* | **1967** | *6* | 1 |
| 3 | Grob, C.A. | *Helv.Chim.Acta* | **1955** | *38* | 594 |
| 4 | Grob, C.A. | *Helv.Chim.Acta* | **1962** | *45* | 1672 |
| 5 | Grob, C.A. | *Angew.Chem.Int.Ed.* | **1969** | *8* | 535 |
| 6 | Ochiai, M. | *J.Org.Chem.* | **1989** | *54* | 4832 |
| 7 | Beugelmans, R. | *Synlett* | **1994** | | 513 |

**5-Benzoyl-1-pentene 2.[6]** BF$_3$·Et$_2$O (0.12 mmol) was added to a solution of DCC (24.7 mg, 0.12 mmol) in CH$_2$Cl$_2$ (0.5 mL) and the mixture was stirred for 1 h at 20°C. The mixture was added to 1-phenyl-3-(tributyltin)cyclohexanol **1** (46.4 mg, 0.1 mmol) and ISB (iodosil benzene PhIO) (26.4 g, 0.12 mmol) in CH$_2$Cl$_2$ (0.5 mL) at 0°C. After 5 h stirring at 0°C the reaction mixture was washed with brine, extracted with CH$_2$Cl$_2$, the solvent evaporated and the product separated by preparative TLC, to afford 14 mg of **2** (81%) isolated yield.

**GROVENSTEIN–ZIMMERMANN** Carbanion Rearrangement

Stereospecific 1,2-sigmatropic rearrangement of 1-halo-2,2-di or 2,2,2-triarylethane with alkali metal derivatives.

$$Ph_3C\text{-}CH_2\text{-}Cl \xrightarrow[\text{i-octane}]{C_5H_{12}Na} Ph_2C\text{=}CHPh$$

**1**                                    **2** (36%)

| 1 | Grovenstein, E. Jr. | *J. Am. Chem. Soc.* | **1957** | *79* | 4895 |
|---|---|---|---|---|---|
| 2 | Zimmermann, H.E. | *J. Am. Chem. Soc.* | **1957** | *79* | 5455 |
| 3 | Grovenstein, E. Jr. | *J. Am. Chem. Soc.* | **1961** | *83* | 412 |
| 4 | Zimmermann, H.E. | *J. Am. Chem. Soc.* | **1961** | *83* | 1196 |
| 5 | Hauser, C.R. | *J. Org. Chem.* | **1966** | *31* | 4273 |
| 6 | Grovenstein, E. Jr. | *J. Am. Chem. Soc.* | **1972** | *94* | 4971 |
| 7 | Grovenstein, E. Jr. | *J. Org. Chem.* | **1989** | *51* | 1671 |

**Triphenylethylene 2.**[2] To a suspension of amylsodium (1.125 g; 9.7 mmol) in isooctane (15 mL) under high speed stirring (12.000 r.p.m.) was added 1,1,1-triphenyl-2-chloroethane **1** (2 g; 6.8 mmol) in $Et_2O$ (30 mL). After 30 min stirring under $N_2$ at 35°C, EtOH (1 mL) was added and the mixture was poured into ice. Extraction with PhH, concentration and chromatography (silica gel, 10-40% $Et_2O$ in hexane) gave 624 mg of **2** (36%), mp 63-65°C.

## **GRUBBS** Olefin Metathesis

Carbon-carbon bond formation by olefin metathesis catalyzed by transition metal ligands (Grubbs (A); Schrock (B); Hermann (C); Tebbe (D); Nugent (E).

| 1 | Calderon, N. | J.Am.Chem.Soc. | **1968** | *90* | 4133 |
|---|---|---|---|---|---|
| 2 | Villemin,D. | Tetrahedron Lett. | **1980** | 21 | 1715 |
| 3 | Grubbs, R.H. | J.Am.Chem.Soc. | **1992** | *11* | 5426;7324; |
|   |   |   |   | *4* | 3974 |
| 4 | Grubbs, R.H. | J.Am.Chem.Soc. | **1993** | *115* | 9856 |
| 5 | Grubbs, R.H. | J.Org.Chem. | **1994** | 59 | 4029 |
| 6 | Grubbs, R.H. | Angew.Chem.Int.Ed.Engl. | **1995** | 34 | 1833 |
| 7 | Wright, D.L. | Curr.Org.Chem. | **1999** | 3 | 211 |

**2-Phenyl-3,4-dihydropyran 2**. To a solution of catalyst A (9.3 mg, 0.01 mmol) in dry PhH was added the acyclic olefin ether **1** (94 mg, 0.5 mmol). The reaction mixture was stirred at 20°C for 5 h. The reaction mixture was quenched by exposure to air, concentrated and purified by flash chromatography to afford 69 mg of **2** (86 %) as a colorless oil.

## GUARESKY-THORPE Pyridone Synthesis

Synthesis of 2-pyridones from β-diketones and activated amides (see 1st edition).

| 1 | Guaresky, A. | *Mem. Real. Accad. Sci. Torino* | **1896** | *46* | 25 (II) |
|---|---|---|---|---|---|
| 2 | Thorpe, J.F. | *J. Chem. Soc.* | **1911** | *99* | 422 |
| 3 | Katritzky, A.R. | *Adv. Heterocycl. Chem.* | **1963** | *1* | 347 |
| 4 | Kellog, R.M. | *J. Org. Chem.* | **1986** | *45* | 2856 |

## GUY-LEMAIRE-GUETTE Reagent

Regioselective chlorination, bromination, nitration by hexachloro-cyclohexadienone reagents or 4-nitro-cyclohexadienone **5** (see 1st edition).

| 1 | Guy, A.; Lemaire, M.; Guette, J.P. | *Tetrahedron* | **1982** | *38* | 2346 |
|---|---|---|---|---|---|
| 2 | Guy, A.; Lemaire, M.; Guette, J.P. | *Tetrahedron* | **1982** | *38* | 2354 |
| 3 | Guy, A.; Lemaire, M.; Guette, J.P. | *Tetrahedron* | **1987** | *43* | 835 |
| 4 | Messmer, A. | *Tetrahedron* | **1986** | *42* | 5415 |
| 5 | Guy, A.; Lemaire, M.; Guette, J.P. | *Jansen Chim. Acta* | **1987** | *5* | 3 |

**1-Hydroxy-4-chloronaphtalene 4.**[2] A solution of α-naphtol **3** (720 mg; 5 mmol) in DMF (10 mL) was treated at 20°C with **2** (301 mg; 5 mmol). After 48 h at 20°C under stirring and after vacuum concentration, the residue was chromatographed on $Al_2O_3$ (heptane:EtOAc 7:3). Purification by chromatography on silica gel with the same solvents gave 0.8 g of **4** (99%), ratio of o-:p- 30:70.

## HADDADIN–ISSIDORIDES Quinoxaline Synthesis

Synthesis of quinoxaline N,N'-dioxides from benzofurazan oxides and ketone enolates or enamines (also known as the Beirut reaction) (see 1st edition).

3 (42%)

48%

| 1 | Haddadin, M.; Issidorides, C.H. | *Tetrahedron Lett.* | **1965** |  | 3253 |
| 2 | Haddadin, M.; Issidorides, C.H. | *J. Org. Chem.* | **1966** | *31* | 4067 |
| 3 | Haddadin, M.; Issidorides, C.H. | *Tetrahedron* | **1974** | *30* | 659 |
| 4 | Haddadin, M.; Issidorides, C.H. | *Heterocycles* | **1978** | *4* | 767 |
| 5 | Haddadin, M.; Issidorides, C.H. | *Heterocycles* | **1993** | *35* | 1503 |
| 6 | Haddadin, M.; Issidorides, C.H. | *Chem. Abstr.* | **1984** | *101* | 171, 227 |
| 7 | Lin, S.K. | *Yonji Huaxue* | **1991** | *11* | 106(1) |

**2-Phenyl-3-benzoylquinoxaline-N,N'-dioxide 3.**[2] A solution of benzofurazan-N-oxide **1** (3.4 g, 25 mmol) and dibenzoyl methane **2** (5.6 g, 26 mmol) in warm Et₃N (25 mL) was allowed to stand at 20°C for 24 h. The mixture was diluted with Et₃N and filtered to give 2.5 g of **3**. The filtrate after another 30 h afforded a second crop of crystals. The total yield of **3** was 3.6 g (42%), mp. 234°C (from MeOH). The benzoyl group can be removed by heating **3** (1 g) in 45 mL of 2% KOH in MeOH until all dissolved, to obtain 0.65 g (95%) of debenzoylated product, mp. 205-206°C.

## H A F N E R  Azulene Synthesis

Synthesis of azulenes by condensation of cyclopentadienes with derivatives of glutaric dialdehydes.

| 1 | Zincke, Th. | *Liebigs Ann.* | **1905** | *338* | 107; 121 |
|---|---|---|---|---|---|
| 2 | König, W. | *J. Prakt. Chem.* | **1904** | *105* | 134 |
| 3 | Ziegler, K.; Hafner, K. | *Angew. Chem.* | **1955** | *67* | 301 |
| 4 | Hafner, K. | *Liebigs Ann.* | **1957** | *606* | 79 |
| 5 | Hafner, K. | *Org. Synth.* | **1984** | *62* | 134 |

**Azulene 5.**[5] A mixture of 1-chloro-2,4-dinitrobenzene **2** (202.6 g; 1 mol) and pyridine **1** (1200 mL) was heated with stirring to 80-90°C for 4 h. To the cooled (0°C) mixture a solution of $Me_2NH$ (100 g; 2.22 mol) in **1** (300 mL) was added dropwise in 30 min and stirred for 12 h at 20°C. Under $N_2$, cyclopentadiene **4** (70 g; 1.06 mol) is added followed by a solution of 2.5M NaOMe (400 mL). Stirring is continued for 4 h, then heated (oil bath) to distill $Me_2NH$ and **1**. After addition of **1** (1000 mL) the mixture was heated to 125°C for 4 days. Evaporation of the solvent, extraction with hexane and chromatography (alumina II) afforded 65-75 g of **5** (51-59%), mp 96-97°C.

## H O U B E N – H O E S C H  Phenol Acylation

Synthesis of ketones (or aldehydes) by Lewis acid catalyzed acylation of phenols with nitriles or ortho formates (see 1st edition).

| 1 | Houben, J. | *Chem. Ber.* | **1913** | *46* | 2447 |
|---|---|---|---|---|---|
| 2 | Hoesch, K. | *Chem. Ber.* | **1915** | *48* | 1122 |
| 3 | Trucare, J. | *J. Org. Chem.* | **1963** | *28* | 3206 |
| 4 | Roger, R. | *Chem. Rev.* | **1961** | *61* | 184 |
| 5 | Spoerri, P.E. | *Org. React.* | **1949** | *5* | 387 |
| 6 | Gross, H. | *Chem. Ber.* | **1963** | *96* | 308 |

## HAJOS–PARRISH Enantioselective Aldol Cyclization

Enantioselective aldol condensation (cyclization) using (S)-proline as catalyst, with high optical yield.

**1**

**2** (97%, 93% ee)

| 1 | Hajos, Z.G., Parrish, D.R. | *J. Org. Chem.* | **1973** | *38* | 3244 |
| 2 | Hajos, Z.G., Parrish, D.R. | *J. Org. Chem.* | **1974** | *39* | 1612, 1615 |
| 3 | Swaminathan, S. | *Tetrahedron Asymm.* | **1996** | *7* | 2189 |

**(+)-(3aS,7aS)-3a,4,7,7a-Tetrahydro-3a-hydroxy-7a-methyl-1,5(6H)-indandione   2.[2]**
2-Methyl-2-(3'-oxobutyl)-cyclopentane-1,3-dione **1** (1.82 g, 10 mmol) and (S)-(-)-proline (1.15 g, 10 mmol) were stirred in MeCN under Ar at 20°C for a period of 6 days. (S)-Proline (1.11 g, 9.65 mmol) was recovered by filtration. After evaporation of the solvent, the residue was dissolved in EtOAc (30 mL) and filtered through silica gel (4 g) by suction, followed by washing the silica gel with EtOAc (60 mL). The combined filtrates gave after evaporation 1.77 g of crude **2** (97%), $\alpha_D^{25}$ = +64.0° (c 1.035, CHCl$_3$). Recrystallization from Et$_2$O gave the pure product, mp 119-119.5°C, $\alpha_D^{25}$ = +60.40° (c 1.06, CHCl$_3$).

## H A L L E R – B A U E R Ketone Cleavage

Cleavage of ketones, lacking α-hydrogens, with sodium amide (see 1st edition).

| 1 | Haller, A., Bauer, E. | *C.R.* | **1909** | *148* | 127 |
|---|---|---|---|---|---|
| 2 | Impastato, F.I. | *J. Am. Chem. Soc.* | **1962** | *84* | 4838 |
| 3 | Kaiser, E.M. | *Synthesis* | **1975** | | 395 |
| 4 | Paquette, L.A. | *J. Org. Chem.* | **1988** | *53* | 704 |
| 5 | Goverdhan, M. | *J. Org. Chem.* | **1955** | *60* | 279 |
| 6 | Paquette, L.A. | *Org. Prep. Proced. Intn.* | **1990** | *22* | 169 |
| 7 | Hamlin, K.E. | *Org. React.* | **1957** | *9* | 1 |

**1-Methyl-2,2-diphenylcyclopropane 2.**[2]   A mixture of NaNH$_2$ (3 g, 75 mmol) and 1-benzoyl-1-methyl-2,2-diphenylcyclopropane 1 (9.3 g, 30 mmol) in PhMe (80 mL) was refluxed for 5 h.  The cooled reaction mixture was treated with cracked ice (50 g) and the separated organic layer, after washing with brine was distilled.  The fraction bp. 106-107°C/2.5 mm was collected.  There were obtained 4.9 g of **2** (79%).

**Benzamide 4.**[3]   To benzophenone 3 (9.1 g, 50 mmol) and DABCO (16.8 g, 0.15 mol) in PhH (200 mL) under N$_2$ was added NaNH$_2$ (5.85 g, 0.15 mol).  After 5 h reflux with stirring, the cooled mixture was treated with 3N HCl (100 mL) and the aqueous layer was extracted with Et$_2$O.  The combined extracts were concentrated and the crystals washed with hexane.  There was obtained 4.4 g of benzamide (73%), mp 126-128°C.

## H A N T S C H Thiazole Synthesis

Condensation of alpha-halo ketones or aldehydes with thioureas in neutral, anhydrous solvents to give 2-amino thiazoles.

| 1 | Hantsch, A. | *Chem. Ber.* | **1887** | 20 | 3118 |
|---|---|---|---|---|---|
| 2 | Sharma, G.M. | *J. Indian. Chem. Soc.* | **1967** | 57 | 44 |
| 3 | Birkinshaw, T.N. | *J. Chem. Soc. Perkin Trans 1* | **1982** | | 939 |
| 4 | Arakawa, K. | *Chem. Pharm. Bull.* | **1972** | 20 | 1041 |
| 5 | Meakins, G.D. | *J. Chem. Soc. Perkin Trans 1* | **1987** | | 639 |
| 6 | Meyers, A.I. | *Tetrahedron Lett.* | **1994** | 35 | 2473 |

**2-(Phenylamino)-4-methylthiazole 3.[5]** To a stirred suspension of anhydrous MgSO$_4$ (1 g) in Me$_2$CO (15 mL) containing N-phenylthiourea 1 (2.5 g, 16.4 mmol), was added dropwise a solution of chloroacetone 2 (1.52 g, 16.4 mmol) in anh. Me$_2$CO (15 mL) under reflux. After 1 h stirring under reflux, the mixture was cooled, poured into brine (80 mL) and basified with 18 M ammonia. Extraction with Et$_2$O and evaporation of the solvent afforded 2.97 g (96%) of crude 3. Recrystallization from MeOH gave 2.1 g (68%) of 3 as a first crop, mp. 86-87°C and 0.72 g (23%) of a second crop of 3, mp. 85-86°C.

## H A N T S C H Pyridine Synthesis

One step synthesis of substituted pyridines from a β-keto ester, an aldehyde and ammonia (see 1st edition).

EtO$_2$C–CH$_2$
|
Me–C=O

**2**

+

NH$_3$

**3**

H$_2$C–CO$_2$Et
|
O=C–Me

**2**

$\xrightarrow[\text{refl. 3 h}]{\text{EtOH}}$

**4 (45%)**

o-Cl–C$_6$H$_4$–CHO  +  Me–CO–CH$_2$–CO$_2$Et

**5**                              **2**

$\xrightarrow{\text{NH}_3}$

**4 (92%)**

| 1 | Hantsch, A. | *Liebigs Ann.* | **1882** | 215 | 172 |
|---|---|---|---|---|---|
| 2 | Phillips, A.P. | *J. Am. Chem. Soc.* | **1949** | 71 | 4003 |
| 3 | Svetlik, J. | *J. Chem. Soc. Perkin 1* | **1987** | | 563 |
| 4 | Osaki, S. | *Synthesis* | **1983** | | 761 |
| 5 | Eisner, U. | *Chem. Rev.* | **1972** | 72 | 1 |

**3,5-Di(ethoxycarbonyl)-1,4-dihydro-2,6-dimethyl-4-(m-nitrophenyl)pyridine   4.**[2]
m-Nitrobenzaldehyde 1 (15.1 g, 0.1 mol), ethyl acetoacetate 2 (28.6 g, 0.22 mol) and conc. NH$_4$OH 3 (8 mL) in EtOH (60 mL) was heated to reflux for 3 h. The hot solution was diluted with water (40 mL), cooled, filtered and washed with 50% EtOH (10 mL) to give 16-18 g of 4 (43-48%), mp. 165-167°C.

**Diethyl 2,6-Dimethyl-4-aryl-1,4-dihydropyridine-3,5-dicarboxylate 6.**[4]   o-Chloro-benzaldehyde 5 (1.405 g, 10 mmol), 2 (2.86 g, 22 mol), EtOH (10 mL) and 3 (28%), were heated in an autoclave for 17 h at 110°C. Evaporation of the solvent and chromatography of the residue (silica gel, EtOAc:hexane) afforded 3.23 g of 6 (92%), mp. 122.5-123°C.

## HASS–BENDER Carbonyl Synthesis

Aldehyde or ketone synthesis by reaction of an alkyl halide with the sodium salt of 2-nitroalkanes (see 1st edition).

| 1 | Hass, H. B.; Bender, M. L. | *J. Am. Chem. Soc.* | **1949** | *71* | 1767 |
| 2 | Bersohn, M. | *J. Am. Chem. Soc.* | **1961** | *83* | 2136 |
| 3 | Epstein, W. W. | *Chem. Rev.* | **1967** | *67* | 247 |

## HASSNER–RUBOTTOM α-Hydroxylation

α-Hydroxylation, iodination, or oximation of carbonyls via silyl enol ethers (see 1st edition).

| 1 | Hassner, A. | *J. Org. Chem.* | **1974** | *39* | 1788,2558 |
| 2 | Rubottom, A. | *Tetrahedron Lett.* | **1974** | | 167 |
| 3 | Hassner, A. | *J. Org. Chem.* | **1975** | *40* | 3427 |
| 4 | Rubottom, A. | *J. Org. Chem.* | **1979** | *44* | 1731 |
| 5 | Ching-Kang, Sho | *J. Org. Chem.* | **1987** | *52* | 3919 |

### HASSNER  Aziridine-Azirine Synthesis

Stereospecific and regioselective addition of IN$_3$ (via iodonium ions) or of BrN$_3$ (ionic or free radical) to olefins and conversion of the adducts to aziridines or azirines (see 1st edition).

| 1 | Hassner, A. | *J. Am. Chem. Soc.* | **1965** | *87* | 4203 |
|---|-------------|---------------------|----------|------|------|
| 2 | Hassner, A. | *J. Am. Chem. Soc.* | **1969** | *91* | 5046 |
| 3 | Hassner, A. | *J. Org. Chem.* | **1968** | *33* | 2686 |
| 4 | Hassner, A. | *J. Am. Chem. Soc.* | **1968** | *90* | 216 |
| 5 | Hassner, A. | *Accts. Chem. Res.* | **1971** | *4* | 9 |
| 6 | Kohn, H. | *J. Org. Chem.* | **1991** | *56* | 4648 |

*trans*-**2-Methyl-3-phenylaziridine 3**.[2] To a slurry (15 g; 0.25 mol) of NaN$_3$ in MeCN (100 mL) below 0°C was added slowly iodine monochloride (18.3 g; 0.113 mol) over 15 min. After 10 min stirring, *E*-1-phenylpropene (0.1 mol) was added and the mixture stirred at 20°C overnight. The slurry was poured into 300 mL of cold 5% sodium thiosulfite and the orange oil extracted with ether, washed with water (5x200 mL), dried and evaporated. Flash chromatography (Woelm neutral alumina, petroleum ether) gave erythro **2** (100%). <u>Note</u>. *Some S-compounds react explosively with IN$_3$.*

To a stirred solution of LAH (2.5 g) in anh. ether (90 mL) was added **2** (10.3 g; 0.035 mol) in ether (10 mL) at 0°C over 20 min. Work up with 20% NaOH (10 mL) stirring, filtration, drying and evaporation gave 4.93 g (85%) of **3** and 5% of **1**.

# H A S S N E R – G H E R A – L I T T L E  MIRC Ring Closure

Ring closure to three, five, six and seven membered rings by Michael Initiated Ring Closure (MIRC) especially of sulfones, stereoselective for (3+2) cycloadditions.

| # | Author | Journal | Year | Vol | Page |
|---|--------|---------|------|-----|------|
| 1 | Ghera, E. | *Tetrahedron Lett.* | **1979** | | 4603 |
| 2 | Little, R.D. | *Tetrahedron Lett.* | **1980** | *21* | 2609 |
| 3 | Ghera, E.; Hassner, A. | *Tetrahedron Lett.* | **1990** | *31* | 3653 |
| 4 | Ghera, E.; Hassner, A. | *J.Org.Chem.* | **1996** | *61* | 4959 |
| 5 | Hassner, A | *Tetrahedron Asymm.* | **1998** | *9* | 2201 |
| 6 | Hassner, A | *Tetrahedron Asymm.* | **1996** | *7* | 2423 |

**Cyclopentane (3).**[3] To a stirred solution of **1** (1 equiv) in THF was added LDA (1.3 equiv) in THF at -78°C. After 15 min the cinnamate ester **2** (1.1 equiv) was added and the reaction mixture was stirred for 45 min. Quenching (aqueous HCl), extraction (Et$_2$O-20% CH$_2$Cl$_2$) and chromatography afforded **3** in 75% yield.

## HAUSER-BEAK  Ortho Lithiation

Ortho-alkylation of benzamides (see 1st edition).

| 1 | Hauser, C.R. | *J. Heterocycl. Chem.* | **1969** | 6 | 475 |
| 2 | Beak, P. | *J. Org. Chem.* | **1977** | 42 | 1823 |
| 3 | Hauser, C.R. | *J. Chem. Eng. Data* | **1978** | 23 | 183 |
| 4 | Beak, P. | *Acc. Chem. Res.* | **1982** | 15 | 306 |
| 5 | Katritzky, A.R. | *Org. Prep. Proceed. Intn.* | **1987** | 19 | 263 |

**2-n-Butylbenzanilide (3).**[5] To benzanilide **1** (1.97 g; 10 mmol) in THF (28.5 mL) and HMPA (1.5 mL) was added 2.5M n-butyllithium (4 mL) dropwise at –70°C. The mixture was warmed to 20°C and $CO_2$ was passed through for 5 min. After removal of the solvent under vacuum, THF (30 mL) was added under Ar and 1.7M tert-butyllithium (6.5 mL) was added slowly at –70°C. The mixture was maintained for 20 min at –20°C and recooled to –70°C. n-Butyl bromide **2** (1.37 g; 10 mmol) was added. After warming to 20°C the mixture was stirred for a few hours. The solvent was removed and 2N HCl was added to the residue at 0°C. The precipitate was collected and recrystallized to give 1.85 g of **3** (73%), mp 72-73°C.

## H A U S E R – K R A U S Annulation

Regioselective annulation of phthalides to naphthalene hydroquinone.

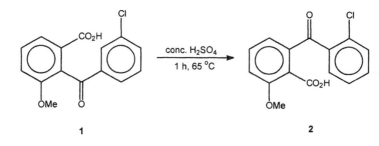

| 1 | Hauser, F. M. | *J. Am. Chem. Soc.* | **1977** | 99 | 4533 |
| 2 | Hauser, F. M. | *J. Org. Chem.* | **1978** | 43 | 178,180 |
| 3 | Kraus, G. A. | *Tetrahedron Lett.* | **1978** | 19 | 2263 |

**Hydroquinone (2).** To LDA(3.3 mmol) in THF (4 mL) and HMPA at −78 °C, was added 3-cyanophthalide **1** (3 mmol) in THF (3 mL) over 2 min. After 10 min stirring at −78 °C, R-CH=CH-CO-R (3 mmol) in THF (3 mL) was added in 1 min. After slow warming to 0 °C, the mixture was quenched with AcOH and diluted with Et$_2$O and water. Work up and chromatography yielded pure hydroquinone **2**.

## H A Y A S Y Rearrangement

Rearrangement of o-benzoylbenzoic acids (see 1st edition).

| 1 | Hayasy, M. | *J. Chem. Soc.* | **1927** | | 2516 |
| 2 | Sandin, R. B. | *J. Am. Chem. Soc.* | **1955** | 78 | 3817 |
| 3 | Caspar, M. L. | *J. Org. Chem.* | **1968** | 33 | 2020 |

**H A Y A S H I – U O Z U M I** Asymmetric Functionalization
Catalytic asymmetric synthesis of optically active alcohols via hydrosilylation of alkenes catalyzed by chiral monophosphine-palladium.

**2a** 83% (93:7)      **2b**

**3 (71%)**

96% ee       90-99% ee

| 1 | Uozumi, Y.; Hayashi, T. | *J. Am. Chem. Soc.* | **1991** | *113* | 9887 |
|---|---|---|---|---|---|
| 2 | Uozumi, Y.; Hayashi, T. | *Tetrahedron Lett.* | **1993** | *34* | 1335 |
| 3 | Uozumi, Y.; Hayashi, T. | *Tetrahedron Asymm.* | **1993** | *4* | 2419 |
| 4 | Hayashi, T. | *Bull. Chem. Soc. Jpn.* | **1995** | *68* | 713 |

**(R)-2-Octanol (3).[5]** To a mixture of PdCl($\eta^3$-C$_3$H$_5$)$_2$ (0.92 mg, 0.0025 mmol), (S)-2-methoxy-2'-diphenylphosphino-1,1'-binaphthyl ((S)-MeO-MOP) (4.68 mg, 0.01 mmol) and 1-octene **1** (560 mg, 5 mmol) was added trichlorosilane (745 mg, 5.5 mmol) at 0 °C and the reaction mixture was stirred for 24 h at 40 °C. Bulb to bulb distillation afforded 1.03 g of a mixture of **2a** and **2b** (83%) in a ratio of 93:7. To a suspension of KF (1.44 g, 24.9 mmol) and KHCO$_3$ (5.0 g, 50 mmol) in THF/MeOH (200 mL) was added **2a** and **2b** (1.03 g, 4.15 mmol) in a ratio of 87:13. To the suspension was added 30% H$_2$O$_2$ (4.15 mL) at 20 °C and the mixture was stirred for 12 h. The excess of H$_2$O$_2$ was reduced with Na$_2$S$_2$O$_3$.5H$_2$O (5 g), and after 1 h stirring, the mixture was filtered through celite. After usual work up 485 g of crude alcohol was obtained. To a solution of a crude mixture of alcohols (3.3 g) in hexane (100 mL) were added EtOH (20 mL) and powdered CaCl$_2$ (2.8 g). After 16 h of vigorous stirring the solid was removed by filtration and the solution concentrated in vacuum and distilled to give 2.5 g of **3** (71% from **2a+2b**), $\alpha_D^{25}$-10.3° (c, 5.59, EtOH).

# HECK–FUJIWARA Coupling

Cross-coupling reactions of aromatic or vinylic halides and olefins catalyzed by palladium derivatives (see 1st edition).

| 1 | Fujiwara, Y. | *Tetrahedron Lett.* | **1967** | | 1119 |
|---|---|---|---|---|---|
| 2 | Heck, R. F. | *J. Am. Chem. Soc.* | **1968** | 90 | 5518 |
| 3 | Heck, R. F. | *J. Am. Chem. Soc.* | **1974** | 96 | 1133 |
| 4 | Hassner, A. | *J. Org. Chem.* | **1984** | 49 | 2546 |
| 5 | Fujiwara, Y. | *Bull. Chem. Soc. Japan.* | **1990** | 63 | 438 |
| 6 | de Meijere, A. | *Synlett.* | **1990** | | 405 |
| 7 | Herrmann, W. A. | *Angew. Chem. Int. Ed.* | **1995** | 34 | 1844 |
| 8 | Halberg, A. | *J. Org. Chem.* | **1977** | 62 | 564 |

| 9  | Beller, M.           | *Tetrahedron Lett.*      | **1997** | *38*  | 2073 |
| 10 | Diaz-Ortiz A.        | *Synlett.*               | **1997** |       | 269  |
| 11 | Heck, R. F.          | *Org. React.*            | **1982** | *27*  | 345  |
| 12 | Hayashi, M.          | *Synthesis*              | **1997** |       | 1339 |
| 13 | Carreira, C. R. D.   | *Synlett.*               | **2000** |       | 1037 |
| 14 | Buono, G.            | *Angew. Chem. Int. Ed.*  | **2000** | *39*  | 1946 |
| 15 | Beletskaya, I. P.    | *Chem. Rev.*             | **2000** | *100* | 3009 |

**Trans-4-(p-di-n-butylaminostyryl) pyridine (3).[4]** A mixture of p-bromo-N,N-dibutylaniline **2** (5.68 g, 20 mmol), 4-vinylpyridine **1** (2.63 g, 25 mmol), Pd (OAc)$_2$ (45 mg, 0.2 mmol), tris o-tolylphosphine (TTP) (120 mg, 4 mmol) and Et$_3$N (10 mL) was heated at 110 °C for 72 h. To the cooled mixture was added water and CHCl$_3$ (all solids dissolved). The water layer was extracted with CHCl$_3$ (2 × 100 mL) and the combined organic solutions were washed, dried and evaporated. The residue recrystallized from cold hexane gave 5.29 g of **3** (86%), mp 80-81 °C.

**Trans-di(μ-acetato)-bis(o-(di-o-tolyphosphino) benzyl) dipalladium (4).** [7] To a red-brown solution of Pd(OAc)$_2$ (4.5 g, 20 mmol) in PhMe (500 mL) was added tris(o-tolyl) phosphine (8 g, 26.3 mmol). The solution was heated to 50 °C for 3 min and then cooled to 25 °C. After concentration in vacuum to a ¼ of its volume, hexane (500 mL) was added and the precipitate was filtered and dried (vacuum) to afford 8.8 g of **4** (93%).

**n-Butyl 4-(formylphenyl) acrylate (7).** A mixture of 4-bromobenzaldehyde **5** (d18.5 g, 100 mmol), n-butyl acrylate **6** (17.5 g, 0.14 mol) catalyst **4** (0.0005 mmol) and anh. NaOAc (9 g) in dimethylacetamide (100 mL) was heated under Ar at 135 °C for 12 h. Usual work up afforded **7** in quantitative yield.

**Trans stilbene (10).** [10] A mixture of styrene **8** (182 mg, 1.75 mmol), bromobenzene **9** (226.8 mg, 1.4 mmol), Pd(OAc)$_2$ (8.4 mg, 0.027 mmol) and triso-tolylphosphine (0.7 mL, 5 mmol) in dry Et$_3$N was charged into a 25 mL teflon vessel, under Ar, and irradiated in a Miele electronic M-720 microwave oven for 22 min. Usual work up and chromatography afforded 252 mg of **10** (100%).

## HELL–VOLHARDT–ZELINSKI Bromination

α-Bromination of carboxylic acids (see 1st edition).

| 1 | Hell, C. | *Chem. Ber.* | **1881** | 14 | 891 |
|---|----------|--------------|----------|-----|------|
| 2 | Volhardt, J. | *Liebigs Ann.* | **1887** | 242 | 141 |
| 3 | Zelinski, Y. | *Chem. Ber.* | **1887** | 20 | 2026 |
| 4 | Gibson, Th. | *J. Org. Chem.* | **1981** | 46 | 1003 |
| 5 | Haworth, C. | *Chem. Rev.* | **1962** | 62 | 99 |

**Methyl 2-(1,5-Dimethylbicyclo[2.1.1] hexanyl-2-bromoacetate)(2).[4]** To a mixture of acid 1(2.92 g, 12.4 mmol) in PBr$_3$ (7.94 g, 29.3 mmol) maintained for 1 h at 20 °C, was added Br$_2$ (7.94 g, 57 mmol) in two batches under Ar. The mixture was heated on a steam bath for 3 h, cooled, quenched with anh.MeOH, diluted with Et$_2$O and the organic layer was washed with 5% NaHCO$_3$ solution. Evaporation of the solvent and distillation of the residue gave 4.0 g of **2** (88%), bp 58-59 °C (0.33 mm).

## HENBEST Iridium Hydride Reagent

Reagent for selective reduction of ketones by means of an iridium hydride (see 1st edition).

| 1 | Henbest, H.B. | *J. Chem.Soc.* | **1962** | | 954 |
|---|---------------|----------------|----------|-----|------|
| 2 | Blicke, T. A. | *Proc. Chem. Soc.* | **1964** | | 361 |
| 3 | Hirschmann, H. | *J. Org. Chem.* | **1966** | 31 | 375 |
| 4 | Hill, J. | *J. Chem. Soc.(C)* | **1967** | | 783 |
| 5 | Kirk, D. M. | *J. Chem. Soc.(C)* | **1969** | | 1653 |

## H E N K E L – R A E C K E Carboxylic Acid Rearrangement

A thermal rearrangement or disproportionation of aromatic alkali metal carboxylates to symmetrical aromatic dicarboxylates.

**1**             **2 (85%)**

| | | | | | |
|---|---|---|---|---|---|
| 1 | Raecke, B. | *Angew. Chem.* | **1958** | *70* | 1 |
| 2 | Raecke, B. | *Brit. Pat.* | **1956** | | 747,204 |
| 3 | Ogata, Y. | *J. Org. Chem.* | **1960** | *25* | 2082 |
| 4 | McNellis, E. | *J. Org. Chem.* | **1963** | *30* | 1209 |
| 5 | Sorm, F. | *Coll. Czech. Commun* | **1959** | *24* | 2553 |

## H E N R Y Nitro Aldol Condensation

Base catalyzed aldol condensation of nitroalkanes with aldehydes (see 1st edition).

**1**        **2**        **3 (75%)**[3]

**4**        **5**        **6 (80%, 92% ee)**[4]

| | | | | | |
|---|---|---|---|---|---|
| 1 | Henry, L. | *C. R.* | **1895** | *120* | 1265 |
| 2 | Barker, R. | *J. Org. Chem.* | **1964** | *29* | 869 |
| 3 | Rosini, G. | *Synthesis* | **1983** | | 1014 |
| 4 | Shibasaki, M. | *Tetrahedron Lett.* | **1993** | *34* | 855 |
| 5 | Ballini. R. | *J. Org. Chem.* | **1994** | *39* | 5466 |
| 6 | Hass, H. B. | *Chem. Rev.* | **1943** | *32* | 406 |
| 7 | Lichtentaler. F. W. | *Angew. Chem. Int. Ed.* | **1964** | *3* | 211 |

# HERBST–ENGEL–KNOOP–OESTERLING Aminoacid Synthesis

Alpha amino acids (and aldehydes) synthesis by reaction of an alpha keto acid with another amino acid (Herbst-Engel) or by reaction of a keto acid with ammonia under reducing conditions (Knoop-Oesterling) (see 1st edition).

| | | | | | | |
|---|---|---|---|---|---|---|
| 1 | Herbst, R. M., Engel, W. | *J. Biol. Chem.* | **1934** | 107 | 505 |
| 2 | Herbst, R. M. | *J. Am. Chem. Soc.* | **1936** | 58 | 2239 |
| 3 | Mix, H. | *Z. Physiol. Chem.* | **1961** | 325 | 106 |
| 4 | Wieland | *Angew. Chem.* | **1942** | 55 | 147 |
| 5 | Knoop, F., Oesterling, H. | *Z. Physiol. Chem.* | **1925** | 148 | 194 |
| 6 | Wieland | *Chem. Ber.* | **1944** | 77 | 34 |

## HERZ Benzothiazole Synthesis

Reaction of aromatic amines with sulfur monochloride and an acyl chloride in the presence of Zn salts to give 1,3-benzothiazoles (see 1st edition).

| | | | | | |
|---|---|---|---|---|---|
| 1 | Herz, R. | *Chem. Zent. Bl.* | **1922** | 4 | 948 |
| 2 | Herz, R. | *U. S. Patent* | | | 1.637.023 |
| 3 | Huestins, L. D. | *J. Org. Chem.* | **1965** | 30 | 2763 |
| 4 | McChenard, B. L. | *J. Org. Chem.* | **1984** | 49 | 1224 |
| 5 | Warburton, W. K. | *Chem. Rev.* | **1957** | 57 | 1011 |

**6-Chloro-2-phenylbenzothiazole (6).**[3] **1** (5.7 g, 0.045 mol) in AcOH (7 mL) was added to $S_2Cl_2$ (42 g, 0.31 mol; 25 mL) stirred for 3 h at 25 °C then for 3 h at 70-80 °C. The cooled mixture was stirred with PhH (50 mL) and filtered to give 9.3 g (93%) of **2**, mp 210-225 °C (dec.). A vigorously stirred suspension of **2** (8.3 g, 37 mmol) in ice-water (500 mL) was made alkaline with 6 N NaOH. Then $NaHSO_3$ (5.0 g, 40 mmol) was added and after 1 h heating, the mixture was treated with Norite and filtered. Excess $ZnSO_4$ was added to precipitate the zinc mercaptide **4** (2.65 g, 38%). To a suspension of **4** (1.3 g, 3.4 mmol) in AcOH(40 mL) was added **5** (2.0 g, 14 mmol). After 30 min reflux, decomposition with water and crystallization from MeOH, gave 1.25 g of **6** (75%), mp 156-157 °C.

# HESSE–SCHMID "Zip" Reaction

Ring expansion of N-aminoalkyl lactams or of some hydroxy ketones by a zip reaction.

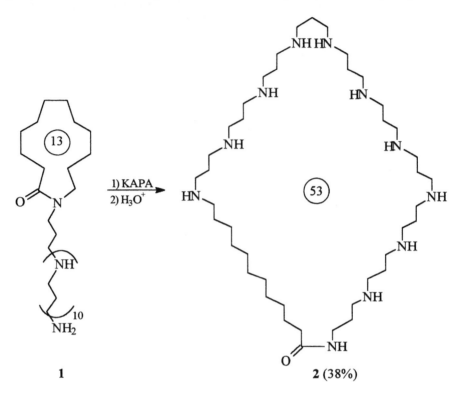

**1**                                                           **2 (38%)**

| 1 | Hesse, M.; Schmid, H. | *Helv. Chim. Acta* | **1968** | *51* | 1813 |
|---|---|---|---|---|---|
| 2 | Hesse, M.; Schmid, H. | *Helv. Chim. Acta* | **1974** | *57* | 414 |
| 3 | Hesse, M.; Schmid, H. | *Angew. Chem. Int. Ed. Engl.* | **1977** | *16* | 861 |
| 4 | Hesse, M.; Schmid, H. | *Angew. Chem. Int. Ed. Engl.* | **1978** | *17* | 200 |
| 5 | Hesse, M.; Schmid, H. | *Chimia* | **1978** | *32* | 58 |
| 6 | Hesse, M.; Schmid, H. | *Tetrahedron* | **1988** | *44* | 1573 |

**1,5,9,13,17,21,25,29,33,37,41-Undecaazatripentacontan-42-one 2.[4]** Treatment of **1** with potassium (3-amino)propylamide in 1,3-diaminopropane (KAPA; 45 min) and acidic work up afforded **2** in 38% yield.

## HILBERT–JOHNSON Nucleoside Synthesis

Nucleoside synthesis from bromosugars and methoxypyrimidines (see also Vorbrueggen) (see 1st edition).

| | | | | | |
|---|---|---|---|---|---|
| 1 | Johnson, T.B. | *Science* | **1929** | *69* | 579 |
| 2 | Hilbert, G.E. | *J. Am. Chem. Soc.* | **1930** | *52* | 2001 |
| 3 | Wolfrom, P.H. | *J. Org. Chem.* | **1965** | *30* | 3058 |
| 4 | Ulbricht, P.L. | *Angew. Chem.* | **1962** | *74* | 767 |
| 5 | Scott, M. | *Chem. Commun.* | **1996** | *26* | 2681 |

## HINSBERG Thiophene Synthesis

Synthesis of thiophenes from α-diketones (see 1st edition).

| | | | | | |
|---|---|---|---|---|---|
| 1 | Hinsberg, O. | *Chem. Ber.* | **1910** | *42* | 901 |
| 2 | Wynberg, N. | *J. Org. Chem.* | **1964** | *29* | 1919 |
| 3 | Wynberg, N. | *J. Am. Chem. Soc.* | **1965** | *87* | 1739 |
| 4 | Chadwick, D.J. | *J. Chem. Soc. Perkin I* | **1972** | | 2079 |

**2-Carbetoxy-3,4-diphenylthiophene-5-carboxylic acid 3.**[3] To a solution of t-BuOK [from 4.2 g K (0.11 g atom) and t-BuOH (100 mL)] was added at 30°C benzil **1** (8.0 g; 38 mmol) and diethyl thioacetate **2** (14.0 g; 68 mmol). After 15 min stirring the mixture was acidified with 15% HCl (20 mL) and the alcohol removed in vacuum. The residue was extracted with $Et_2O$ and the organic layer was extracted with 2N ammonia (20 mL portions) until the aqueous layer gave no precipitation upon acidification. The combined ammonia extracts were heated to remove $Et_2O$ and acidified to give 12.4 g of **3** (93%), mp 205-210°C.

### HIYAMA  Aminoacrylate Synthesis

Synthesis of 3-aminoacrylic acids or derivatives from nitriles and enolates by an aldol type condensation (see 1st edition).

| 1 | Hiyama, T. | *Tetrahedron Lett.* | **1982** | 23 | 1597 |
|---|-----------|---------------------|----------|----|------|
| 2 | Hiyama, T. | *Tetrahedron Lett.* | **1983** | 24 | 3509 |
| 3 | Hiyama, T. | *Bull. Chem. Soc. Jpn.* | **1987** | 60 | 2127, 2131, 2139 |

### HINSBERG–STOLLÉ  Indole-Oxindole Synthesis

Indole synthesis from anilines and glyoxal (Hinsberg), oxindole synthesis from anilines and α-haloacyl halides (Stollé) (see 1st edition).

| 1 | Hinsberg, O. | *Chem. Ber.* | **1888** | 21 | 110 |
|---|--------------|--------------|----------|----|-----|
| 2 | Burton, H. | *J. Chem. Soc.* | **1932** | | 546 |
| 3 | Stollé, R. | *Chem. Ber.* | **1913** | 46 | 3915 |
| 4 | Stollé, R. | *J. Prakt. Chem.* | **1930** | 128 | 1 |
| 5 | Julian, P.L. | *J. Am. Chem. Soc.* | **1935** | 57 | 563, 2026 |
| 6 | Sumter, W. | *Chem. Rev.* | **1944** | 34 | 396 |

### HIYAMA–HEATHCOCK Stereoselective Allylation

Stereoselective synthesis of anti homoallylic alcohols by $Cr^{2+}$ promoted allylation of aldehydes or ketones (see 1st edition).

| # | | | | | |
|---|---|---|---|---|---|
| 1 | Hiyama, T. | *J. Am. Chem. Soc.* | **1977** | *99* | 3179 |
| 2 | Heathcock, C.H. | *J. Am. Chem. Soc.* | **1977** | *99* | 247; 9109 |
| 3 | Heathcock, C.H. | *Tetrahedron Lett.* | **1978** | | 1185 |
| 4 | Hiyama, T. | *Bull. Chem. Soc. Jpn.* | **1982** | *55* | 562 |
| 5 | Hiyama, T. | *Tetrahedron Lett.* | **1983** | *24* | 5281 |
| 6 | Mulzer, J. | *Angew. Chem. Int. Ed.* | **1990** | *29* | 679 |
| 7 | Cintas, P. | *Synthesis* | **1992** | | 248 |

**1-Allylcyclohexanol 6.**[1] To a suspension of $CrCl_3$ (370 mg; 2.3 mmol) in THF (5 mL) at 0°C was added $LiAlH_4$ (44 mg; 1.2 mmol) under Ar, followed by cyclohexanone **4** (84.3 mg; 0.86 mmol) and then by allyl bromide **5** (145 mg; 1.2 mmol). After 2 h stirring at 20°C and work up there were obtained after distillation 93.9 mg of **6** (78%).

## H O C H – C A M P B E L L Aziridine Synthesis

Aziridines from oximes or from α-haloimines via azirines (see 1st edition).

| 1 | Hoch, J. | C.R. | **1934** | 198 | 1865 |
|---|----------|------|----------|-----|------|
| 2 | Campbell, K.N. | J. Org. Chem. | **1943** | 8 | 103 |
| 3 | De Kimpe, N. | J. Org. Chem. | **1980** | 45 | 5319 |
| 4 | Laurent, A. | Bull. Soc. Chim. Fr. | **1973** | | 2680 |
| 5 | Kotera, K. | Tetrahedron | **1968** | 24 | 3681, 5677 |

## HOFMANN Amide Degradation

Degradation of amides to amines by means of hypohalides or NBS (see 1st edition).

| 1 | Hofmann, A.W. | Chem. Ber. | **1881** | 14 | 2725 |
|---|---------------|------------|----------|-----|------|
| 2 | Magnieri, E. | J. Org. Chem. | **1958** | 23 | 2029 |
| 3 | Cohen, L.A. | Angew. Chem. | **1961** | 73 | 260 |
| 4 | Wawzoneck, S. | Org. Prep. Proced. Intn. | **1985** | 17 | 65 |
| 5 | Keillor, J.W. | J. Org. Chem. | **1997** | 62 | 7495 |
| 6 | Applequist, J. | Chem. Rev. | **1954** | 54 | 1083 |
| 7 | Wallis, E.S. | Org. React. | **1946** | 3 | 268 |

**Carbamate 6.**[5]   Amide **5** (76 mg, 0.5 mmol), NBS (90 mg, 0.5 mmol) and DBU (230 mL) in MeOH, were refluxed for 15 min. A second portion of NBS (90 mg) was added, reflux continued for 10 min, the solvent evaporated and the residue dissolved in EtOAc (50 mL). Flash chromatography (silica gel, 5% EtOAc in $CH_2Cl_2$) gave 86 mg of **6** (95%), mp. 87-89°C.

## H O F M A N N  Isonitrile Synthesis

Isonitrile synthesis from primary amines and dichlorocarbene (from chloroform) or dibromocarbene (see 1st edition).

$$EtNH_2 + CHBr_3 \xrightarrow[20°]{R_4\overset{+}{N}\overset{-}{O}H} Et-N\equiv C$$

$$\underset{1}{PhNH_2} + \underset{2}{CHCl_3} \xrightarrow[50\% \text{ NaOH}]{PhCH_2\overset{+}{N}Me_3Cl^-} \underset{3\ (57\%)}{Ph-N\equiv C}$$

| | | | | | |
|---|---|---|---|---|---|
| 1 | Hofmann, A.W. | *Liebigs Ann.* | **1868** | *146* | 107 |
| 2 | Smith, P.A.S. | *J. Org. Chem.* | **1958** | *23* | 1599 |
| 3 | Ugi, J.K. | *Angew. Chem. Int. Ed.* | **1972** | *11* | 530 |
| 4 | Weber, W.P. | *Tetrahedron Lett.* | **1972** | | 1637 |

**Phenylisocyanide 3.[4]**  To PhNH₂ 1 (18.6 g, 0.2 mol), alcohol free 2 (24 g, 0.2 mol) and benzyltrimethylammonium chloride (0.5 g) in CH₂Cl₂ (60 mL) was added at once 50% NaOH (60 mL).  After 10 min. induction, reflux began.  Reflux and stirring was continued for 1 h and work up gave 12 g of 3 (57%), bp. 50–52°C/11 torr.

## H O F M A N N  Elimination

Olefins by elimination from quaternary ammonium salts to form preferentially the less substituted olefin (see 1st edition).

| | | | | | |
|---|---|---|---|---|---|
| 1 | Hofmann, A.W. | *Chem. Ber.* | **1881** | *14* | 659 |
| 2 | Hinskey, R.G. | *J. Org. Chem.* | **1964** | *29* | 3678 |
| 3 | Cope, A.C. | *J. Org. Chem.* | **1965** | *30* | 2163 |
| 4 | Francke, H. | *Angew. Chem.* | **1960** | *72* | 397 |
| 5 | Horvath, A. | *Synthesis* | **1994** | | 102 |
| 6 | Brewster, J.H. | *Org. React.* | **1953** | *7* | 137 |
| 7 | Cope, A.C. | *Org. React.* | **1960** | *11* | 317 |

## **HOFMANN–LOEFFLER–FREYTAG**  Pyrrolidine Synthesis

Synthesis of pyrrolidines and piperidines from N-haloamines via a free radical reaction (see 1st edition).

| 1 | Hofmann, A.W. | *Chem. Ber.* | **1883** | *16* | 558 |
|---|---|---|---|---|---|
| 2 | Loeffler, K.; Freytag, C. | *Chem. Ber.* | **1909** | *42* | 3427 |
| 3 | Kimura, M. | *Synthesis* | **1976** | | 201 |
| 4 | Corey, E.J. | *J. Am. Chem. Soc.* | **1980** | *82* | |
| 5 | Wolff, M.E. | *Chem. Rev.* | **1963** | *63* | 55 |

## **HOFMANN – MARTIUS – REILLY – HICKINBOTTOM**  Aniline Rearrangement

Thermal or Lewis acid catalyzed rearrangement of N-alkylanilines to o-(p) alkylated anilines (see 1st edition).

| 1 | Hofmann, A.W.; Martius, C.A. | *Chem. Ber.* | **1871** | *4* | 742 |
|---|---|---|---|---|---|
| 2 | Hart, H. | *J. Org. Chem.* | **1962** | *27* | 116 |
| 3 | Ogatta, Y. | *Tetrahedron* | **1964** | *20* | 2717 |
| 4 | Reilly, J., Hickinbottom, W. | *J. Chem. Soc.* | **1920** | *117* | 103 |
| 5 | Cripps, R.W. | *J. Chem. Soc.* | **1943** | | 14 |
| 6 | Fischer, A. | *J. Org. Chem.* | **1960** | *25* | 463 |

## H O F F M A N – Y A M A M O T O Stereoselective Allylations

Synthesis of syn or anti homoallylic alcohols from Z or E crotylboronate and aldehydes (Hoffman) or of syn homoallylic alcohols from crotylstannanes, $BF_3$ and aldehydes – (Yamamoto).

| | | | | | |
|---|---|---|---|---|---|
| 1 | Hoffman, R.W. | *Angew. Chem. Int. Ed.* | **1979** | *18* | 326 |
| 2 | Hoffman, R.W. | *J. Org. Chem.* | **1981** | *46* | 1309 |
| 3 | Yamamoto, Y. | *J. Am. Chem. Soc.* | **1980** | *102* | 7107 |
| 4 | Yamamoto, Y. | *Aldrichimica Acta.* | **1987** | *20* | 45 |
| 5 | Brown, H.C. | *J. Am. Chem. Soc.* | **1985** | *107* | 2564 |
| 6 | Roush, W.R. | *Tetrahedron Lett.* | **1988** | *29* | 5579 |

**Homoallyl alcohol (3).**[6] Metalation of (E)-butene (1.05 equiv.) with n BuLi (1 equiv) and KOtBu (1 equiv.) in THF at –50°C for 15 min. followed by treatment of the (E)-crotyl potassium salt with $B(OiPr)_3$ at –78°C gave, after quenching with 1 N HCl and extraction with $Et_2O$ containing 1 equiv. of diisopropyl tartarate, the crotyl boronate **2**. A solution of decanal **1** (156 mg, 1 mmol) was added to a toluene solution of **2** (1.1-1.5 equiv.) (0.2 M) at –78°C containing 4Å molecular sieves (15-20 mg/L). After 3 h at –78°C, 1N NaOH was added, followed by extraction and chromatography to afford 208 mg of **3** (90%), anti:syn 99:1.

## H O L L E M A N N  Pinacol Synthesis

Dimerization of ketones to 1,2-diols by means of Mg-Hg or other metals (see 1st edition).

$$2 \quad \text{cyclohexanone} \xrightarrow[\text{-10 to 0°}]{\text{Mg-Hg, TiCl}_4} \text{2 (93%)}$$

**1**

$$\text{C}_6\text{H}_5\text{-CHO} \xrightarrow[\text{THF, -78°}]{\text{Cp}_2\text{Ti}^{\text{III}}\text{ Cl}_2} \text{4 (98% d, l)}$$

**3**

| 1 | Hollemann, M.A.F. | *Rec. Trav. Chim.* | **1906** | *25* | 206 |
|---|---|---|---|---|---|
| 2 | Goth, H. | *Helv. Chim. Acta* | **1965** | *48* | 1395 |
| 3 | Corey, E.J. | *J. Org. Chem.* | **1976** | *41* | 260 |
| 4 | Olah, G.E. | *Synthesis* | **1978** | | 358 |
| 5 | Zimmermann, H.E. | *J. Org. Chem.* | **1986** | *51* | 4644 |
| 6 | Pierce, K.G. | *J. Org. Chem.* | **1995** | *60* | 11 |
| 7 | Schwartz, J. | *J. Am. Chem. Soc.* | **1996** | *118* | 5480 |
| 8 | Gausauer, A. | *J. Chem. Soc. Chem. Commun.* | **1997** | | 4579 |

## H O N Z L – R U D I N G E R  Peptide Synthesis

Peptide synthesis by coupling of acyl azides with amino esters (see 1st edition).

$$\text{A-NH-CH-C-NH-NH}_2 \xrightarrow[\text{HCl, -10°}]{\text{NaNO}_2} \text{A-NH-CH-C-N}_3 \xrightarrow[\text{H}_2\text{N-CH-COO-B}]{\text{iPr}_2\text{NEt}}$$
$$\underset{R_1 \ \ O}{} \qquad \qquad \underset{R_1 \ \ O}{} \qquad \underset{R_2}{}$$

**1**

$$\text{A-NH-CH-C-NH-CH-COO-B}$$
$$\underset{R_1 \ \ O \qquad \ R_2}{}$$

**3**

| 1 | Honzl, I.; Rudinger, I. | *Coll. Czech. Chem. Comm.* | **1961** | *26* | 2333 |
|---|---|---|---|---|---|
| 2 | Siebel, F. | *Helv. Chim. Acta.* | **1970** | *53* | 2134 |
| 3 | Medzihradsky, K. | *Acta. Chim. Acad. Sci. Hung.* | **1962** | *30* | 105 |
| 4 | Ondetti, M.A. | *J. Am. Chem. Soc.* | **1968** | *90* | 4711 |
| 5 | Klausner, Y.S. | *Synthesis* | **1974** | | 554 |

## H O O K E R Quinone Oxidation - Rearrangement

Oxidation of 2-alkyl-3-hydroxy-1,4-quinones with $KMnO_4/NaOH$ or $H_2O_2/Na_2CO_3$ and $CuSO_4/NaOH$ with shortening of the alkyl group by one C and regiochemical rearrangement of the alkyl and hydroxy substituents.

| 1 | Hooker, S.C. | *J. Am. Chem. Soc.* | **1936** | 58 | 1168, 1179 |
| 2 | Fieser, L.F. | *J. Am. Chem. Soc.* | **1948** | 70 | 3215 |
| 3 | Moore, H.W. | *J. Org. Chem.* | **1995** | 60 | 461 |

**2,3-Dihydroxy-2-ethyl-4,6-dimethoxy-1-oxoindan-3-carboxylic       acid       2.[3]**
2-Ethyl-3-hydroxy-5,7-dimethoxy-1,4-naphthoquinone 1 (262 mg, 1 mmol) was added to dioxane (5 mL) and $H_2O$ (5 mL) containing $Na_2CO_3$ (120 mg, 1.13 mmol). The mixture was treated with 30% $H_2O_2$ (0.2 mL) and heated at 70°C for 1.5 h. The cooled mixture (ice bath) was treated with 36% HCl (5 drops) followed by a sat. solution of $SO_2$ in water. Remaining $SO_2$ was purged with $N_2$ (0.5 h). Extraction with EtOAc (3 x 30 mL) and washing of the organic extract was followed by drying. Evaporation of the solvent afforded 224 mg of 2 (76%) as an oil. White plates from $CHCl_3$ mp. 148-150°C.

**2-Hydroxy-3-methyl-5,7-dimethoxy-1,4-naphthoquinone 3.** 2 (87 mg, 0.29 mmol) in $H_2O$ (2 mL) was treated with 25% NaOH solution (0.8 mL) The pale yellow solution was treated with $CuSO_4$ (277 mg, 1.7 mmol) in water (1.5 mL) and heated to 70°C for 10 min. Filtration over Celite, acidification of the filtrate (HCl pH = 1-2) was followed by extraction with $CHCl_3$ (3 x 25 and 2 x 25 mL). The organic extract was washed, dried and the solvent was evaporated in vacuum. Flash chromatography (3:1 hexane: EtOAc) afforded 89 mg of 3 (72%), mp. 223-225°C.

## HOPPE Enantioselective Homoaldol Reaction

Enantioselective homoaldol reaction induced by sparteine and Ti catalyzed, also asymmetric deprotonation of allyl carbamates.

| 1 | Hoppe, D. | *Angew.Chem.Int.Ed.Engl.* | **1990** | *29* | 1422 |
|---|-----------|---------------------------|----------|------|------|
| 2 | Hoppe, D. | *Tetrahedron* | **1992** | *48* | 5657, 5667 |
| 3 | Hoppe, D. | *Angew.Chem.Int.Ed.Engl.* | **1989** | *28* | 69 |
| 4 | Hoppe, D. | *Pure.Appl.Chem.* | **1990** | *62* | 1999 |
| 5 | Hoppe, D. | *Synthesis* | **1996** | *62* | 141 |
| 6 | Kocienski, P.J. | *Synlett* | **1996** | | 652 |
| 7 | Hoppe, D. | *Synlett* | **2000** | | 1067 |
| 8 | Hoppe, D. | *Angew.Chem.Int.Ed.* | **1997** | *36* | 2282 |

**Z (3S,4R)-4-Hydroxy-3,5-dimethyl-1-hexenyl N, N-diisopropylcarbamate 5.[2]** (E)-Butenyl carbamate **1** (408 mg, 2 mmol) diluted with pentane (2 mL) was added slowly to a solution of sparteine (514.8 mg, 2.2 mmol) and BuLi (2.5 mmol in hexane 1.6 N) in pentane/cyclohexane (7 mL + 1.5 mL) and stirred vigorously. To this precooled suspension precooled Ti isopropoxide (4-10 mmol) was added very quickly at –70°C and stirring was continued for 15 min. 2-Methylpropanal **4** (360 mg, 5 mmol) was injected and the reaction mixture was allowed to warm to 20°C. Quenching with 2N HCl (10 mL) was followed by extraction with Et₂O. The residue obtained after evaporation of the solvent was purified by LC (silica gel Et₂O:pentane) to provide 488 mg of **5** (90%).

## H O R N E R - W A D S W O R T H - E M M O N S  Olefination

Wittig type reaction of phosphonate stabilized carbanions with aldehydes or ketones to form olefins (mainly *E*) (see 1st edition).

| 1 | Horner, L. | *Chem. Ber.* | **1958** | *83* | 733 |
|---|---|---|---|---|---|
| 2 | Wadsworth, W.S; Emmons, W.D. | *J. Am. Chem. Soc.* | **1961** | *83* | 1733 |
| 3 | Berkowitz, W.F. | *J. Org. Chem.* | **1982** | *47* | 824 |
| 4 | Tamizawa, T.K. | *Synthesis* | **1985** | | 887 |
| 5 | Sampson, C.R. | *J. Org. Chem.* | **1986** | *52* | 2525 |
| 6 | Tsuge, O. | *Bull. Chem. Soc. Jpn.* | **1987** | *60* | 4091 |
| 7 | Boutagy, J. | *Chem. Rev.* | **1974** | *79* | 87 |
| 8 | Paterson, I. | *Synlett.* | **1993** | | 774 |
| 9 | Endo, K. | *J. Org. Chem.* | **1997** | *62* | 1934 |

**Unsaturated ketone 3.**[2] To NaH (21.4 mg; 0.883 mmol) in DME (4 mL), under $N_2$ was injected **1** (210 mg; 0.95 mmol) in DME (1 mL). After stirring for 1 h (voluminous precipitate) and ice cooling, aldehyde **2** (100 mg; 0.442 mmol) in DME (1 mL) was injected. Stirring was continued for 30 min under ice cooling followed by 2.5 h at 20°C. The mixture was neutralized with AcOH (0.12 mL) and concentrated. Chromatography on silica gel (45 g) and elution with EtOAc:hexane 1:1, gave 125 mg of **3** (87%).

# HUNSDIECKER-BORODIN

## CRISTOL-FIRTH-KOCHI Halogenation

Substitution of carboxylic groups by halogen via Ag salts (Hunsdiecker-Borodin), Hg salts (Cristol-Firth) or Pb salts (Kochi) (see 1st edition).

| 1 | Borodin, A. | *Liebigs Ann.* | **1861** | *119* | 121 |
|---|---|---|---|---|---|
| 2 | Hunsdiecker, H.C. | *Chem. Ber.* | **1942** | *75* | 291 |
| 3 | Cristol, S.; Firth, W. | *J. Org. Chem.* | **1961** | *26* | 280 |
| 4 | Kochi, J.K. | *J. Org. Chem.* | **1965** | *30* | 3265 |
| 5 | Roy, S. | *J. Org. Chem.* | **1997** | *62* | 199 |
| 6 | Meyers, A.I. | *J. Org. Chem.* | **1979** | *44* | 3405 |
| 7 | Johnson, R.G. | *Chem. Rev.* | **1956** | *56* | 219 |
| 8 | Wilson, C.V. | *Org. React.* | **1957** | *9* | 332 |
| 9 | Roy, S. | *Tetrahedron* | **2000** | *56* | 1364 |

**1-Bromo-2-(p-methoxyphenyl)ethene 2.**[5] To LiOAc (0.2 mmol) in MeCN:$H_2O$ (97:3; 4.5 mL) was added p-methoxycinnamic acid **1** (336 mg; 2 mmol). After 5 min stirring at 20°C NBS (365 mg; 2.1 mmol) was added as a solid. The mixture was stirred for 10 min. Work up and chromatography (silica gel, hexane:EtOAc 3:2) afforded 386 mg of **2** (91%).

**Cyclobutyl chloride 4.**[4] To a solution of cyclobutanecarboxylic acid **3** (100 mg; 10 mmol) in PhH (10 mL) was added Pb(OAc)₄ (2 g; 4.5 eq) and the mixture was stirred at 20°C until it became homogeneous. Anhydrous LiCl (240.4 mg; 6.2 mmol) was added under $N_2$ and the mixture was heated at 81°C. Work up afforded 90.5 mg of **4** (100%).

**4-Chlorobromobenzene 6.**[6] A solution of p-chlorobenzoic acid **1** (1.56 g; 10 mmol) in CCl₄ (50 mL) and HgO (15 mmol) was refluxed and irradiated (100W bulb). Bromine (15 mmol) was added via a syringe. After 3 h the mixture was cooled to 20°C, washed (NaHCO₃ aq, 30 mL). Usual work up afforded 1.5 g of **6** (80%).

## H U I S G E N  Tetrazole Rearrangement

Rearrangement of 5-substituted (aryl or akyl) tetrazoles to 1,3,4-oxadiazoles by acylation.

| 1 | Huisgen, R. | *Chem. Ber.* | **1960** | *93* | 2106; 2885 |
|---|---|---|---|---|---|
| 2 | Huisgen, R. | *Tetrahedron* | **1960** | *11* | 241 |
| 3 | Huisgen, R. | *Chem. Ber.* | **1961** | *94* | 2509 |
| 4 | Huisgen, R. | *Liebigs Ann.* | **1962** | *654* | 146 |
| 5 | Huisgen, R. | *Chem. Ber.* | **1965** | *98* | 2966 |
| 6 | Brown, H.C. | *J. Org. Chem.* | **1967** | *32* | 1871 |
| 7 | Marchand, A.P. | *Heterocycles* | **1995** | *40* | 223 |

**2-Phenyl-5-(3-cyanophenyl)-1,3,4-oxadiazole 3.**[5] A mixture of 5-phenyltetrazole **1** (5.11 g; 35 mmol) and 3-cyanobenzoyl chloride **2** (8.25 g, 49.8 mmol) in pyridine (50 mL) was heated (water bath) for 30 min (850 mL of $N_2$ evolved). Hydrolysis with HCl, filtration of the product and washing with water gave after drying ($P_2O_5$) 8.4 g of **3** (97%), mp 143-146°C; from EtOH mp 147-148°C.

**5,5'-Bis(perfluoropropyl)-2,2'-bi-1,3,4-oxadiazole 6.**[6] A mixture of 5-perfluoro propyltetrazole **4** (2.5 g; 10.5 mmol) and oxalyl chloride **5** (667 mg; 5.3 mmol) in $CH_2Cl_2$ (2 mL) was heated for 5 h at 100°C. Removal of the solvent and sublimation in vacuum afforded 1.37 g of **6** (55%), mp 165-165.8°C.

**I V A N O V** Grignard Reagent

Formation of a polyfunctional organomagnesium reagent useful in the synthesis of β-hydroxy acids, diacids.

$$Ph\text{-}CH_2\text{-}COOH + RMgX \longrightarrow Ph\text{-}\underset{\underset{MgBr}{|}}{CH}\text{-}COOMgBr \xrightarrow{CO_2} Ph\text{-}\underset{\underset{COOH}{|}}{CH}\text{-}COOH$$

| **1** | | **2** | |

$$2 + CH_2O \longrightarrow Ph\text{-}\underset{\underset{CH_2OH}{|}}{CH}\text{-}COOH \qquad 2 + FeCl_3 \longrightarrow Ph\text{-}\underset{\underset{Ph\text{-}CH\text{-}COOH}{|}}{CH}\text{-}COOH$$

| **3** | **4** |

| 1 | Grignard, V. | *Bull. Soc. Chim. Fr.* | **1904** | *31* | 651 |
|---|---|---|---|---|---|
| 2 | Ivanov, D. | *Bull. Soc. Chim. Fr.* | **1931** | *49* | 19; 371 |
| 3 | Blicke, F.F. | *J. Am. Chem. Soc.* | **1952** | *74* | 253 |
| 4 | Blicke, F.F. | *J. Org. Chem.* | **1964** | *29* | 2036 |
| 5 | Ivanov, D. | *Synthesis* | **1970** | | 615 |

**Reagent 2.**[3] Mg (9.7 g; 0.4 atg) in Et$_2$O (200 mL) containing EtBr (0.5 mL) and iPrCl (1 mL) was warmed to initiate the reaction. iPrCl (33 g; 0.42 mol) was added to maintain reflux. After further 30 min reflux a solution of phenylacetic acid **1** (24 g; 0.176 mol) in Et$_2$O (200 mL) was added to maintain reflux. The mixture was refluxed until gas evolution ceased (ca 12 h).

**Tropic acid 4.** To the ice cooled solution of **2** was added a flow of CH$_2$O (**3**) by heating of paraformaldehyde (14 g; 0.466 mol) under a flow of N$_2$. After introduction of **3** (ca 30 min) the mixture was poured into ice (300 g) and 98% H$_2$SO$_4$ (30 mL). Work up afforded 20.7-24.5 g of **4** (71-83%), mp 116-117°C.


**K E I N A N** Silane Reagent

Diiodosilane (DIS) reagent for mild hydrolysis of ketals, acetals or reductive iodination of ketones and aldehydes (see 1st edition).

| 1 | Keinan, E. | *J. Org. Chem.* | **1987** | *52* | 4846 |
|---|---|---|---|---|---|
| 2 | Keinan, E. | *J. Org. Chem.* | **1990** | *55* | 2927 |
| 3 | Keinan, E. | *Synthesis* | **1990** | | 641 |

## **J A C O B S E N** Asymmetric Epoxidation

Asymmetric olefin epoxidation (also conjugated olefins) with NaOCl, catalyzed by chiral Mn(III) salene complex **2** (compare Sharpless asymmetric epoxidation)

**1** (95%)

**2** (95%)

**3**

NaOCl
(R,R) **2** (6mol%)
4-PPNO (0.25 equiv.)

**4** (95-97% ee)
(56% yield +13% trans isomer)

**2**
ionic luiq.

86%, 96% ee

| 1 Jacobsen, E. N. | *J. Am. Chem. Soc.* | **1990** | *112* | 2801 |
|---|---|---|---|---|
| 2 Jacobsen, E. N. | *J. Am. Chem. Soc.* | **1991** | *113* | 7063 |
| 3 Jacobsen, E. N. | *J. Am. Chem. Soc.* | **1991** | *56* | 2296 |
| 4 Jacobsen, E. N. | *Tetrahedron Lett.* | **1991** | *32* | 5055 |
| 5 Jacobsen, E. N. | *J. Org. Chem.* | **1992** | *57* | 4320 |
| 6 Jacobsen, E. N. | *J. Org. Chem.* | **1994** | *59* | 1939 |
| 7 Hughes, D. L. | *J. Org. Chem.* | **1997** | *62* | 2222 |
| 8 Houk, K. H. | *Org. Lett.* | **1999** | *1* | 419 |
| 9 Pozzi, G. | *Chem. Commun.* | **1998** | | 877 |
| 10 Song, C. E. | *Chem. Commun.* | **2000** | | 837 |

**(2R,3R)-Ethyl-3-phenylglycidate (4).[5]** To cis-ethyl cinnamate **3** (5g, 25.5 mmol) and 4-phenylpyridine-N-oxide (4-PPNO) (1.16 g, 6.78 mmol) in $CH_2Cl_2$ (60 mL) was added **2** (1.08 g, 6.08 mol%). Cooled buffered bleach (160 mL, pH=11.25) was added at $4^0$ C. After 12 h extraction with tert-butyl methyl ether (500 mL) and distillation afforded 4 g of a mixture 70% cis **4** (56%) and 13% trans **4** (10%). The e,e of the cis epoxide was 95-97% (NMR, Eu(HFC) )

## J A P P  Oxazole Synthesis

Oxazole synthesis from benzoin and nitriles or ammonium formate (see 1st edition).

| 1 | Japp, F. R. | J. Chem. Soc. | 1893 | 63 | 469 |
| 2 | Bredereck, H. I. | Chem. Ber. | 1954 | 87 | 726 |
| 3 | Willey, R. H. | Chem. Rev. | 1945 | 37 | 420 |

## J A P P – K L I N G E M A N N  Hydrazone Synthesis

Synthesis of hydrazones from diazonium salts and an activated methylene group (or enamine) (see 1st edition).

| 1 | Japp, F. R.; Klingemann, F. | Chem. Ber. | 1887 | 20 | 2492 |
| 2 | Frank, R. L. | J. Am. Chem. Soc. | 1949 | 71 | 2804 |
| 3 | Jackman, A. | Chem. Commun. | 1967 | | 456 |
| 4 | Philips, R.R. | Org. React. | 1959 | 10 | 143 |
| 5 | Robinson, R. | Chem. Rev. | 1969 | 69 | 233 |

## JAROUSSE–MAKOSZA Phase Transfer Reaction

Phase transfer (PT) catalysis by quaternary ammonium salts of substitution, addition, carbonyl formation, oxidation, reduction (see 1st edition).

$$CH_3-CO-CH_2COOCH_3 \; + \; CH_3I \xrightarrow[\text{NaOH - CHCl}_3]{R_4N^+HSO_4^-} CH_3-CO-\underset{\underset{CH_3}{|}}{CH}-COOCH_3$$

$$\textbf{1} \qquad\qquad \textbf{2} \qquad\qquad\qquad\qquad\qquad \textbf{3 (80\%)}$$

$$C_6H_5OH \; + \; n-C_4H_9Br \xrightarrow[\text{NaOH}]{PT} C_6H_5-O-C_4H_9 \; (n)$$

$$\textbf{4} \qquad \textbf{5} \qquad\qquad\qquad\qquad \textbf{6 (90\%)}$$

| | | | | | |
|---|---|---|---|---|---|
| 1 | Jarousse, M.I. | *C. R. Ser. C.* | **1951** | 232 | 1424 |
| 2 | Makosza, M. | *Rocz. Chem.* | **1965** | 39 | 1977 |
| 3 | Dockx, J. | *Synthesis* | **1973** | | 411 |
| 4 | Weber, W. P. | *Angew. Chem. Int. Ed.* | **1972** | 11 | 530 |
| 5 | Dehmlow, E. V. | *Angew. Chem. Int. Ed.* | **1974** | 13 | 170 |
| 6 | Harris, J. M. | *J. Org. Chem.* | **1985** | 50 | 5230 |
| 7 | Wang, J. X. | *J. Org. Chem.* | **1986** | 51 | 275 |

**Methyl acetopropanoate (3).**[3] (C-alkylation). Tetrabutylammonium hydrogen sulfate (PT catalyst) (34.6 g, 0.1 mol) and NaOH (8.0 g, 0.2 mol) in water (75 mL) was added to well stirred **1** (11.6 g, 0.1 mol) and MeI **2** (28.4 g, 0.2 mol) in CHCl₃ (75 mL). The reaction is exothermic and becomes neutral after a few min. The CHCl₃ layer was evaporated. Et₂O was added to filter the PT catalyst. Evaporation of the Et₂O gave a mixture of **3** (80%) (monoalkylated) and 20% dialkylated product.

## JEGER Tetrahydrofuran Synthesis

Free radical ring closure of alcohols with Pb(AcO)₄ to tetrahydrofurans (see 1st edition).

| | | | | | |
|---|---|---|---|---|---|
| 1 | Jeger, O. | *Helv. Chim. Acta.* | **1959** | 42 | 1124 |
| 2 | Micovic, V. N. | *Tetrahedron* | **1964** | 20 | 2279 |
| 3 | Moon Sung | *J. Org. Chem.* | **1969** | 34 | 288 |
| 4 | Michailovici, M. L. | *Synthesis* | **1970** | | 209 |
| 5 | Jeger, O. | *Helv. Chim. Acta.* | **1964** | 47 | 1883 |

## JONES-SARETT Oxidizing Reagent

Oxidation of alcohols to aldehydes or ketones with $CrO_3$-$H_2SO_4$ in $Me_2CO$ (Jones) or $CrO_3$ in pyridine (Sarett) (see 1st edition).

| 1 | Jones, E.R.H. | J.Chem.Soc. | **1946** | | 39 |
|---|---|---|---|---|---|
| 2 | Burgstahler, A.W. | J.Org.Chem. | **1969** | *34* | 1562 |
| 3 | Dauben, W.C. | J.Org.Chem. | **1980** | *45* | 4413 |
| 4 | Liotta, D. | J.Org.Chem. | **1983** | *48* | 2932 |
| 5 | Sarett, L.H. | J.Am.Chem.Soc. | **1953** | *75* | 422 |
| 6 | Holum, J.R. | J.Org.Chem. | **1961** | *26* | 4814 |
| 7 | Gassmann, P.G. | J.Org.Chem. | **1964** | *28* | 323 |
| 8 | Collins, J.C. | Tetrahedron Lett. | **1968** | | 3363 |

**Phenyl ethynyl ketone 2.**[1] To 1 (342 g, 2.59 mol) in $Me_2CO$ was added slowly $CrO_3$ (175 g, 1.75 mol) in water (500 mL) and 98% $H_2SO_4$ (158 mL) under stirring and $N_2$ at 5°C over 4-5 h. After stirring for a further 30 min, dilution (water), extraction ($Et_2O$), evaporation of the solvent and recrystallization (MeOH) gave 258 g of **2** (68%), mp 50-51°C.

**4b-Methyl-7-etheylenedioxy-1,2,3,4,4a,5,6,,7,8,10,10b-dodecahydrophenananthrene 1,4-dione (6).** [5] A solution of **3**(3.12 g, 10 mmol) in pyridine (30 mL) was maintained with $CrO_3$ (3.1 g) in Pyridine (30 mL). After 24 h at 20°C usual work up and recrystallization ($Et_2O$) gave 2.76 g of **6** (89%), mp 117-120°C.

## J U L I A – B R U Y L A N T S  Cyclopropyl Carbinol Rearrangement

Synthesis of homoallyl halides (usually E) by acid catalyzed rearrangement of cyclopropyl carbinols (see 1st edition).

| | | | | | |
|---|---|---|---|---|---|
| 1 | Bruylants, P. | *Bull. Acad. Royal Belge* | **1928** | *28* | 160 |
| 2 | Julia, M. | *Bull. Soc. Chim. Fr.* | **1960** | | 1072 |
| 3 | Julia, M. | *Bull. Soc. Chim. Fr.* | **1961** | | 1849 |
| 4 | Julia, M. | *Tetrahedron Suppl.* | **1966** | | 443 |
| 5 | Corey. E. J. | *J. Org. Chem.* | **1967** | *32* | 4160 |
| 6 | Kociensky, P. J. | *Chem.& Ind.* | **1981** | | 549 |
| 7 | Nakamura, H. | *Tetrahedron Lett.* | **1973** | | 111 |
| 8 | Johnson, W. S. | *J. Am. Chem. Soc.* | **1968** | *90* | 2882 |
| 9 | Faulkner, J. | *Synthesis* | **1971** | | 175 |

## J U N G – O L A H – V O R O N K O V  Ether Cleavage

Cleavage of ethers or esters, carbamates, phosphonates with trimethylsilyl iodide. Deoxygenation of sulfoxides (see 1st edition).

| | | | | | |
|---|---|---|---|---|---|
| 1 | Jung, M. E. | *J. Am. Chem. Soc.* | **1977** | 99 | 968 |
| 2 | Jung, M. E. | *J. Org. Chem.* | **1977** | 42 | 3761 |
| 3 | Olah, G. A. | *Angew. Chem. Int. Ed.* | **1976** | 15 | 774 |
| 4 | Olah, G. A. | *Synthesis* | **1977** | | 581 |
| 5 | Jung, M. E. | *Chem. Commun* | **1978** | | 315 |
| 6 | Olah, G. A. | *Tetrahedron* | **1982** | 38 | 2225 |
| 7 | Voronkov, M. G. | *Zh. Obshch. Khim.* | **1976** | 46 | 1908 |

**Benzoic acid (3a).** Methyl benzoate 1 (136 mg, 1 mmol) trimethylsilyl iodide 2 (0.16 mL, 1.2 mmol) in $CCl_4$ (0.5 mL) was heated to 50 °C for 35 h. (NMR yield of 3 100%). The reaction mixture was stirred with 10% $NaHCO_3$ (10 mL) for 30 min. Acidification of the aq.layer and extraction with $Et_2O$, followed by evaporation of the solvent afforded 104 mg of **3a** (95%), mp 118-119 °C.

## J U L I A – C O L O N N A Asymmetric Epoxidation

Asymmetric epoxidation of electron-poor olefins catalyzed by poly-α amino acids.

**1**  **2**  **3 (92%, 96% ee)**

**4**  1) 1-PPL, UHP, DBU  2) mCPBA, KF  **5 (70%, 96% ee)**

| 1 | Julia, S.; Colonna, S. | *Angew. Chem. Int. Ed.* | **1980** | *19* | 929 |
|---|---|---|---|---|---|
| 2 | Julia, S.; Colonna, S. | *J. Chem. Soc. Perkin 1* | **1982** | | 1317 |
| 3 | Julia, S.; Colonna, S. | *Tetrahedron* | **1983** | *39* | 1655 |
| 4 | Julia, S.; Colonna, S. | *Tetrahedron* | **1984** | *40* | 5207 |
| 5 | Roberts, S. M. | *J. Chem. Soc. Perkin1* | **1997** | | 3501 |
| 6 | Geller, T. | *J. Chem. Soc. Perkin1* | **1999** | | 1397 |

**Catalyst (1).** [4] To a solution of N-carboxy-L-alanine anhydride (2.5 g, 21.7 mmol) in MeCN anh. (50 mL) was added MeCN (20 mL with 0.43 mmol $H_2O$). After 4 days stirring at 20 °C the solvent was removed in vacuum and the residue stirred 24 h in $Et_2O$, filtered and dried.

**Epoxide (3).** To a solution of chalcone **2** (500 mg, 2.4 mmol) in PhMe (6 mL) was added **1** (400 mg) and all was stirred for 30 min at 20 °C. The mixture was added to a solution fo NaOH in $H_2O_2$ (0.08 g/mL) (4.4 mL) and stirred for 24 h. The reaction was monitored by TLC (silica gel, petroleum ether:$Et_2O$ 9:1) Usual work up and chromatography afforded 494 mg of **3** (92%, 96% ee).

**Epoxide (5).** [5] TO I-PLL(immobilized poly-L-leucine, 7 g) in THF (50 mL) was added DBU (4.1 mL, 27.48 mmol) and urea-hydrogen peroxide (UHP) (2.07 g, 21.98 mmol). Under stirring was added **4** (4.01 g, 18.37 mmol), followed after 3 h by a second quantity of DBU and UHP. Separation of the epoxide and oxidation with m-CPBA afforded 3.20 g of **5** (70% yield from 4), mp 53-55 °C, 96% ee.

## JULIA - LYTHGOE Olefination

Synthesis of olefins by reductive elimination of $\alpha$ -substituted sulfones.

| 1 | Julia, M. | *Bull.Soc.Chim.Fr.* | 1973 | | 743 |
|---|---|---|---|---|---|
| 2 | Julia, M. | *Tetrahedron Lett.* | 1973 | | 4833 |
| 3 | Lythgoe, B. | *J.Chem.Soc.Perkin 1* | 1978 | | 834 |
| 4 | Kocienski, P.J. | *Chem. and Ind.* | 1981 | | 548 |
| 5 | Julia, M. | *Tetrahedron Lett.* | 1982 | 23 | 2465 |
| 6 | Seebach, D. | *Helv.Chim.Acta* | 1982 | 65 | 385 |
| 7 | Kende, A.S. | *Tetrahedron Lett.* | 1990 | 31 | 7105 |
| 8 | Fukumoto, R. | *Synlett* | 1994 | | 859 |
| 9 | Keck, G.E. | *J.Org.Chem.* | 1995 | 60 | 3194 |
| 10 | Ferezou, J.P. | *Synlett* | 1998 | | 1223 |

**Sulfone 3.**[9] To sulfone **1** (1g, 3.85 mmol) in THF (35 mL) cooled to −78°C was added n-BuLi (1.88 mL, 2.25 M, 4.24 mmol). After 30 min stirring, benzaldehyde (429 mg, 4.04 mmol) in THF (4 mL) was added. The mixture was stirred for 3 h at −78°C, Ac$_2$O was added after 1h at −78°C, this was slowly warmed to 20°C. Usual work up and RPLC (4 mm plate) afforded 1.325 g of **3** (85%).

**1,4-Diphenyl-2-(phenylsulfonyl)-1-butene 4.** To **3** (1.47 g, 3.6 mmol) in THF (50 mL) was added DBU (3.3 g, 21.65 mmol) dropwise. After 18 h (TLC), the mixture was quenched with Et$_2$O and brine. Washing, drying, filtration through Celite and silica gel and purification by RPLC afforded 1.26 g of **4** (97%).

**1,4-Diphenylbut-1-ene 5.** To samarium (249 mg, 1.66 mmol) in THF (15 mL) was added iodine (373 mg, 1.47 mmol). The mixture was heated to 65°C (bath temp.) for 90 min, cooled to 20°C and **4** (64.03 mg, 0.184 mmol) was added followed by DMPU (1,3-dimethyl-3,4,5,6-tetrahydro-2(H)-pyrimidone) (236 mg, 1.84 mmol) in THF (2 mL). After 30 min the mixture was worked up to afford 32.5 mg of **5** (85%).

## **K A B E** Chromanone Synthesis

Synthesis of 4-chromanones by condensation of salicylaldehydes or o-hydroxyaryl ketones with enamines or ketones (see 1st edition).

| 1 | Kabe, H. J. | *Liebigs Ann.* | **1976** | *511* | 511 |
|---|-------------|----------------|----------|-------|-----|
| 2 | Kabe, H. J. | *Synthesis* | **1978** | | 887,888 |
| 3 | Kabe, H. J. | *Angew. Chem. Int. Ed.* | **1982** | *21* | 247 |
| 4 | Kelly, E. S. | *J. Org. Chem.* | **1991** | *56* | 1325 |

## **K A I S E R – J O H N S O N – M I D D L E T O N** Dinitrile cyclization

Synthesis of heterocycles by cyclization of dinitriles by means of HBr(see 1st edition).

| 1 | Johnson, F. | *J. Org. Chem.* | **1962** | 27 | 2241, 2473, 3953 |
|---|-------------|----------------|----------|-----|------------------|
| 2 | Kaiser, A.M. | *U. S.Patent* | **1953** | | 2,630,433; 2,658,893 |
| 3 | Middleton, W. J. | *J. Am. Chem. Soc.* | **1958** | 80 | 2822, 2832 |
| 4 | Johnson, F. | *J. Org. Chem.* | **1964** | 29 | 153 |

**2-Amino-6-bromopyridine (2).**[1] Glutacononitrile 1 (12.5 g, 27 mmol), was added dropwise to a solution of HBr in AcOH (30 g of 30%) in 5 min with cooling and stirring. The yellow precipitate after filtration and washing (NaHCO₃ sol) was extracted with Et₂O and recrystallized from Et₂O-petroleum ether to give 2.7 g of **2** (60%), mp 88-89 °C.

## KAGAN–MODENA Asymmetric Oxidation

Asymmetric oxidation of sulfides to chiral sulfoxides by chiral titanium complexes and hydroperoxide.

| 1 | Kagan, H. B. | *Tetrahedron Lett.* | **1984** | | 1049 |
|---|---|---|---|---|---|
| 2 | Modena, G. | *Synthesis* | **1984** | | 325 |
| 3 | Kagan, H. B. | *J. Am. Chem. Soc.* | **1984** | *106* | 8188 |
| 4 | Kagan, H. B. | *Tetrahedron* | **1987** | *43* | 5135 |
| 5 | Kagan, H. B. | *Pure Appl. Chem.* | **1985** | *57* | 1911 |
| 6 | Kagan, H. B. | *Tetrahedron Lett.* | **1989** | *30* | 3659 |
| 7 | Kagan, H. B. | *Synlett* | **1990** | | 643 |
| 8 | Kagan, H. B. | *J. Am. Chem. Soc.* | **1994** | *116* | 9430 |
| 9 | Potvin, P. G. | *Can. J. Chem.* | **1992** | *70* | 2256 |
| 10 | Potvin, P. G. | *Tetrahedron Asymm.* | **1999** | *10* | 1661 |

**(R)-Methyl p-tolyl sulfoxide (2).** [4] To a solution of (R,R)-diethyl tartarate (DET) (1.71 mL, 10 mmol) in $CH_2Cl_2$ (50 mL) under Ar are added $Ti(OiPr)_4$ (1.49 mL, 5 mmol) and water (0.09 mL, 5 mmol). The solution became homogeneous after 20 min stirring. Methyl p-tolyl sulfide **1** (0.69 g, 5 mmol) was added, the mixture was cooled to –30 °C, followed by dropwise addition of a 3.6 M toluene solution of tert.butyl hydroperoxide (1.52 mL, 5.5 mmol). After 18 h at –23 °C, water (2 mL) was added and the mixture was stirred for 1 h at 20 °C. Usual work up and flash chromatography (silica gel EtOAc) gave 0.7 g of **2** (90%, 89% ee).

## KAGAN - HORNER - KNOWLES Asymmetric Hydrogenation

Enantioselective hydrogenation of prochiral olefins such as conjugated acids or enamides (also asymmetric hydroboration) with chiral Rh phosphine catalysts (also Ti-catalysts)[9] (see 1st edition).

$$\text{Ph} \underset{\text{CO}_2\text{H}}{\overset{\|}{\bigwedge}} + H_2 \ (1.1 \ \text{bar}) \xrightarrow[\text{0.1 equiv NEt}_3]{\substack{\text{0.03 equiv Rh} \\ \text{(R,R)-DIOP}}} \underset{\text{Ph} \qquad \text{CO}_2\text{H}}{\bigwedge}$$

95%, 63% e,e

$$\underset{\substack{\text{AcNH} \qquad \text{CO}_2\text{H} \\ \mathbf{1}}}{\overset{\text{Ph}\diagdown \qquad \diagup \text{H}}{\diagup\diagdown}} + H_2 \ (1.1 \ \text{bar}) \xrightarrow[\text{PhH : EtOH}]{\substack{[\text{RhCl(cyclooctene)}_2]_2 \\ \text{(R,R)-DIOP}}} \underset{\substack{\text{AcNH} \qquad \text{CO}_2\text{H} \\ \mathbf{2} \ (90\%, \ 82\% \ \text{e,e})}}{\overset{\text{CH}_2\text{Ph}}{\big|}}$$

(R,R) DIOP

| 1 | Horner, L. | *Angew.Chem.Int.Ed.* | 1968 | 7 | 942 |
|----|--------------|---------------------------|------|-----|-------|
| 2 | Knowles, W.S. | *J.Chem.Soc.Chem.Commun.* | 1968 | | 1445 |
| 3 | Kagan, H.B. | *J.Chem.Soc.Chem.Commun.* | 1971 | | 481 |
| 4 | Kagan, H.B. | *J.Am.Chem.Soc.* | 1972 | 94 | 6429 |
| 5 | Kagan, H.B. | *J.Organomet.Chem.* | 1975 | 90 | 353 |
| 6 | Kagan, H.B. | *Pure Appl.Chem.* | 1975 | 43 | 401 |
| 7 | James, B.R. | *J.Organomet.Chem.* | 1985 | 279 | 31 |
| 8 | Burgess, K. | *Tetrahedron Asym.* | 1991 | 2 | 613 |
| 9 | Buchwald, S.L. | *J.Am.Chem.Soc.* | 1993 | 115 | 12569 |
| 10 | Kagan, H.B. | *Bull.Soc.Chim.Fr.* | 1988 | | 846 |
| 11 | Kagan, H.B. | *C.R.Acad.Sci., Serie IIb* | 1996 | 322 | 131 |

**N-Acetyl-(R)-phenylalanine 2.**[4] The rhodium catalyst was obtained by adding (R,R)-DIOP (from diethyl tartarate) to a benzene solution of [RhCl (cyclooctene)$_2$]$_2$ under Ar and stirring for 15 min. A solution of the catalyst (1 mmol in EtOH:PhH 4:1) was introduced under H$_2$ to a solution of $\alpha$-N-acetylamino-$\beta$-phenylacrylic acid 1 (molar ratio catalyst:substrate 1:540). After hydrogenation at 1.1 bar, the solvent was evaporated, the residue was dissolved in 0.5N NaOH, the catalyst was filtered and the solution acidified and concentarted to dryness to afford 2 in 90% yield and 82% e,e.

## KAGAN - MOLANDER Samarium reagent

Lantanides and $SmI_2$ specifically in carbon-carbon bond formation or for functional group transformation (cyclization, Barbier type reaction, intramolecular coupling, aldol, Evans, Tishenco).

| | | | | | |
|---|---|---|---|---|---|
| 1 | Kagan, H.B. | *J.Am.Chem.Soc.* | **1980** | *102* | 2693 |
| 2 | Kagan, H.B. | *Tetrahedron* | **1986** | *42* | 6573 |
| 3 | Molander, G.A. | *J.Org.Chem.* | **1986** | *51* | 5259 |
| 4 | Molander, G.A. | *J.Org.Chem.* | **1987** | *52* | 3943 |
| 5 | Molander, G.A. | *J.Org.Chem.* | **1989** | *54* | 3525 |
| 6 | Molander, G.A. | *J.Org.Chem.* | **1991** | *56* | 4112 |
| 7 | Fukuzawa, G. | *Synlett* | **1993** | | 803 |
| 8 | Skrydstrup, T. | *Angew.Chem.Int.Ed.* | **1997** | *36* | 345 |
| 9 | Fang, J.M. | *J.Org.Chem.* | **1999** | *64* | 843 |
| 10 | Molander, G.A. | *Chem.Rev.* | **1992** | *92* | 29 |
| 11 | Krief, A. | *Chem.Rev.* | **1999** | *99* | 745 |

**Cyclopropanation of 1.**[5] To samarium metal (316 mg, 2.1 mmol) under Ar was added THF (5 mL), followed by a solution of $HgCl_2$ (54 mg, 0.2 mmol) in THF (5 mL). After 10 min stirring the allyl alcohol **1** (64 mg, 0.5 mmol) was added. The mixture was cooled to -78°C and chloroiodomethane (353 mg, 2 mmol) was added dropwise. The mixture was allowed to warm to 20°C and stirred for an additional 1-2h. The reaction mixture was quenched with aq. sat $K_2CO_3$ solution and extracted with $Et_2O$. Chromatography afforded 71 mg of a mixture of 2:3 in ratio 200:1, yield 99%.

## KAKIS-KIKUCHI  Oxidative Aryl Rearrangement

Formation of ketones by bromination (chlorination)-rearrangement of aryl substituted ethylenes (Kakis) (see 1st edition). Conversion of 1-arylalkenes to 2-arylaldehydes with $I_2$ and $Ag_2O$ at room temperature, via aryl migration (Kikuchi).

| 1 | Kakis, F.J. | *J. Org. Chem.* | **1971** | *36* | 4117 |
| 2 | Kakis, F.J. | *J. Org. Chem.* | **1973** | *38* | 1733 |
| 3 | Kikuchi, H. | *Chem. Lett.* | **1984** | | 341 |
| 4 | Koreeda, M. | *Synlett* | **1993** | | 207 |

**Ketone 2.**[1] **1** (3.3 g; 10 mmol) in $CHCl_3$ (250 mL) ice cooled was saturated with $Cl_2$ (yellow color). The mixture was treated with 9:1 $MeOH:H_2O$ saturated with $AgNO_3$ and stirred for 20 h. The salts were filtered off and the filtrate diluted with water. The organic layer was washed and dried ($MgSO_4$) and the solvent removed in vacuum to give a residue which crystallized spontaneously, 3.27 g of **2** (94%), mp 183-184°C.

**2-(4-Methoxyphenyl)propionaldehyde 4.**[3] **3** (740 mg; 5 mmol) in 5:1 dioxane:water (30 mL) was treated with iodine (1.98 g; 7.8 mmol) and $Ag_2O$ (2.04 g; 7.8 mmol) at 20°C for 3 h. The mixture was filtered and the filtrate was extracted with $Et_2O$. The organic layer after washing and drying was chromatographed to give 803 mg of **4** (98%).

## K A L U Z A  Isothiocyanate Synthesis

Formation of isothiocyanates from amines and $CS_2$ (see 1st edition).

| 1 | Kaluza, H. | *Monatsh.* | **1912** | *33* | 363 |
|---|---|---|---|---|---|
| 2 | Hodgkins, J. E. | *J. Org. Chem.* | **1956** | *21* | 404 |
| 3 | Hodgkins, J. E. | *J. Am. Chem. Soc.* | **1961** | *83* | 2532 |
| 4 | Hodgkins, J. E. | *J. Am. Chem. Soc.* | **1964** | *29* | 3098 |

## K A M E T A N I  Amine Oxidation to Nitriles

Oxidation of primary amines to nitriles by $Cu(I)Cl$-$O_2$-pyridine (see 1st edition).

| 1 | Kametani, T. | *Synthesis* | **1977** | *245* |
|---|---|---|---|---|
| 2 | Capdevielle, P. | *Synthesis* | **1989** | *451* |

**p-Methoxybenzonitrile (2).**[2] p-Methoxybenzylamine **1** (0.137 g, 1 mmol), 4 Å molecular sieves (8 g) and $Cu_2Cl_2$ (0.6 Cu(I) equiv) in dry pyridine (50 mL) were stirred at 60 °C for 4 h under $O_2$ atm. More Cu catalyst was added and the reaction continued 20 h. The mixture was poured on ice (100 g) and 36% HCl (60 mL) and extracted with $CH_2Cl_2$(3×50 mL). The extract was washed with aqueous $NaHCO_3$, dried and evaporated to give 0.131 g of **2** (99%), mp 61 °C.

## KATRITZKY Amine Displacement

Nucleophilic replacement of aliphatic primary amino groups by H, halogen-, O-, S-, Se-, N-, P- and C-linked substituents via pyrylium salts.

| 1 | Katritzky, A.R. | *J. Chem. Soc. Perkin I* | **1979** |    | 430 |
| 2 | Katritzky, A.R. | *J. Chem. Soc. Perkin I* | **1979** |    | 442 |
| 3 | Katritzky, A.R. | *Tetrahedron* | **1980** | *36* | 679 |
| 4 | Katritzky, A.R. | *J. Chem. Soc. Perkin I* | **1980** |    | 849 |
| 5 | Katritzky, A.R. | *J. Chem. Soc. Perkin I* | **1980** |    | 2901 |
| 6 | Balaban, A.T. | *Adv. Heterocyclic Chem.* | **1969** | *10* | 241 |
| 7 | Katritzky, A.R. | *Angew. Chem. Int. Ed.* | **1984** | *23* | 420 |
| 8 | Katritzky, A.R. | *J. Org. Chem.* | **1998** | *63* | 6704 |

**Pyridinium tetrafluoroborate 3.**[1] Benzylamine 1 (2.0 g; 18.7 mmol) and a suspension of 2,4,6-triphenylpyrylium tetrafluoroborate 2 (2.0 g; 15 mmol) in EtOH (50 mL) were stirred for 12 h. The clear solution was evaporated in vacuum (60°C; 20 mm), the residue was washed with Et₂O and recrystallized from anh. EtOH to yield 4.2 g of 3 (85%), mp 196-197°C.

**Benzyl acetate 4.** A mixture of 3 (1.99 g; 5 mmol), 2,4,6-triphenyl-pyridine (460 mg; 1.5 mmol) and anhydrous NaOAc (820 mg; 10 mmol) was heated to 100°C at 0.1-0.2 mm for 4 h to remove the water, then to 210°C when 525 mg of 4 (70%) was collected in a liquid nitrogen trap.

## K A T R I T Z K Y Stereoselective Ester Olefination

Stereoselective olefination of carboxylic eaters or synthesis of allylamines from α-amino acid esters mediated by benzotriazole(Bt) derivatives.

| | Katritzky, A. R. | *J. Org. Chem.* | **1997** | 62 | 238 |
|---|---|---|---|---|---|
| 1 | Katritzky, A. R. | *J. Org. Chem.* | **1997** | 62 | 238 |
| 2 | Katritzky, A. R. | *Synthesis* | **1994** | | 597 |
| 3 | Katritzky, A. R. | *J. Org. Chem.* | **1997** | 62 | 721 |
| 4 | Katritzky, A. R. | *J. Org. Chem.* | **1998** | 63 | 3438 |

α-(4-(Dimethylanilino)- α-(benzotriazol-1-yl)-acetophenone (3).[4] To a solution of 1 (2.016 g, 8 mmol) in THF cooled at -78 °C was added BuLi under Ar. After 15 min a solution of ester 2 (1.26 g, 8.4 mmol) in THF (5 mL) was added dropwise. After the dark color disappeared, NH₄Cl solution (20 mL) was added. Usual work up followed by recrystallization (hexane:EtOAc 1:1) afforded 2.62 g of 3 (92%).

**Dimethylaminostilbene. 5.** A solution of 3 (1.78 g, 5 mmol) in EtOH was treated with NaBH₄(0.5 g), heated to 50 °C for 15 min and cooled to 20 °C. Quenching, extraction (CH₂Cl₂) and evaporation of the solvent gave 4 (mixture of diastereoisomers). A solution of 4 in DME (20 mL) was treated with a low-valent titanium mixture (from (Zn-Cu(5.4 g) and TiCl₃(3.85 g, 15 mmol) see ref 1). After overnight refluxing, filtration, extraction (CH₂Cl₂) and evaporation of the solvent, chromatography (CH₂Cl₂:hexane 1:1) afforded 981 mg of 5 (88%) trans only.

## **K A U F F M A N N** Dimerisation

Synthesis of polyheteroarenes, heteroprotophanes, ketazines by Cu catalyzed dimerization of magnesium or Li derivatives.

| 1 | Kauffmann, Th. | *Angew.Chem.Int.Ed.* | **1967** | *6* | 633 |
| 2 | Kauffmann, Th. | *Chem.Ber.* | **1968** | *101* | 3022 |
| 3 | Kauffmann, Th. | *Chimia* | **1972** | *26* | 511 |
| 4 | Kauffmann, Th. | *Angew.Chem.Int.Ed.* | **1974** | *13* | 291 |

**Benzophenonazine 3.**[2] A solution of phenylmagnesium bromide **1** (from brombenzene 20.4 g, 0.13 mol and Mg 3.6 g, 0.15 g At) in Et$_2$O (50 mL) was treated with benzonitrile **2** (10.3 g, 0.1 mol) in Et$_2$O (50 mL) under stirring at 20°C. Benzophenoniminium-magnesium bromide appeared as a colorless crystalline product. After 12 h dry Cu$_2$Cl$_2$ (0.5 g, 5 mmol) and THF (50 mL) were added and the mixture was heated to 35°C under stirring. After 3 h stirring at 20°C dry O$_2$ was bubbled through the reaction mixture for 1-2 h. Dilution with PhH (100 mL) quenching with water (10 mL) and evaporation of the solvent gave after recrystallization from EtOH 17 g of **3** (94%).

## K A W A S E  N-Acyl Rearrangement

Rearrangement of N-acylprolines or N-acyl-1,2,3,4-tetrahydroisoquinoline- 1-carboxylic acids with trifluoroacetic anhydride to 5-trifluoromethyl oxazoles.

| 1 | Kawase, M. | *J.Chem.Soc.Chem.Commun.* | **1990** | | 1382 |
| 2 | Kawase, M. | *Heterocycles* | **1993** | *36* | 2441 |
| 3 | Kawase, M. | *Tetrahedron Lett.* | **1993** | *34* | 859 |
| 4 | Kawase, M. | *Tetrahedron Lett..* | **1994** | *35* | 149 |
| 5 | Kawase, M. | *Heterocycles.* | **1998** | *48* | 285 |
| 6 | Kawase, M. | *Chem.Pharm.Bull.* | **1998** | *46* | 749 |
| 7 | Kawase, M. | *J.Chem.Soc.Chem.Commun.* | **1998** | | *641* |

**2-t-Butyl-4-(3-hydroxypropyl)-5-trifluoromethyloxazole (3).**[3] To a stirred solution of N-pivaloylproline **1** (298.5 mg, 1.5 mmol), pyridine (0.73 mL, 9 mmol) and DMAP (28 mg, 0.23 mmol) in PhH (5 mL) at 0°C under $N_2$ was added trifluoroacetic anhydride (0.64 mL, 4.5 mmol). After 3 h stirring at 25°C, the reaction mixture was refluxed for 5 h. The residue **2** obtained after evaporation in vacuum was stirred with a mixture of 10% HCl and dioxane (3 mL/ 2 mL) for 3 h at 60°C. After usual work up and column chromatography (silica gel, EtOAc: hexane 1:4) there were obtained 328.4 mg of **3** (87%).

## K E C K Allylation

Replacement of halogen by an allyl moiety via thermal or photochemical free radical reaction with trialkylallylstannanes.

| 1 | Kosugi, M.J. | *J. Organomet. Chem.* | **1973** | *56* | C11 |
|---|---|---|---|---|---|
| 2 | Grignon, J. | *J. Organomet. Chem.* | **1973** | *61* | C33 |
| 3 | Grignon, J. | *J. Organomet. Chem.* | **1975** | *96* | 225 |
| 4 | Seyferth, D. | *J. Org. Chem.* | **1961** | *26* | 4797 |
| 5 | Keck, G.E. | *J. Org. Chem.* | **1982** | *47* | 3590 |
| 6 | Keck, G.E. | *J. Am. Chem. Soc.* | **1982** | *104* | 5829 |

**1-Methyl-1-allylcyclohexane 3.**[6] **1** (177 mg; 1 mmol) in degased PhMe (1 mL) and **2** (661 mg; 2 mmol) was treated with AIBN (24.5 mg; 0.15 mmol) and heated for 8 h at 80°C to afford 110 mg of **3** (80%).

## KENNEDY Oxidative Cyclization

Stereoselective rhenium heptoxide-periodate induced oxidative cyclization to tetra-hydrofurans (syn addition).

| 1 | Kennedy, R.M. | Tetrahedron Lett. | 1992 | 33 | 3729; 5299; 5303 |
| 2 | Kennedy, R.M. | Tetrahedron Lett. | 1994 | 35 | 5133 |
| 3 | Keinan, E. | J. Am. Chem. Soc. | 1995 | 117 | 1447 |

**Bis-perhydrofuran 2.**[3] To a solution of **1** (337 mg; 1 mmol) in dry $CH_2Cl_2$ was added $Re_2O_7$ (726 mg; 1.5 mmol) and $H_5IO_6$ (447.8 mg; 2 mmol). After 35 min stirring at 20°C, the mixture was quenched with aqueous $NaHSO_3$ and extracted with $CH_2Cl_2$. Evaporation of the solvent and chromatography (silica gel, EtOAc:hexane 1:1) afforded 265.5 mg of **2** (75%).

**Spirane 4.**[1] To a solution of **3** (142 mg; 1 mmol) in dry $CH_2Cl_2$ (5 mL) at 0°C under an Ar atmosphere was added 2,6-lutidine (963 mg; 9 mmol) and $Re_2O_7$ (1.452 g; 3 mmol). The mixture was stirred for 12 h at 20°C. A solution of NaOOH (2M; 13 mL) was added dropwise under stirring. Extraction with EtOAc, evaporation of the solvent and acetylation in $CH_2Cl_2$ with $Ac_2O$ (204 mg; 2 mmol), $Et_3N$ (25.3 mg; 4 mmol) and DMAP (12.2 mg; 0.1 mmol) followed by chromatography gave 119 mg of **4** (56%).

## KHARASH-LIPSHUTZ-POSNER Cuprate Reagents

Organocuprate reagents as active intermediates in 1,4-addition to unsaturated carbonyls, in substitutions and epoxide opening.

| 1 | Kharash, M.S. | *J. Am. Chem. Soc.* | **1941** | *63* | 2308 |
| 2 | Parham, W.E. | *J. Org. Chem.* | **1969** | *34* | 1899 |
| 3 | Posner, G.H. | *J. Am. Chem. Soc.* | **1972** | *94* | 5106 |
| 4 | Lipshutz, B.H. | *J. Am. Chem. Soc.* | **1981** | *103* | 7672 |
| 5 | Lipshutz, B.H. | *Tetrahedron* | **1986** | *42* | 2873 |
| 6 | Posner, G.H. | *Org. React.* | **1977** | *19* | 1 |
| 7 | Lipshutz, B.H. | *Org. React.* | **1992** | *41* | 1 |

**Hexahydrofluoren-9-one 2.**[2] MeLi in Et$_2$O (1.4M; 34 mL; 0.048 mmol) was added to a slurry of CuI (4.76 g; 25 mmol) in Et$_2$O at 0°C under N$_2$. After 30 min stirring **1** (2 g; 10.9 mmol) in Et$_2$O (40 mL) was added dropwise and after another 30 min, usual work up gave 2.21 g of crude **2** (100%). Short-path distillation afforded 2.09 g of **2** (95%), bp 94-98°C/0.2 mm.

**Ketone 5.**[5] CuCN (102 mg; 1.14 mmol) in THF (1 mL) under Ar was cooled to -78°C. 2-Thienyllithium (Aldrich or from thiophene, 1.14 mmol) in THF (1 mL) at -30°C and 1.14 mmol t-BuLi (0.47 mL; 2.44 mmol in hexane) were stirred at 0°C for 30 min. All was added to CuCN at -78°C over 30 min. Grignard reagent **4** (80 mL; 1.42M in THF; 1.14 mmol) cooled to -78°C was added dropwise and the mixture was warmed up to 0°C for 2 min and cooled back to -78°C. Cyclohexenone **3** (100 µL; 1.03 mmol) was added for 2.25 h at -78°C and quenched with 5 mL NH$_4$OH/NH$_4$Cl. Normal work up and chromatography (Et$_2$O, Skellysolve) gave 186 mg of **5** (85%).

## KHARASH-SOSNOVSKY Allylic Oxidation

Cu catalyzed allylic or propargylic oxidation with t-butyl peresters (see 1st edition).

$$Et-C\equiv C-Et \xrightarrow[\text{CuCl; 140°; 14 h}]{\underset{2}{\overset{O}{\overset{\|}{Ph-C-O-O-tBu}}}} Et-C\equiv C-\underset{\underset{Me}{|}}{\overset{H}{\underset{|}{C}}}-O\text{-COPh}$$

1                                           3 (57%)          4

$$\xrightarrow[\text{Cu(OTf)}_2]{\text{2; DBU; 20°}} \quad \text{5 (80%)} \quad \text{OCOPh}$$

| | | | | | |
|---|---|---|---|---|---|
| 1 | Kharash, M.; Sosnovsky, G. | *J. Am. Chem. Soc.* | **1958** | *80* | 756 |
| 2 | Kochi, J.K. | *Tetrahedron* | **1968** | *24* | 5099 |
| 3 | Walling, C. | *J. Am. Chem. Soc.* | **1967** | *85* | 2084 |
| 4 | Sosnovsky, G. | *Synthesis* | **1972** | | 1 |
| 5 | Julia, M. | *Tetrahedron Lett.* | **1976** | | 2141 |
| 6 | Kropf, H. | *Synthesis* | **1977** | | 894 |
| 7 | Sing, V.K. | *Tetrahedron Lett.* | **1996** | | 8435 |

**2-Cyclohexenol benzoate 5.**[7] A solution of DBU (18.24 mg; 0.12 mmol) and Cu(OTf)$_2$ (36.1 mg; 0.1 mmol) in Me$_2$CO (4 mL) was stirred for 15 min at 20°C. Cyclohexene **4** (820 mg; 10 mmol) was added followed by dropwise addition of t-butyl perbenzoate (194 mg; 1 mmol). After consumation of perbenzoate (TLC), usual work up and purification gave **5** in 80% yield.

# KHUN-WINTERSTEIN

## GAREGG-SAMUELSSON Olefin Synthesis

Conversion of vic-diols into alkenes by $P_2I_4$ (Khun-Winterstein) or by $I_2$-Ph$_3$P-imidazol (Garegg-Samuelsson) (see 1st edition).

| 1 | Khun, R.; Winterstein, A. | *Helv. Chim. Acta* | **1928** | *11* | 87 |
| 2 | Khun, R.; Winterstein, A. | *Helv. Chim. Acta* | **1955** | *27* | 309 |
| 3 | Inhoffen, C. | *Liebigs Ann.* | **1965** | *684* | 24 |
| 4 | Mitchel, R.H. | *Can. J. Chem.* | **1977** | *55* | 1480 |
| 5 | Block, A. | *Org. React.* | **1984** | *30* | 452 |
| 6 | Garegg, P.J.; Samuelsson, B. | *Synthesis* | **1979** | | 469;813 |
| 7 | Garegg, P.J.; Samuelsson, B. | *J. Carbohydr. Chem.* | **1984** | *3* | 189 |
| 8 | Zamojski, A. | *Carbohydrate Res.* | **1990** | *205* | 410 |

**Olefin 4.**[4] The diol **3** (500 mg; 1.75 mmol) and $P_2I_4$ (500 mg; 0.87 mmol) were stirred for 12 h in Et$_2$O (100 mL) and THF (80 mL), followed by reflux of the orange solution for 4 h. Washing with aqueous Na$_2$S$_2$O$_3$ solution (to remove the iodine), evaporation of the solvent and chromatography (silica gel, PhH) gave 325 mg of **4** (80%).

## KILIANI-FISCHER Sugar Homologation

Synthesis of $C_{n+1}$ sugars from $C_n$ sugars (see 1st edition).

| 1 | Kiliani, H. | *Chem. Ber.* | **1885** | *18* | 3066 |
| 2 | Fischer, E. | *Chem. Ber.* | **1889** | *22* | 2204 |
| 3 | Wood, H.B. | *J. Org. Chem.* | **1961** | *26* | 1969 |
| 4 | Mowry, D.T. | *Chem. Rev.* | **1948** | *42* | 239 |

**D-Ribose 3.**[3] **2** (1.1 g; 6.8 mmol) in water (50 mL) and sodium acid oxalate (2.0 g) at 0°C was treated with NaBH₄ (0.5 g; 13 mmol) in water (10 mL). The pH was kept at 4.5-4. After dilution with MeOH to precipitate the salts, the solution was deionized by Amberlite IR-120-H⁺ and Duolite A-4 and the concentrate was treated with anhydrous EtOH. After several days at 5°C, crystals were filtered, 0.42 g of **3** (38%), mp 102-104°C.

## K N O C H E L  Zinc Vinyl Coupling

Copper-zinc mediated coupling of vinyl halides with alkyl or aryl iodides.

| 1 | Knochel, P.  | *J. Org. Chem.*      | **1988** | 53 | 2390 |
|---|--------------|----------------------|----------|----|------|
| 2 | Knochel, P.  | *Tetrahedron. Lett.* | **1989** | 30 | 4795 |
| 3 | Knochel, P.  | *Synlett*            | **1994** |    | 849  |
| 4 | Knochel, P.  | *Pure Appl. Chem.*   | **1992** | 64 | 361  |
| 5 | Knochel, P.  | *Chem. Rev.*         | **1993** | 93 | 2117 |
| 6 | Knochel, P   | *Tetrahedron*        | **1998** | 54 | 8275 |
| 7 | Knochel, P   | *J. Org. Chem.*      | **1999** | 64 | 186  |
| 8 | Erdik, E.    | *Tetrahedron*        | **1992** | 48 | 9577 |

**(E)-10-Pivaloxy-5-decenitrile 3.**[3] Zinc dust (1.3 g, 20 mmol) in THF (3 mL) was activated with 1,2-dibromoethane (112 mg) and Me$_3$SiCl (10.8 mg), then 4-iodobutyl pivalate **1** (2.84 g, 10 mmol) in THF (1 mL) was added. After 4 h stirring at 25-35 °C, THF ( 3 mL) was added, the excess zinc was allowed to settle and the supernatant (the alkylzinc iodide intermediate) was transferred to a solution of CuCN(0.89 g, 10 mmol) and  LiCl (0.85 g, 20 mmol) in N-methylpyrrolidone (NMP) (10 mL). After 5 min at 0 °C, 6-iodo-5-hexenenitrile **2** (1.1 g, 5 mmol) was added. After 18 h at 60 °C, the solution was poured into Et$_2$O and aq. NH$_4$Cl, followed by usual work up. Chromatography (hexane: Et$_2$O 3:1) afforded 1.09 g of **3** (87%), 100% E.

## K N O E V E N A G E L - D O E B N E R - S T O B B E Condensation

Base catalyzed aldol condensation of aldehydes or ketones with an activated methylene group of a malonic ester (Knoevenagel-Doebner) (see also Laszlo) or a succinic ester (Stobbe) (see 1st edition).

$$Ph-CH=O \ + \ NC-CH_2-CO_2R \xrightarrow[\text{PhH; refl. 12 h}]{\text{NH}_4\text{OAc; HOAc}} Ph-CH=C-CO_2R$$

$$\underset{1}{\phantom{Ph-CH=O}} \qquad \underset{2}{\phantom{NC-CH_2-CO_2R}} \qquad \underset{|}{\overset{|}{\phantom{x}}}$$

$$\underset{3 \ (93\%)^4}{CN}$$

$$\underset{4}{\underset{Ph}{\overset{Ph}{>}}C=O} \ + \ \underset{5}{\overset{CH_2-COOEt}{\underset{CH_2-COOEt}{|}}} \xrightarrow[20°]{NaH} \underset{6 \ (97\%)^8}{\underset{Ph}{\overset{Ph}{>}}C=\underset{CH_2COOEt}{\overset{|}{C}}-COOEt}$$

| | | | | | |
|---|---|---|---|---|---|
| 1 | Knoevenagel, E. | *Chem. Ber.* | **1896** | *29* | 172 |
| 2 | Doebner, O. | *Chem. Ber.* | **1900** | *33* | 2140 |
| 3 | Rapoport, H. | *J. Org. Chem.* | **1981** | *46* | 5064 |
| 4 | Cativiela, C. | *Synth. Commun.* | **1990** | *20* | 3145 |
| 5 | Emden, D. | *Chem. Ber.* | **1987** | *120* | 2717 |
| 6 | Jones, G. | *Org. React.* | **1967** | *15* | 204 |
| 7 | Stobbe, H. | *Chem. Ber.* | **1893** | *26* | 2312 |
| 8 | Daub, G.R. | *J. Am. Chem. Soc.* | **1948** | *70* | 418 |
| 9 | Johnson, W.S. | *Org. React.* | **1951** | *6* | 1 |

## K N U N Y A N T S Fluoroalkylation

Fluoroalkylation of aromatics using hexafluoroacetone (see 1st edition).

| | | | | | |
|---|---|---|---|---|---|
| 1 | Knunyants, I.L. | *Zh. Vses. Chim. Obsh.* | **1960** | *4* | 114 |
| 2 | Simmons, H.E. | *J. Am. Chem. Soc.* | **1960** | *82* | 2288 |
| 3 | Gilbert, E.E. | *J. Org. Chem.* | **1965** | *30* | 998; 1001 |
| 4 | Knunyants, I.L. | *Zh. Akad. Nauk. SSSR* | **1962** | *4* | 682 |

**Bis(Trifluoromethyl)phenylcarbinol (3)**.[3] To a suspension of AlCl₃ (5.0 g; 37 mmol) in PhH **1** (880 g; 11.3 mol) cooled externally, was bubbled hexafluoroacetone **2** (bp = -28°C) until was absorbed 115 g (6.72 mol; ca 6 h). The mixture was washed, dried and distilled to give 541 g of **3** (94%).

## **K N O R R** Pyrazole Synthesis

Pyrazole synthesis from a β-dicarbonyl compound and a hydrazine (see 1st edition).

| 1 | Knorr, L.       | *Chem. Ber.*   | **1883** | 16 | 2587 |
| 2 | Seidel, F.      | *Chem. Ber.*   | **1935** | 68 | 1922 |
| 3 | Mosley, M. S.   | *J. Chem. Soc.* | **1957** |    | 3997 |
| 4 | Katritzky, A. R. | *Tetrahedron*  | **1964** | 20 | 299  |

## **K N O R R** Quinoline Synthesis

Quinoline synthesis by acid catalyzed cyclization of acetoacetanilides (see 1st edition).

| 1 | Knorr, L.        | *Liebigs Ann.*  | **1886** | 236 | 69   |
| 2 | Hodgkinson, A.   | *J. Org. Chem.* | **1969** | 34  | 1709 |
| 3 | Bergstrom, F. W. | *Chem. Rev.*    | **1944** | 35  | 157  |
| 4 | Bergstrom, F. W. | *Chem. Rev.*    | **1948** | 48  | 47   |

**3-Chloro-4-methyl-2-quinoxolone (2).**[2] 2-Chloroacetoacetanilide **1** (1.0 g, 4.7 mmol) was heated in 98% $H_2SO_4$ (2 mL) at 95 °C for 1 h . Usual work up afforded 761 mg of **2** (83%), mp 272-274 °C.

## KOCH-HAAF Carboxylation

Carboxylation of alcohols or olefins with $HCO_2H$ or with CO (super-saturated solution) in conc. sulfuric acid via carbocations, usually with rearrangement.

| 1 | Koch, H., Haaf, W. | *Liebigs Ann.* | **1958** | *618* | 251 |
|---|---|---|---|---|---|
| 2 | Koch, H., Haaf, W. | *Angew. Chem.* | **1958** | *70* | 311 |
| 3 | Haaf, W. | *Chem. Ber.* | **1966** | *99* | 1149 |
| 4 | Takahashi, Y. | *Synth. Commun.* | **1989** | *19* | 1945 |

**2,2-Dimethylnonanoic acid (2).** [4] To 98% sulfuric acid (12 mL) cooled to 0-5 °C was added drowise 100% formic acid (3.9 mL) in 3 min under stirring. A solution of 2-methyl-2-nonanol 1 (795 mg, 5 mmol) in $CCl_4$ (9 mL) was added during 3-5 h under stirring (100 rpm) at the same temperature. Stirring was continued for an additional 5 min at 5 °C. Quenching with ice (100 g) and extraction ($Et_2O$) was followed by washing the extracts with 5% $Na_2CO_3$. The alkaline solution was acidified and extracted with $Et_2O$. Removal of the solvent afforded a sufficiently pure residue, 935 mg (100%). Distillation (Kugelrohr) afforded 888 mg of 2 (95%), of 95-100% purity.

## K O C H I   Cross Coupling

Cross coupling of organometallics with vinyl halides catalyzed by iron (III) or by Fe(III)-dibenzoylmethane (FeDBM$_3$) (see 1st edition).

| 1 | Kochi, J. | *Synthesis* | **1971** | | 303 |
|---|-----------|-------------|----------|-----|------|
| 2 | Kochi, J. | *J. Org. Chem.* | **1975** | *40* | 599 |
| 3 | Kochi, J. | *J. Org. Chem.* | **1976** | *41* | 502 |
| 4 | Molander, G.A. | *Tetrahedron Lett.* | **1983** | *24* | 5449 |

## K O E N I G S – K N O R R  Glycosidation

Synthesis of glycosides from halosugars or acetoxysugars in the presence of Ag⁺, Hg²⁺ or base (e.g. tetrabutylammonium bromide(TBAB)-NaOH) (see 1st edition).

| 1 | Koenigs, W.; Knoor, E. | *Chem.Ber.* | **1901** | *34* | 957 |
|---|---|---|---|---|---|
| 2 | Ice, C.H. | *J.Am.Chem.Soc.* | **1952** | *74* | 4606 |
| 3 | Knochel, A. | *Tetrahedron Lett.* | **1974** | | 551 |
| 4 | Israel, M. | *J.Med.Chem.* | **1982** | | 28 |
| 5 | Gabrey, S. | *Synthesis* | **1992** | | 1078 |

## K R E S Z E  Amination Agent

Regiospecific allylic amination of alkenes by bis (methoxycarbonyl) sulfur diimide (3).

| 1 | Kresze, G. | *Lielbigs Ann.* | **1975** | | 1725 |
|---|---|---|---|---|---|
| 2 | Kresze, G. | *Lielbigs Ann.* | **1980** | | 629 |
| 3 | Kresze, G. | *J.Org.Chem.* | **1983** | *48* | 3561 |

**Bis(methoxycarbonyl)sulfur diimide 3.**[3] N,N-dichlorocarbamate 1 (144 g, 1 mol), pyridine (0.5 mL) and SCl₂ (5 g) were heated (50-60°C) under stirring until a vigorous evolution of Cl₂ (5-10 min). The heating is removed and SCl₂ (5 g) is added to maintain a rapid evolution of Cl₂ at 35°C. The mixture is heated at 60°C (10 mbar) for 10 min followed by removal of volatiles (20°C/0.01 mbar-1 h) to give a yellow oil (moisture sensitive) (quantitative yield).

**Methyl N-(2-alkenyl)carbamate 5.** Alkene 4 (13.2 g, 0.1 mol) was added dropwise to reagent 3 (17.8 g, 0.1 mol) under stirring in CHCl₃ (15 mL) at 0°C. After 20 h stirring at 20°C, the solvent was removed in vacuum and the residue after usual work up and vacuum distillation (70°C/0.01 mbar) afforded 10.45 g of 5 (51%).

**Alkenyl amine 6.** A mixture of 5 (20.5 g, 0.1 mol), KOH (28 g, 0.5 mol), MeOH (70 mL) and water (50 mL) was refluxed for 30 h. Evaporation of the solvent, the residue basified (KOH), extraction (Et₂O), evaporation of the solvent and distillation gave 11 g of 6 (75%), bp 81°C/3mbar.

## K O L B E  Electrolysis

Electrochemical decarboxylation-dimerization (via free radicals) (see 1st edition).

$$2\ Ph_2CH\text{-}COOH \xrightarrow[\text{DMF; 17 h}]{\text{electrolysis}} Ph_2CH\text{-}CHPh_2$$

**1**                    **2** (24%)[3]

**3**                                        **4**

| 1 | Kolbe, H. | *Liebigs Ann.* | **1849** | *69* | 257 |
|---|---|---|---|---|---|
| 2 | Crum Brown, A.; Walker, A. | *Proc. Roy. Soc. Edinburgh* | **1890** | *17* | 292 |
| 3 | Finkelstein, M. | *J. Org. Chem.* | **1969** | *25* | 156 |
| 4 | Rabson, M. | *J. Org. Chem.* | **1981** | *46* | 4082 |
| 5 | Marquet, B. | *Bull. Soc. Chim. Fr.* | **1988** | | 571 |
| 6 | Vijh, A.K. | *Chem. Rev.* | **1967** | *67* | 625 |
| 7 | Schaefer, H.J. | *Angew. Chem. Int. Ed.* | **1981** | *20* | 911 |
| 8 | Steckhan, E. | *Synthesis* | **1996** | | 71 |
| 9 | Renault, P. | *Synlett* | **1997** | | 181 |

**Diester 4.**[8] Bicyclo[2.2.2]monomethyl-octane-1,4-dicarboxylate **3** (2.5 g; 11.8 mmol) in MeOH (4 mL) was electrolyzed (Pd foil electrode, each 0.24 cm$^2$; distance anode-catode 1 cm; voltage 300-400V; 0.7 A/cm$^2$). In 5 h there were obtained 589 mg of **4** (30%), mp 229°C.

## K O L B E - S C H M I D T  Salicylic Acid Synthesis

Carboxylation (usually ortho) of phenols. Industrial method to obtain salicylic acid derivatives (see 1st edition).

**1**                    **2** (48%)

| 1 | Kolbe, H. | *Liebigs Ann.* | **1860** | *113* | 125 |
|---|---|---|---|---|---|
| 2 | Schmidt, R. | *J. Prakt. Chem.* | **1885** | *31* | 397 |
| 3 | Doub, L. | *J. Org. Chem.* | **1958** | *23* | 1422 |
| 4 | Lindsey, A.S. | *Chem. Rev.* | **1957** | *57* | 583 |
| 5 | Raecke, B. | *Angew. Chem.* | **1958** | *70* | 1 |
| 6 | Ota, K. | *Bull. Soc. Chim. Jpn.* | **1974** | *47* | 2343 |

## K O N A K A   Nickel Oxidizing Agent

Oxidation of alcohols to carboxylic acids (or ketones) with nickel peroxide (see 1st edition).

$$NiSO_4 \cdot 6H_2O + NaOCl \xrightarrow[0.5\,h]{20°} \text{Ni peroxide} \xrightarrow[2\% \text{ NaOH; } 30°]{Ph\frown OH} Ph\text{-}CO_2H \qquad Ph_2CHOH \xrightarrow{1} Ph_2C=O$$

1

2 (86%)

| | | | | | |
|---|---|---|---|---|---|
| 1 | Konaka, | *J. Org. Chem.* | **1962** | *27* | 1660 |
| 2 | Konaka, | *J. Org. Chem.* | **1969** | *34* | 1334 |

**Benzoic acid (2).** Benzyl alcohol (2.16 g; 20 mmol) and NaOH (1.0 g; 25 mmol) in water (50 mL) was treated with **1** (16.0 g; 1.5 equiv) under stirring at 30°C. After 3 h the solution was filtered and the filtrate was acidified. The dried precipitate afforded 2.1 g of **2** (86%), mp 122.5°C.

## K Ö N I G   Benzoxazine Synthesis

Benzoxazine synthesis from quinones and aminoalkyl halides (see 1st edition).

| | | | | | |
|---|---|---|---|---|---|
| 1 | König, K.H. | *Chem. Ber.* | **1959** | *92* | 257 |
| 2 | König, K.H. | *Z. Anal. Chem.* | **1959** | *166* | 92 |
| 3 | Flemming, J. | *Z. Phys. Chem. (Leipzig)* | **1964** | *223* | 106 |
| 4 | Day, J.H. | *J. Org. Chem.* | **1965** | *30* | 4107 |
| 5 | McMurtrey, K.D. | *J. Org. Chem.* | **1970** | *35* | 4252 |

**3,4-Dihydro-4-methyl-2H-1,4-benzoxazine-6-ol (4).**[5] To **1** (30.0 g; 0.277 mol) and **2** (30.0 g; 0.137 mol) in 50% water-MeOH (2000 mL) at 0°C, was added dropwise 0.2N NaOH (500 mL). After 2 h, filtration and trituration with Me$_2$CO, gave from the acetone fraction 6 g of **3** (18%), mp 140-144°C.

**3** (1.0 g; 4 mmol) in CHCl$_3$ (100 mL) was shaken with aqueous sodium dithionite until colorless. The residue after evaporation was dissolved in dioxane:TEA (1:1) (100 mL) by heating 12 h on a steam bath. Evaporation and chromatography gave 0.6 g of **4** (89%), mp 77-78.3°C.

### K O R N B L U M  Aldehyde Synthesis

Synthesis of aldehydes from primary alkyl halides or tosylates using dimethylsulfoxide (DMSO) (see 1st edition).

$$p\text{-Br-}C_6H_4\text{-CO-CH}_2\text{Br} \xrightarrow[9\,h;\;20°]{DMSO} p\text{-Br-}C_6H_4\text{-CO-CHO} \qquad C_7H_{15}\text{-I} \xrightarrow{AgTs} \xrightarrow[150°]{DMSO} C_6H_{13}\text{-CHO}$$

| | | **1** | | **2 (91%)**[1] | **3** | | **4 (70%)** |

| | | | | | | |
|---|---|---|---|---|---|---|
| 1 | Kornblum, N. | *J. Am. Chem. Soc.* | **1957** | 79 | 6562 |
| 2 | Kornblum, N. | *J. Am. Chem. Soc.* | **1959** | 81 | 4113 |
| 3 | Kornblum, N. | *Angew. Chem. Int. Ed.* | **1975** | 14 | 734 |
| 4 | Chandrasekar, S. | *Tetrahedron Lett.* | **2000** | 41 | 5423 |

**Heptanal (4).**[2] To silver tosylate (11.0 g; 38 mmol) in MeCN (100 mL) was added **3** (7.0 g; 30 mmol). The light protected mixture was kept 24 h at 20°C, poured on ice, extracted with $Et_2O$, evaporated and the residue poured into $Na_2CO_3$ (20 g) in DMSO (150 mL). After heating 5 min at 150°C under $N_2$, the aldehyde was separated as its 2,4-dinitrophenylhydrazone (DNPH), 6.9 g of **4** DNPH (70%), mp 106-107°C.

### K O S E R  Tosylation

Vic-bis tosylation of alkenes by means of hydroxytosyloxyiodobenzene (see 1st edition).

| | | | **2** | | **1** | | **3 (44%)** |

| | | | | | | |
|---|---|---|---|---|---|---|
| 1 | Koser, G.F. | *J. Org. Chem.* | **1977** | 42 | 1476 |
| 2 | Neiland, O. | *J. Org. Chem. USSR (Engl).* | **1970** | 6 | 889 |
| 3 | Koser, G.F. | *J. Org. Chem.* | **1980** | 45 | 1542 |
| 4 | Koser, G.F. | *J. Org. Chem.* | **1984** | 49 | 2462 |

**Erythro(dl)-2,3-bis(tosyloxy)pentane (3).**[4] Hydroxy(tosyloxy)iodobenzene **1** (3.92 g; 10 mmol), **2** (2.5 mL; 1.6 g; 23 mmol) and $CH_2Cl_2$ (20 mL) was kept for 28 h at 3°C. The yellow solution and scum was washed (water) and concentrated (vacuum). The residue after washing with pentane (15 mL) and recrystallization from MeOH (6 mL) and pentane (3 mL) at -20°C gave 827 mg of **3** (40%), mp 82-83°C.

## **K R A P C H O** Dealkoxycarbonylation

Dealkoxycarbonylation of malonate esters, β-keto esters and α-cyano esters or other activated esters in dipolar aprotic solvents in the presence of an equiv. of water or of water with added salts.

| 1 | Krapcho, A.P. | *Tetrahedron Lett.* | **1967** | | 215 |
|---|---|---|---|---|---|
| 2 | Krapcho, A.P. | *Tetrahedron Lett.* | **1974** | | 1091 |
| 3 | Krapcho, A.P. | *J. Org. Chem.* | **1978** | *43* | 138 |
| 4 | Klemmensen, P.D. | *J. Org. Chem.* | **1979** | *44* | 416 |
| 5 | Krapcho, A.P. | *Synthesis* | **1982** | | 805; 893 |
| 6 | Krapcho, A.P. | *J. Org. Chem.* | **1987** | *52* | 1880 |
| 7 | Loupy, A. | *J. Chem. Res. (S)* | **1993** | | 36 |

**4-(2,2-Dichloroethenyl)-5,5-dimethyltetrahydrofuran-2-one (2).**[4] Lactone 1 (267 g; 1 mol) in DMF (600 mL) and water (27 mL; 1.5 mol) was heated to reflux for 4-12 h. Water and DMF were removed in vacuum, the residue dissolved in MeOH (500 mL) and precipitated with water (100 mL), to yield a total of 198 g of 2 (95%), mp 116-119°C.

**1,5-Dicyano-3-acetyl pentane (4).**[6] 3 (60 g; 0.227 mol) in DMSO (300 mL), water (4.5 g; 0.25 mol) and LiCl (10.6 g; 0.25 mol) were heated at reflux for 5 h. The cooled mixture was diluted with 200 mL water, extracted with $CH_2Cl_2$ (3x150 mL). The dried extract was evaporated and distillation gave 33.5 g of 4 (90%), bp 173-175°C/0.5 mm.

## K R Ö H N K E - O R T O L E V A   Keto Pyridinium Salts

Synthesis of aryl carboxylic acids by base catalyzed cleavage of keto pyridinium salts formed by reaction of α-haloketone derivatives with pyridine (see 1st edition).

| 1 | Kröhnke, F. | *Chem. Ber.* | **1933** | *66* | 604 |
|---|---|---|---|---|---|
| 2 | Ortoleva, G. | *Gazz. Chim. Ital.* | **1899** | *25 I* | 503 |
| 3 | King, I.C. | *J. Am. Chem. Soc.* | **1944** | *66* | 894; 1612 |
| 4 | Kröhnke, F. | *Angew. Chem. Int. Ed.* | **1963** | *2* | 380 |
| 5 | Alvarez, S.I. | *Tetrahedron* | **1986** | *42* | 699 |

## K U C H E R O V - D E N I G E S   Mercuric Catalyzed Hydration

Water addition to a triple bond (Kucherov) or to a double bond (Deniges) under mercury salt catalysis, sometimes with carbocation rearrangement (see 1st edition).

| 1 | Kucherov, M. | *Chem. Ber.* | **1881** | *14* | 1540 |
|---|---|---|---|---|---|
| 2 | Thomas, R.J. | *J. Am. Chem. Soc.* | **1938** | *60* | 718 |
| 3 | Deniges, G. | *Bull. Soc. Chim. Fr.* | **1898** | *19* | 494 (3) |
| 4 | Shearer, D.A. | *Can. J. Chem.* | **1955** | *33* | 1002 |
| 5 | Arzoumanian, N. | *Synthesis* | **1971** | | 527 |

**1,2-Dihydroxy-2-methyl-1-phenylpropane (4).**[5] To a suspension of HgO (21.7 g; 0.1 mol) in 1N $H_2SO_4$ (200 mL; 0.2 mol) was added **3** (6.61 g; 50 mmol). The mixture was shaken for 2 days, filtered and the residue washed with MeOH and $Et_2O$. The filtrate was extracted with $Et_2O$ and $CHCl_3$ and the solvent evaporated to yield 5.8 g of **4** (70%), mp 54-62°C.

## **KULINKOVICH** Hydroxycyclopropanation

Synthesis of 1-substituted cyclopropanols from esters and a Grignard reagent or by reductive coupling of carboxylic esters with terminal olefins, catalyzed by $Ti(OiPr)_4$.

$$C_4H_9\text{-}COOCH_3 \xrightarrow[20°]{EtMgBr/Ti(OiPr)_4}$$

**1**

**2** (90%)

$$\text{CO}_2\text{Me} \xrightarrow{nBuMgCl/Ti(OiPr)_4}$$

$$EtMgBr + Ti(OiPr)_4 \longrightarrow \|\text{---}Ti(OiPr)_2 \xrightarrow[MeCO_2Et]{Ph} (iPrO)_2Ti \longrightarrow (iPrO)_2Ti \longrightarrow$$

**3** (42%)

| | | | | | |
|---|---|---|---|---|---|
| 1 | Kulinkovich, O.G. | *Zh. Org. Khim.* | **1989** | 25 | 2245 |
| 2 | Kulinkovich, O.G. | *Synthesis* | **1991** | | 234 |
| 3 | Kulinkovich, O.G. | *Mendeleev Commun.* | **1993** | | 230 |
| 4 | Corey, E.J. | *J. Am. Chem. Soc.* | **1994** | 116 | 9345 |
| 5 | Cha, J.K. | *J. Am. Chem. Soc.* | **1996** | 118 | 4198 |
| 6 | Kulinkovich, O.G. | *Tetrahedron Lett.* | **1998** | 39 | 1823 |
| 7 | Kulinkovich, O.G. | *Tetrahedron Lett.* | **1999** | 40 | 000 |

**1-Butylcyclopropanol (2).**[2] To a stirred solution of methyl valerate **1** (2.9 g; 25 mmol) and $Ti(OiPr)_4$ (1.7 mL; 2.5 mmol) in $Et_2O$ (80 mL) was added over a period of 1 h a solution of EtMgBr (53 mmol) in $Et_2O$ (60 mL) at 20°C, under stirring. After further stirring (ca 15 min) at the same temperature, the mixture was poured into a cooled (5°C) solution of 10% $H_2SO_4$ (ca 250 mL). Extraction with $Et_2O$, evaporation of the solvent and distillation of the residue afforded 2.1 g of **2** (90%), bp 67-69°C/19 mm.

***cis*-1-Methyl-2-phenylcyclopropanol (3).**[5] Dropwise addition of ethylmagnesium bromide (2 equiv.) in ether to a boiling solution of ethyl acetate (1 equiv.), styrene (2 equiv.) and titanium isopropoxide (0.05 equiv.) gave, in addition to a small amount of 1-methylcyclopropanol, *cis*-1-methyl-2-phenylcyclopropanol **3** in 42% yield.

## K U R S A N O V – P A R N E S Ionic Hydrogenation

A non-catalytic hydrogenation of C=C, C=O, C=N bonds and hydrogenolysis of C-OH, C-Hal, under the action of an acid and a silyl hydride ion donor (see 1st edition).

$$\xrightarrow[20°,\ 1\ h]{Ph_2SiH_2,\ TFA}$$

**1**

**2 (80%)**[3]

$$\xrightarrow[BF_3 \cdot Et_2O]{Et_3SiH}$$

**3**

**4 (70%)**

$$\xrightarrow[BF_3 \cdot Et_2O]{Et_3SiH}$$

**5**

**6 (96%)**[5]

| | | | | | |
|---|---|---|---|---|---|
| 1 | Parnes, Z. N. | *Dokl. Akad. Nauk. SSSR* | **1966** | *166* | 122 |
| 2 | Parnes, Z. N. | *Tetrahedron* | **1967** | 23 | 2235 |
| 3 | Kursanov, D. N. | *Synthesis* | **1974** | | 633 |
| 4 | Rouzaud, D. | *J. Chem. Soc. Chem. Commun.* | **1983** | | 1325 |
| 5 | Horikawa, H. | *Chem. Pharm. Bull.* | **1990** | 38 | 2024 |

**5-(2-Tetrahydrothienyl)valeric (4).** To a mixture of **3** (5.52 g, 30 mmol) and Et₃SiH (7.19 g, 62 mmol) cooled at 0 °C was added dropwise a solution of BF₃Et₂O (1.15 g, 8 mmol) in TFA (30.78 g, 270 mmol). After 20 min stirring at 20 °C, the volatiles were removed by distillation and the residue recrystallized (hexane) to give 3.95 g fo **4** (70%), mp 50-51 °C.

**N-Acetyl-2-methylbutyl amine (6).**[5] A solution of **5** (2 g, 2 mmol) and Et₃SiH (278 mg, 2.4 mmol) in CH₂Cl₂ (3 mL) was treated with BF₃. Et2O (2.4 mmol) at 5 °C. After 2 h stirring at 5 °C, the mixture was diluted with CH₂Cl₂, washed (aq NaHCO₃), the solvent evaporated and the residue chromatographed (silica gel, CHCl₃:Me₂CO 5:1) to give **6** in 96% yield.

## L A P W O R T H   ( B E N Z O I N )   Condensation

Condensation of two molecules of aryl aldehydes to an α-hydroxy ketone catalyzed by CN ions (via cyanohydrins) (see 1st edition).

| 1 | Lapworth, A. | J.Chem.Soc. | **1903** | 83 | 995 |
|---|---|---|---|---|---|
| 2 | Buck, J.S. | J.Am.Chem.Soc. | **1931** | 53 | 2351 |
| 3 | Hensel, A. | Angew.Chem. | **1953** | 65 | 491 |
| 4 | Dahn, H. | Helv.Chim.Acta. | **1954** | 37 | 309 ;1612 |
| 5 | Solodar, C. | Tetrahedron Lett. | **1971** | | 287 |
| 6 | Enders, D. | Helv.Chim.Acta. | **1996** | 79 | 1217 |
| 7 | Ide, V.S. | Org.React. | **1948** | 4 | 269 |
| 8 | Hassner, A. | Compreh.Org.Synthesis | **1991** | 1 | 541 |

**p-Dimethylaminobenzpiperoin (3).**[2] A solution of piperonal **1** (6 g, 40 mmol) and p-dimethylaminobenzaldehyde **2** (5.96 g, 40 mmol) in EtOH (30 mL) was treated with a saturated solution of KCN (4 g, 61 mmol) in water. After 2 h reflux and 3 days at 20°C, the crystals were filtered and recrystallized from EtOH to give 9.18 g of **3** (76.7%), mp 132°C.

**Benzoin (5).**[6] To a stirred solution of Ph-CHO **4** (4.664 g, 44 mmol) and catalyst (4S,5S)-4-(2,2'-Dimethyl-4-phenyl-1,3-dioxan-5-yl)-1-phenyl-4H-1,2,4-triazoline perchlorate **6** (240 mg, 0.55 mmol) in THF was added $K_2CO_3$ (35 mg, 0.25 mmol) at 20°C. After 60 h the mixture was poured into water, extracted with $CH_2Cl_2$, the solvent evaporated and the residue chromatographed (silica gel, $Et_2O$/ pentane), to afford 3.07 g of **5** (66%), 75% ee, $\alpha_D^{20}$ = -108.4 (R).

## L A R O C K   Annulation

Carbo and heteroannulation of 1,2-, 1,3-, 1,4-dienes, vinyl cyclopropanes, vinyl cyclobutanes catalyzed by arylmercury, thallium or palladium.

| 1 | Larock, R.C. | *J.Org.Chem* | **1984** | *49* | 3663 |
|---|---|---|---|---|---|
| 2 | Larock, R.C. | *J.Am.Chem.Soc.* | **1984** | *106* | 5281 |
| 3 | Larock, R.C. | *Tetrahedron Lett.* | **1987** | *28* | 5291 |
| 4 | Larock, R.C | *Synth.Commun.* | **1989** | *19* | 1463 |
| 5 | Larock, R.C. | *J.Org.Chem* | **1990** | *55* | 3447 |
| 6 | Larock, R.C. | *Synlett* | **1990** | | 529 |

**2-Allyldihydrobenzofuran (3).[6]**   A mixture of Pd(OAc)$_2$ (28 mg, 0.0125 mmol), n-Bu$_4$NCl (0.25 mmol), KOAc (98 mg, 1 mmol), 2-iodophenol 1 (55 mg, 0.25 mmol) and allylcyclopropane 2 (85 mg, 1.25 mmol) and DMF (1 mL) were heated under stirring for 3 days at 80$^0$C. Usual work up and flash chromatography afforded 27.8 mg of 3 (70%).

## L A S Z L O　Clay Catalyst

Modified clays (e.g. montmorilonite K-10) as mild Lewis acid catalysts in Knoevenagel, Michael, Diels-Alder reactions, aromatic chlorination and nitration.

| 1 | Laszlo, P. | *Synthesis* | **1880** | | 849 |
|---|-----------|-------------|----------|-----|-----|
| 2 | Laszlo, P. | *J. Org. Chem.* | **1983** | 48 | 4771 |
| 3 | Laszlo, P. | *Tetrahedron Lett.* | **1984** | 25 | 1567 |
| 4 | Laszlo, P. | *Synthesis* | **1986** | | 655 |
| 5 | Laszlo, P. | *Synlett* | **1994** | | 155 |

**Cyclohexyl 1-tetrahydropyranyl ether 3.**[4] To a solution of cyclohexanol **1** (2.0 g; 20 mmol) in dry $CH_2Cl_2$ (25 mL) containing K-10 clay (500 mg) was added, under stirring at 20°C, a solution of dihydro-4H-pyran **2** (2.52 g; 30 mmol) in dry $CH_2Cl_2$ over a period of 5 min. After 30 min the completion of the reaction was tested by TLC (Merck Kieselgel E, EtOAc:hexane 1:3). The catalyst was removed by filtration and the solvent evaporated in vacuum. Chromatography of the residue (silica gel, hexane:$CHCl_3$ 1:1) afforded 3.23 g of **3** (95%).

**1-Methyl-3,3,5,5-tetramethyl-4-isopropenylcyclohexene 5.**[3] A stirred solution of dimethylpentadiene **4** (19.2 g; 0.2 mol) in $CH_2Cl_2$ at 0°C in the presence of acidic montmorilonite (K-10), doped with $Fe^{3+}$ and 4-t-butylphenol (1.38 g; 10 mmol) afforded 17.2 g of **5** (90%).

## **L A W E S S O N**  Thiacarbonylation Reagent

2,4-Bis(4-methoxyphenyl)-1,3-dithia-2,4-diphosphetane-2,4-disulfide  **1**  reagent  for thiacarbonylation and synthesis of thia heterocycles (see 1st edition).

| 1 | Lecher, H.Z. | *J. Am. Chem. Soc.* | **1956** | 78 | 5018 |
|---|---|---|---|---|---|
| 2 | Lawesson, S.O.; Scheibe, S. | *Bull. Soc. Chim. Belge* | **1978** | 87 | 293 |
| 3 | Lawesson, S.O. | *Bull. Soc. Chim. Belge* | **1979** | 88 | 305 |
| 4 | Lawesson, S.O. | *Tetrahedron* | **1979** | 35 | 1339 |
| 5 | Heimgartner, H. | *Helv. Chim. Acta* | **1987** | 70 | 1001 |
| 6 | Hoffmann, R.W. | *Angew. Chem.* | **1980** | 42 | 559 |
| 7 | Kaneko, K. | *Synthesis* | **1988** |  | 152 |
| 8 | Moriya, T. | *J. Med. Chem.* | **1988** | 31 | 1197 |
| 9 | Sandstrom, J. | *J. Chem. Soc. Perkin I* | **1988** |  | 2085 |
| 10 | L'abbe, G. | *Bull. Soc. Chim. Belge* | **1979** | 88 | 737 |
| 11 | Nishio, T. | *J. Org. Chem.* | **1997** | 62 | 1106 |
| 12 | Cava, M.P.; | *Tetrahedron* | **1985** | 41 | 5061 |

**Synthesis of reagent 1.[2]** A mixture of anisole and $P_4S_{10}$ in the molar ratio of 10:1 was heated to reflux under stirring. After 6 h at 155°C, the solid dissolved, accompanied by evolution of $H_2S$. On cooling **1** crystallized. Filtration, washing ($CH_2Cl_2$:$Et_2O$ 1:1) and drying afforded **1** in 80% yield, mp 228-229.5°C.

**Thiouracyl 3.[8]** To a suspension of uracyl **2** (1.121 g; 10 mmol) in HMPA (10 mL) was added **1** (2.225 g; 5.5 mmol). After 1 h heating (120°C) under Ar, the mixture was cooled, water (100 mL) and charcoal were added and the mixture was heated to reflux. After hot filtration, the filtrate was cooled and the precipitate collected, washed and dried to afford 988 mg of **3** (78%).

## L E B E D E V  Methoxymethylation

Methoxymethyl methyl sulfate **1** as an electrophilic reagent for methoxymethylation of alkenes (see 1st edition).

| 1 | Lebedev, M. Yu. | *Zh. Org. Khim.* | **1987** | 23 | 960 |
|---|---|---|---|---|---|
| 2 | Lebedev, M. Yu. | *Zh. Org. Khim. USSR (Eng. trans.)* | **1989** | 25 | 391 |
| 3 | Kalyan, Yu. B. | *Izv. Akad. Nauk. SSSR Ser. Khim.* | **1985** | 9 | 2082 |

## L E H M S T E D – T A N A S E S C U  Acridone Synthesis

Acridone synthesis from *o*-nitrobenzaldehyde and aryls (see 1st edition).

| 1 | Tanasescu, I. | *Bull. Soc. Chim. Fr.* | **1927** | 41 | 528 |
|---|---|---|---|---|---|
| 2 | Lehmsted, K. | *Chem. Ber.* | **1932** | 65 | 834 |
| 3 | Spalding, D.P. | *J. Am. Chem. Soc.* | **1946** | 68 | 1596 |
| 4 | Silberg, I. | *Rev. Roum. Chim.* | **1965** | 10 | 1035 |

**3,6-Dichloroacridone (3).**[3] A mixture of 2-nitro-4-chlorobenzaldehide **1** (18.5 g; 0.1 mol), chlorobenzene **2** (78.7 g; 0.7 mol), conc. $H_2SO_4$ (37.5 mL) and $NaNO_2$ (0.35 g) was alternatively shaken for 9 h and allowed to stand 15 h, for a total of 6 days. At the end of each two-day period a mixture of $H_2SO_4$ (10 mL) and $NaNO_2$ (0.1 g) was added. The mixture was poured into water (500 mL) and steam distilled until no further aldehyde solidified in the condenser. The residue from steam distillation was filtered and digested with PhH, leaving 14 g of **3** (53%).

## L E H N  Cryptand Synthesis

Synthesis of diaza-polyoxa-macrobicyclic compounds (cryptands) and spherical macrotricycles ligands (supercryptands) (see 1st edition).

(40%)[2]

3, n = 1 (34%)          4, m = 2 (33%)

| 1  | Lehn, J.M.        | *J. Chem. Soc. Chem. Commun.* | **1972** |      | 1100 |
|----|-------------------|-------------------------------|----------|------|------|
| 2  | Lehn, J.M.        | *Tetrahedron*                 | **1973** | 29   | 1624 |
| 3  | Lehn, J.M.        | *Acc. Chem. Res.*             | **1978** | 11   | 49   |
| 4  | Schmidtchen, F.P. | *Chem. Ber.*                  | **1980** | 113  | 864  |
| 5  | Lehn, J.M.        | *Angew. Chem. Int. Ed.*       | **1990** | 29   | 1304 |
| 6  | Echegoyen, L.     | *J. Org. Chem.*               | **1991** | 56   | 1524 |
| 7  | Lochhard, J.C.    | *Polyhedron*                  | **1993** | 12   | 2315 |
| 8  | Dietrich, B.      | *Pure. Appl. Chem.*           | **1993** | 65   | 1457 |
| 9  | Krakowiak, K.E.   | *J. Org. Chem.*               | **1995** | 60   | 7070 |
| 10 | Gibson, H.W.      | *Org. Lett.*                  | **1999** |      | 1001 |

**Diazacryptand 3.**[9] To $K_2CO_3$ (13.8 g; 0.1 mol) in MeCN (400 mL) under reflux and stirring were added with syringe pumps diamine **1** (1.48 g; 10 mmol) and **2** (6.9 g; 20 mmol) followed by 6 days of reflux. Evaporation, chromatography ($Al_2O_3$ then silica gel) was followed by treatment with LAH (5.6 g; 0.15 mol) in THF and chromatography (silica gel) to give 1.54 g of **3** (34%).

**Supercryptand 4.** To a suspension of $Na_2CO_3$ (15 g; 0.14 mol) in $C_3H_7CN$ (200 mL) were added **3** (0.6 g; 1.5 mmol) and diiodoether compound **5** (592 mg; 1.6 mmol). Work up afforded 285 mg of **4** (33%).

## LEIMGRUBER–BATCHO Indole Synthesis

Synthesis of indoles by Al-Hg reduction of o-nitro-β-dimethylaminostyrenes, obtainable from o-nitrotoluenes.

| 1 | Batcho, A.D.; Leimgruber, W. | U.S. Pat. 3,976,639; cf. C.A., **1977**, *86*, 29624t | | | |
|---|---|---|---|---|---|
| 2 | Clark, R.D. | *Heterocycles* | **1984** | 22 | 195 |
| 3 | Somei, M. | *Chem. Pharm. Bull.* | **1981** | 29 | 726 |
| 4 | Clark, R.D. | *J. Heterocyclic Chem.* | **1985** | 22 | 121 |
| 5 | Gilmore, J. | *Synlett.* | **1992** | | 79 |
| 6 | Still, I.W.J. | *Org. Prep. Proced. Int.* | **1995** | 27 | 576 |

**6-Aminoindole (4).**[6] To **3** (2.23 g; 8.43 mmol) in THF (80 mL) was added freshly prepared aluminium amalgam (2.23 g; 85 mat/g) and distilled water (2 mL). After gas evolution (15 min) the mixture was maintained in a sonicator for 5 h. Filtration through celite, concentration, chromatography, $R_f$ = 0.45 and recrystallization from PhH/hexane gave 0.73 g of **4** (64%), mp 67-69°C.


## LEUCKART Thiophenol Synthesis

Formation of thiophenols from diazonium salts and xanthates (see 1st edition).

| 1 | Leuckart, R. | *J. Prakt. Chem.* | **1890** | 41 | 187 (2) |
|---|---|---|---|---|---|
| 2 | Bourgeoise, E. | *Rec. Trav. Chim.* | **1899** | 18 | 447 |
| 3 | Tarbel, D.S. | *J. Am. Chem. Soc.* | **1952** | 74 | 48 |

## **L E U C K A R T – P I C T E T – H U B E R T**  Phenanthridine Synthesis

Amidation of aryls by isocyanates (Leuckart) or by amides (Pictet-Hubert), catalyzed by Lewis acids and leading to phenanthridines (see 1st edition).

| | | | | | |
|---|---|---|---|---|---|
| 1 | Leuckart, R. | *Chem. Ber.* | **1885** | *18* | 873 |
| 2 | Buttler, J.M. | *J. Am. Chem. Soc.* | **1949** | *71* | 2578 |
| 3 | Schmutz, I. | *Helv. Chim. Acta.* | **1965** | *48* | 336 |
| 4 | Pictet, A; Hubert, A. | *Chem. Ber.* | **1896** | *29* | 1182 |
| 5 | Boyer, J.H. | *Synthesis* | **1978** | | 205 |
| 6 | Eisch, J. | *Chem. Rev.* | **1957** | *57* | 525 |

## **L E U C K A R T – W A L L A C H**  Reductive Amination

Reductive amination of aldehydes or ketones with amines and formic acid or $H_2$-Ni (Miquonac) or $NaBH_4$ (see Borch), see also Eschneiler-Clarke (see 1st edition).

| | | | | | |
|---|---|---|---|---|---|
| 1 | Leuckart, R. | *Chem. Ber.* | **1885** | *18* | 2341 |
| 2 | Wallach, O. | *Liebigs Ann.* | **1892** | *272* | 100 |
| 3 | Miquonac, G. | *C.R.* | **1921** | *172* | 223 |
| 4 | Raudvere, F. | *Ann. farm. bio. (Buenos Aires)* | **1943** | *18* | 81 |
| 5 | Marcus, E. | *J. Org. Chem.* | **1960** | *25* | 199 |
| 6 | Bhattacharyya, S. | *Tetrahedron Lett.* | **1994** | *35* | 2401 |
| 7 | Moore, M.I. | *Org. React.* | **1949** | *5* | 301 |

## LEY–GRIFFITH Ru Oxidation Reagent

Tetrapropylammonium perruthenate $Pr_4N^+RuO_4^-$ and N-methylmorpholine-N-oxide (NMO) as catalytic oxidants of primary, secondary, allylic and benzylic alcohols to carbonyl derivatives. The same catalyst polymer supported perruthenate (PSP) used as efficient oxidant (see 1st edition).

1     2 (70%)

90%

3     4 (95%)

| | | | | |
|---|---|---|---|---|
| 1 Ley, S.V; Griffith, W.P. | *J. Chem. Soc. Chem. Commun.* | **1987** | | 1625 |
| 2 Ley, S.V; Griffith, W.P. | *Tetrahedron Lett.* | **1989** | 30 | 3204 |
| 3 Mehta, G. | *Tetrahedron Lett.* | **1991** | 32 | 3215 |
| 4 Dubois, L. | *Tetrahedron* | **1993** | 49 | 901 |
| 5 Ley, S.V; Griffith, W.P. | *Synthesis* | **1994** | | 639 |
| 6 Ley, S.V. | *J. Chem. Soc. Perkin I* | **1997** | | 3291; 1907 |
| 7 Ley, S.V. | *Synthesis* | **1998** | | 977 |
| 8 Ley, S.V. | *J. Chem. Soc. Perkin I* | **1998** | | 3907 |

**Oxirane aldehyde 2.**[1] Alcohol **1** (TBDPS = tert-butyldiphenylsilyl) (192 mg; 0.5 mmol) in $CH_2Cl_2$ (5 mL) containing molecular sieves (4Å) and NMO (0.1 g; 0.75 mmol) was stirred for 10 min. $Pr_4N^+RuO_4^-$ (TPAP) (8.3 mg; 0.025 mmol) was added and the reaction was followed by TLC until complete. Usual work up afforded 134.4 mg of **2** (70%).

**3-Dimethylaminopropanal 4.**[7] To a solution of 3-dimethylaminopropanol **3** (20.6 mg; 0.2 mmol) in PhMe (2 mL), PSP (200 mg; 0.02 mmol) was added and the mixture was stirred at 85°C under $O_2$ ($O_2$ balloon) for 8 h. The mixture was filtered through cotton wool and the residue washed with PhMe. Evaporation of the solvent in vacuum afforded 195.7 mg of **4** (95%).

## L I E B E N  Hypohalide Oxidation

Oxidation of methyl ketones with hypochloride (or hypobromide) to carboxylic acids and chloroform; with NaOH and iodine, iodoform is formed (see 1st edition).

| 1 | Lieben, A. | *Liebigs Ann. Suppl.* | **1870** | *7* | *218* |
|---|---|---|---|---|---|
| 2 | Fieser, L.F. | *J. Am. Chem. Soc.* | **1936** | *58* | *1055* |
| 3 | Farrart, M.V. | *J. Am. Chem. Soc.* | **1949** | *71* | *1946* |
| 4 | Sasson, Y. | *Tetrahedron* | **1996** | *37* | *2063* |
| 5 | Fuson, R.C. | *Chem. Rev.* | **1934** | *15* | *275* |

## L I E B I G  Benzylic Acid Rearrangement

Rearrangement of diketones (also α-ketols) to benzylic acid[5] or in general to α-hydroxyacids (see 1st edition).

| 1 | Liebig, v.J. | *Liebigs Ann.* | **1858** | *25* | *27* |
|---|---|---|---|---|---|
| 2 | Warren, K.S. | *J. Org. Chem.* | **1963** | *28* | *2152* |
| 3 | Houber, G. | *Angew. Chem.* | **1951** | *63* | *501* |
| 4 | Eastham, J.F. | *Quart. Rev. Chem. Soc.* | **1960** | *14* | *221* |
| 5 | Guthor, S | *Org. Synth. Coll. Vol.* | | *I* | *89* |

## L I E P A  Phenanthrene Synthesis

Conversion of stilbene derivatives to phenanthrenes with $VOF_3$ in trifluoroacetic acid (TFA).

| 1 | Liepa, A.J. | *J.Am.Chem.Soc.* | **1973** | 95 | 6861 |
| 2 | Liepa, A.J. | *J.Chem.Soc.Chem.Commun.* | **1977** | | 826 |
| 3 | Ciufolini, M.A. | *J.Am.Chem.Soc.* | **1996** | *118* | 12082 |

**(+) Tylophorine (2).**[3] A cold ($0^{\circ}C$) solution of (+)-depticine **1** (54mg, 0.14 mmol) and $VOF_3$ (84 mg, 0.6 mmol) in $CH_2Cl_2$ (3 mL) was stirred for 15 min, and TFA (136 μL) was slowly added and stirring was continued at $0^{\circ}C$ for an additional 15 min. The mixture was poured into 10% NaOH and extracted with $CH_2Cl_2$. Evaporation of the solvent and recrystallization (MeCN) afforded 40 mg of **2** (74%), mp $272\text{-}274^{\circ}C$.

## L O S S E N  Rearrangement

Rearrangement of O-acyl hydroxamic acid derivatives with base or heat to amines or urea derivatives (via isocyanates), or rearrangement of carboxylic acids via their hydroxamic acids to amines (see 1st edition).

$$\underset{1}{\text{Ph-}\underset{O}{\overset{O}{\underset{\|}{C}}}\text{-}\underset{K}{N}\text{-O-}\underset{\|}{\overset{O}{C}}\text{-Ph}} \xrightarrow[\text{DMSO, 30°}]{\overset{\overset{\text{H}_2\text{N-C}_4\text{H}_9}{2}}{}} \text{Ph-N=C=O} \xrightarrow{2} \underset{3\,(83\%)}{\text{Ph-NH-}\underset{O}{\overset{O}{\underset{\|}{C}}}\text{-NH-C}_4\text{H}_9}$$

$$\text{Naphth-CO}_2\text{H} + \text{H}_2\text{N-OH} \xrightarrow{\text{PPA}} \underset{(82\%)}{\text{Naphth-NH}_2}$$

$$\underset{4}{(\text{iPrO})_2\overset{O}{\underset{\|}{P}}\text{-}\overset{O}{\underset{\|}{C}}\text{-SEt}} + \text{NH}_2\text{OH·HCl} \xrightarrow[\text{72 h}]{\text{Py; 20°}} \underset{5}{(\text{iPrO})_2\overset{O}{\underset{\|}{P}}\text{-NH}_2}$$

| 1 | Lossen, W. | *Liebigs Ann. Chem.* | **1869** | *150* | 314 |
|---|---|---|---|---|---|
| 2 | Brend, D.C. | *J. Org. Chem.* | **1966** | *31* | 976 |
| 3 | Popp, F.V. | *Chem. Rev.* | **1958** | *58* | 374 |
| 4 | Cohen, L.A. | *Angew. Chem.* | **1961** | *73* | 260 |
| 5 | Snyder, H.R. | *J. Am. Chem. Soc.* | **1953** | *75* | 2014 |
| 6 | Ulrich, H. | *J. Org. Chem.* | **1978** | *43* | 1544 |
| 7 | Brener, E. | *J. Org. Chem.* | **1997** | *62* | 3858 |

## L U C H E  Ce Reducing Agent

Selective 1,2-reduction of conjugated ketones with $NaBH_4$-$CeCl_3$, usually in MeOH (in the absence of $CeCl_3$ double bond reduction often occurs). Also ketone reduction in the presence of an aldehyde.

|  |  |  |  | | |
|---|--------|---------|------|------|------|
| 1 | Luche, J.L. | *J. Am. Chem. Soc.* | **1978** | *100* | 2226 |
| 2 | Luche, J.L. | *J. Am. Chem. Soc.* | **1979** | *101* | 5848 |
| 3 | Luche, J.L. | *J. Am. Chem. Soc.* | **1981** | *103* | 5454 |
| 4 | Krieff, A. | *Synlett* | **1991** | | 273 |
| 5 | Toda, F. | *J. Org. Chem.* | **1991** | *56* | 4334 |

**Cyclopentenol 2.**[1] To a solution of cyclopentenone **1** (82 mg; 1 mmol) and $CeCl_3 \cdot 7H_2O$ (372 mg; 1 mmol) in MeOH (2.5 mL) was added in one portion $NaBH_4$ (38 mg; 1 mmol). After gas evolution ceased, stirring was continued for another few minutes and the pH was adjusted to neutral with dil. HCl. Extraction with $Et_2O$, evaporation of the solvent and chromatography afforded practically pure cyclopentenol **2** in 96% yield.

*cis*-**Pulegol 4.**[3] In the same manner as above, pulegone **3** (150 mg; 0.98 mmol) afforded 150 mg of alcohol **4** (100%) as an oil, which crystallized on standing. The product, washed with pentane, showed mp 29-30°C and $[\alpha]_D = -104°$ (EtOH:$H_2O$ 95:5).

## L U C H E   Zn Allylation

Addition of allylic halides to ketones or aldehydes in the presence of Zn in aqueous media, analogous to Barbier reaction or in the absence of solvents (see Toda).

$$C_6H_{13}\text{-CHO} + \quad \text{\raisebox{0pt}{Br}} \quad \xrightarrow[\text{18 h; }20°]{\text{Zn/C}_{18}} \quad C_6H_{13} \overset{\text{OH}}{\diagdown}$$

| | | |
|---|---|---|
| **1** | **2** | **3** (98%) |

$$\text{cyclohexanone} =O \; + \; 2 \quad \xrightarrow[20°]{\text{Zn; NH}_4\text{Cl(s)}} \quad \text{cyclohexanol with allyl}$$

**4** (90%)

| 1 | Luche, J.L. | *J. Org. Chem.* | **1985** | *50* | 91 |
|---|---|---|---|---|---|
| 2 | Luche, J.L. | *Tetrahedron Lett.* | **1985** | *26* | 1449 |
| 3 | Luche, J.L. | *J. Organomet. Chem.* | **1987** | *322* | 177 |
| 4 | Wilson, S.R. | *J. Org. Chem.* | **1989** | *54* | 3087 |
| 5 | Toda, F. | *J. Org. Chem.* | **1991** | *56* | 4333 |

**1-Decen-4-ol (3).**[4] A mixture of heptaldehyde **1** (119.5 mg; 1.05 mmol), saturated aqueous NH$_4$Cl (1 mL), reverse phase resin (C$_{18}$) (200 mg), allyl bromide (0.1 mL) and zinc dust (78 mg; 1.2 mmol) was stirred overnight at 20°C open to air. Filtration, washing with Et$_2$O and the solvent evaporation afforded 160 mg of **3** (98%).
**1-Allylcyclohexanol (5).**[5] **4** (500 mg; 5.1 mmol), **2** (3.09 g; 25.5 mmol), Zn powder (5 g) and NH$_4$Cl (2g) was grounded in an agate mortar and pestle and mixture was kept for 2 h at 20°C. Work up and evaporation gave 642 mg of **5** (90%).

## M A N N   Ether Dealkylation

Dealkylation of alkyl aryl ethers and sulphides by diaryl-posphide or arsenide ions.

$$\text{Ph-O-Me} + \text{Ph}_2\text{PH-BuLi} \xrightarrow[\text{refl. 4 h}]{\text{THF}} \text{Ph-O} \overset{\text{Me}}{\underset{\text{Li}}{\diagup}} \text{PPh}_2 \longrightarrow \text{PhOH} + \text{Ph}_2\text{PMe}$$

| | | | |
|---|---|---|---|
| **1** | **2** | **3** (83%) | **4** (87%) |

$$\text{Ph-S-CH}_2\text{-Ph} \xrightarrow{\text{Ph}_2\text{PLi}} \text{PhSH} + \text{Ph}_2\text{P-CH}_2\text{-Ph} \qquad \textbf{1} \xrightarrow{\text{Ph}_2\text{As}^-} \textbf{3} + \text{Ph}_2\text{As-Me}$$

$$(79\%) \qquad\qquad\qquad\qquad (85\%)$$

| 1 | Mann, F.G. | *J. Chem. Soc.* | **1963** | | 1155 |
|---|---|---|---|---|---|
| 2 | Mann, F.G. | *Chem. and Ind.* | **1963** | | 1558 |
| 3 | Mann, F.G. | *Chem. and Ind.* | **1964** | | 1386 |
| 4 | Mann, F.G. | *J. Chem. Soc.* | **1965** | | 4120 |
| 5 | Veriott, G. | *Across Organics Acta* | **1995** | *1* | 40 |

**Phenol (3).**[4] An ice-cooled **2** (9.1 g; 50 mmol) in THF (110 mL) was treated with n-BuLi (1.24M; 45 mL). **1** (5.8 g; 53 mmol) was added and the red solution was refluxed for 4 h. Evaporation, addition of Et$_2$O and water and distillation afforded 8.5 g of **4** (87%), bp 87-90°C/0.2 mm from Et$_2$O and 3.81 g of **3** (83%) from the aqueous layer.

## **M A C D O N A L D**  Porphyrine Synthesis

Porphyrine synthesis from dipyrrolemethanes (see 1st edition).

| 1 | MacDonald, S.P. | *J.Am.Chem.Soc.* | **1960** | *82* | 4384 |
| 2 | Clesy, P.S. | *Austr.J.Chem.* | **1965** | *18* | 1835 |
| 3 | Chang, C.K. | *J.Org.Chem.* | **1981** | *46* | 4610 |

## **M A D E L U N G**  Indole Synthesis

Indole synthesis by cyclization of N-acyl-o-toluidines, (see 1st edition).

| 1 | Madelung, W. | *Chem.Ber.* | **1912** | *45* | 1128 |
| 2 | Pichat, L. | *Bull.Soc.Chim.Fr.* | **1954** | | 85 |
| 3 | Hertz, W. | *J.Org.Chem.* | **1960** | *25* | 2242 |
| 4 | Houlihan, W.J. | *J.Org.Chem.* | **1981** | *46* | 4511 |

## MAKOSZA  Vicarious Nucleophilic Substitution

Introduction of functionalized alkyls, OH or $NH_2$ groups into electrophilic aromatic rings (e.g. nitrobenzenes), via replacement of hydrogen (see also 1st edition).

| 1 | Makosza, M. | *J.Org.Chem.* | **1983** | *48* | 3860 |
|---|---|---|---|---|---|
| 2 | Makosza, M. | *J.Org.Chem.* | **1989** | *54* | 5094 |
| 3 | Makosza, M. | *Russian Chem.Rev.* | **1989** | *58* | 747 |
| 4 | Makosza, M. | *Acc.Chem.Rev.* | **1987** | *20* | 282 |
| 5 | Makosza, M. | *Synthesis* | **1991** | | 103 |
| 6 | Makosza, M. | *Liebigs Ann.* | **1997** | | 1805 |
| 7 | Nilsson, M. | *Synthesis* | **1994** | | 242 |
| 8 | McCluskey, J.G. | *J.Org.Chem.* | **1998** | *63* | 4199 |

**3-Dichloromethyl-5-nitropyridine (3).**[2]  A solution of 3-nitropyridine 1 (372 mg, 3 mmol) and $CHCl_3$ (395 mg, 3.3 mmol) in DMF (2 ml) was added dropwise to a vigorously stirred mixture of NaOMe (650 mg, 12 mmol) in liq. $NH_3$ (10 mL) at -70°C. After 1 min stirring $NH_4Cl$ (1.5 g) was added, ammonia was evaporated, water (50 mL) was added to the residue and usual work up afforded 447 mg of 3 (72%).

**Diethyl 2,6-dinitrophenyl malonate (6).**[7]  To t-BuOK (393 mg, 3.5 mmol) in DME (15 mL) was added CuCl (248 mg, 2.5 mmol) at 0°C and all was stirred for 30 min. Pyridine (1 mL) and 1,3-dinitrobenzene (168 mg, 1 mmol) was added and after cooling at -20°C diethyl bromomalonate 5 (211 g, 1 mmol) in DME (5 mL) was added. After 2h stirring at -20°C and 30 min at 0°C, quenching and usual work up afforded (chromatography 20% EtOAc in hexane) 250mg of 6 (83%), mp 159-160°C (MeOH).

# MALAPRADE-LEMIEUX-JOHNSON

## Olefin (diol) cleavage

Oxidative cleavage of 1,2-glycols to two carbonyls (Malaprade) or direct oxidation of olefins by $IO_4^-$ and $OsO_4$ catalyst (Lemieux-Johnson) (see 1st edition).

| | | | | |
|---|---|---|---|---|
| 1 | Malaprade, L. | *Bull.Soc.Chim.Fr.* | **1828** | 43 | 683 |
| 2 | Baddiley, J. | *J.Chem.Soc.* | **1954** | | 3826 |
| 3 | Fatiatide, A.J. | *Synthesis* | **1974** | | 229,255 |
| 4 | Jackson, E.I. | *Org.React.* | **1944** | 2 | 341 |
| 5 | Lemieux, R.U. | *Anal Chem.* | **1954** | 26 | 920 |
| 6 | Lemieux, R.U.; Johnson, W.S. | *J.Org.Chem.* | **1956** | 21 | 478 |
| 7 | Rapoport, H. | *J.Am.Chem.Soc.* | **1958** | 80 | 5767 |
| 8 | Djerassi, C. | *J.Am.Chem.Soc.* | **1962** | 84 | 2990 |
| 9 | Henbest, N.R. | *Chem.Commun.* | **1968** | | 1036 |

**Dialdehyde (2).**[2] Glucoside **1** (2.8 g, 10 mmol) in water (750 mL) was treated with NalO$_4$ (2.14 g, 10 mmol) and kept for a week at 20$^0$C. Filtration gave 2.9 g of **2** (100%), mp 142$^0$C. **Undecanal (4).** Water (5 mL), dioxane (15 mL), dodecene-1 **3** (0.71 g, 4.2 mmol) and OsO$_4$ (11.3 mg, 0.044 mmol) were stirred for 5 min. Powdered NalO$_4$ (2.06 g, 9.6 mmol) was added over 30 min and the slurry stirred for 90 min. The mixture was extracted with Et$_2$O and **4** was isolated as the 2,4-DNPH, 0.96 g, mp 102-106$^0$C, second crop 0.14 g, total yield 68%.

## M A N D E R Methoxycarbonylation Reagent

Methyl cyanoformate, agent for regioselective methoxycarbonylation of carbanions, can function as dienophile, dipolarophile or radical cyanating agent.

| 1 | Mander, L.N. | *Tetrahedron Lett.* | **1983** | 24 | 5425 |
| 2 | Padwa, A. | *J.Am.Chem.Soc.* | **1982** | 104 | 286 |
| 3 | Akiyama,Y | *Chem.Lett.* | **1983** | | 1231 |
| 4 | Mander, L.N. | *Synlett* | **1990** | | 169 |
| 5 | Padwa, A. | *J.Org.Chem.* | **1991** | 56 | 3271 |
| 6 | Potthoff, B. | *Synthesis* | **1986** | | 584 |
| 7 | Krebs, A. | *Tetrahedron Lett.* | **1981** | 22 | 1675 |

**β-Ketoester 2.[4]** To a solution of 8-methoxy-4α-methyl-4,4α,9,10-tetrahydro phenanthren-2(3H)-one (512 mg, 5 mmol) in $NH_3 \cdot Et_2O$ and t-BuOH (!40 mg, 1.9 mmol) under $N_2$, was added Li (35 mg, 5 mmol) under stirring at –33°C. After 45 min isoprene was added, then $NH_3$ was evaporated under a stream of $N_2$. The residue was dried under high vacuum for 5 min then $Et_2O$ (20 mL) was added, the mixture was cooled to –78°C and methyl cyanoformate (187 mg, 2.2 mmol) was added dropwise. After 20 min at –78°C the mixture was allowed to warm to 0°C, EtOAc was added, followed by water. Usual work up and chromatography (silica gel) afforded 449 mg of **2** (71%), mp 143-145°C.

## M A N N I C H  Aminomethylation

Aminomethylation of activated methyl or methylene groups by in situ formed imminium species $Me_2N^+=CH-R$ (see also 1st edition).

| 1 | Mannich, C. | Arch.Pharm. | **1912** | *250* | 647 |
|---|---|---|---|---|---|
| 2 | House, H.O. | J.Org.Chem. | **1964** | *29* | 1339 |
| 3 | Tramontini, M. | Synthesis | **1973** | | 703 |
| 4 | Dimnoch, D. | Die Pharmazie | **1986** | *91* | 284 |
| 5 | Krawczyk, H. | Syn.Commun. | **1995** | *25* | 3357 |
| 6 | Jnhua, Zon | Org.Prep.Proced.Int. | **1996** | *28* | 618 |
| 7 | Kobayashi, S. | J.Am.Chem.Soc. | **1997** | *119* | 7153 |
| 8 | Blicke, F.F. | Org.React. | **1942** | *1* | 303 |

**1-Phenyl-1-(p-chloroanilino)-3-hexanone (4).**[6]   To a mixture of p-chloro-aniline **1** (637 mg, 5 mmol), 2-pentanone **2** (450 mg, 5 mmol) and PhCHO **3** (450 mg, 5 mmol) in EtOH (5 mL) under cooling (ice bath), 35% HCl (0.2 mL) was added. After 12 h stirring at 14°C and 10 h at 0°C, the mixture was neutralized with 10% NaHCO₃ (pH=7) and the product filtered. Recrystallization from EtOH gave 1.197 g of **4** (90%), mp 84-86°C.

**2-(Morpholinomethyl)acrylonitrile (8).**[5]   To cyanoacetic acid **5** (25.5 g, 0.3 mmol), paraformaldehyde **6** (21.6 g, 0.72 mmol) in PhH (150 mL) was added morpholine **7** (26.1 g, 0.3 mmol). After 6 h reflux with a Dean-Stark water separator, the solvent was evaporated, the residue was dissolved in CHCl₃ and the organic phase was washed with water. Evaporation of the solvent and distillation afforded 36.5 g of **8** (80%), bp 142°C/25 Torr.

## M A R K O V N I K O V　Regioselectivity

Description of selectivity during addition of unsymmetrical reagents to unsymmetrical olefins. H-X adds selectively with H forming a bond to the less substituted olefin carbon (Markovnikov). Now supplanted by the general term **regioselectivity** introduced by A. Hassner, denoting selectivity in bond making between an unsymmetrical reagent X-Y and an unsymmetrical substrate, now includes regioselective (o,m,p)-substitution and also applied to bond breaking reactions (regioselective elimination) (see 1st edition).

| 1 | Markovnikov, W. | *Liebigs.Ann.* | **1870** | *153* | 256 |
| 2 | Hassner, A. | *J.Org.Chem.* | **1969** | *34* | 2628 |
| 3 | Stasey, F.M. | *Org.React* | **1963** | *13* | 155 |
| 4 | Hassner, A. | *Acc.Chem.Res.* | **1971** | *4* | 9 |

## M A R S C H A L C K　Aromatic alkylation

Alkylation of quinones or aminoquinones with aldehydes (see 1st edition).

| 1 | Marschalck, 0. | *Bull.Soc.Chim.Fr.* | **1936** | *3* | 1545 |
| 2 | Marschalck, 0. | *Bull.Soc.Chim.Fr.* | **1939** | *6* | 655 |
| 3 | Brockmann, H. | *Chem.Ber.* | **1958** | *91* | 1920 |
| 4 | Havlincova, L. | *J.Chem.Soc.* | **1970** | | 657 |
| 5 | Krohn, E. | *Angew.Chem.Int.Ed.* | **1979** | *18* | 621 |

# M A R T I N  Dehydrating Reagent

Sulfurane reagent for conversion of trans diols to epoxides, generally for dehydration of diols to olefins or cyclic ethers, or as an oxidizing agent (see 1st edition).

| 1 | Martin, J.C. | J.Am.Chem,Soc. | **1971** | 93 | 4327 |
|---|---|---|---|---|---|
| 2 | Martin, J.C. | J.Am.Chem,Soc. | **1974** | 96 | 4604 |
| 3 | Martin, J.C. | J.Am.Chem,Soc. | **1977** | 99 | 3511 |
| 4 | Bartlett, P.D. | J.Am.Chem,Soc. | **1980** | 102 | 3515 |
| 5 | Eschenmoser, W | Helv. Chim. Acta. | **1982** | 65 | 353 |
| 6 | Burnett, D.A. | J.Am.Chem,Soc. | **1984** | 106 | 8201 |
| 7 | Martin, J.C. | Organic Synthesis | **1977** | 57 | 22 |

# M A S C A R E L L I  Fluorene Synthesis

Synthesis of fluorenes from 2-amino-2'-alkylbiphenyls via diazonium ions (see 1st edition).

| 1 | Mascarelli, L. | Gazz.Chim.Ital. | **1936** | 66 | 843 |
|---|---|---|---|---|---|
| 2 | Mascarelli, L. | Gazz.Chim.Ital. | **1937** | 67 | 812 |
| 3 | Mascarelli, L. | Gazz.Chim.Ital. | **1938** | 68 | 4565 |
| 4 | Cohen, T. | J.Am.Chem.Soc. | **1964** | 86 | 2514 |
| 5 | Puskas, I. | J.Org.Chem. | **1968** | 3 | 4237 |

## M A T T E S O N  Boronic Esters

Asymmetric synthesis by means of α–halo boronic esters intermediates leading to drial aldehydes.

2
(Cy = cyclohexyl)

3

4

5

6 (55%)

| 1 | Matteson, D.S. | *J.Am.Chem.Soc.* | **1963** | *85* | 2599 |
|---|---|---|---|---|---|
| 2 | Matteson, D.S. | *J.Am.Chem.Soc.* | **1980** | *102* | 7590; 7588 |
| 3 | Rathke, M.W. | *J.Organomet.Chem.* | **1976** | *122* | 145 |
| 4 | Matteson, D.S. | *J.Am.Chem.Soc.* | **1996** | *118* | 4560 |
| 5 | Matteson, D.S. | *Tetrahedron* | **1998** | *54* | 10555 |

**[4R-[2(R*), 4α, 5β]]-4,5-Dicyclohexyl-2-[1-(phenylmethoxy)propyl]-1,3,2-dioxaboro lane 3.[4]** To a solution of **1** (54 g, 204 mmol) and CH₂Cl₂ (52 g, 610 mmol) in THF (300 mL) was added LDA (120 mL, 2 M, 240 mmol) at –40°C. After 10 min , ZnCl₂ (55.5 g, 408 mmol) was added to the solution. After 30 min the mixture was allowed to warm to 20°C and was kept for 2 h to give **2** (NMR analysis). The solution was evaporated in vacuum to remove CH₂Cl₂, THF (300 mL) was added and this solution was added dropwise to PhCH₂OMe (from PhCH₂OH 26 g and NaH 9 g in THF/DMSO). After 48 h stirring at 20°C, hexane (1000 mL) and aqueous NH₄Cl (500 mL) was added followed by HCl (to acid). Usual work up and evaporation of the solvent afforded 75 g of crude **3** which was used in the next step without further purification.

## M A T T O X - K E N D A L L   Dehydrohalogenation

Dehydrohalogenation of α-haloketones with 2,4-dinitrophenylhydrazine or LiCl-DMF (see 1st edition).

| 1 | Mattox, V.R.; Kendall, E.C. | *J.Am.Chem.Soc.* | **1948** | *70* | 882 |
|---|---|---|---|---|---|
| 2 | Djerassi, C. | *J.Am.Chem.Soc.* | **1953** | *75* | 3500 |
| 3 | Warnhof, E.W. | *J.Org.Chem.* | **1963** | *28* | 887 |

## M c C O R M A C K – K U C H T I N – R A M I R E Z   Phosphole Synthesis

Formation of phospholes from butadienes (McCormack) or of dioxaphospholes from 1,2-diketones (Kuchtin-Ramirez), (see 1st edition).

| 1 | McCormack, W.B. | *U.S. Pat.* | 2.663.736, | | 2.663.737 |
|---|---|---|---|---|---|
| 2 | Hajos, A.G. | *J.Org.Chem.* | **1956** | *30* | 1213 |
| 3 | Quin, L.D. | *J.Org.Chem.* | **1981** | *46* | 461 |
| 4 | Kuchtin, V.A. | *Doklad.Akad.Nauk.USSR* | **1958** | *121* | 466 |
| 5 | Ramirez, F. | *J.Am.Chem.Soc.* | **1960** | *82* | 2651 |
| 6 | Mitsuo, S. | *J.Org.Chem.* | **1981** | *46* | 4030 |

## M c F A D Y E N - S T E V E N S   Ester Reduction

Reduction of esters to aldehyde via hydrazides (see 1st edition).

| 1 | McFadyen, J.S.; Stevens, T.S. | *J.Chem.Soc.* | **1936** | | 584 |
|---|---|---|---|---|---|
| 2 | Nieman, C. | *J.Am.Chem.Soc.* | **1942** | 62 | 1681 |
| 3 | Martin, C.B. | *J.Org.Chem.* | **1974** | 39 | 2285 |
| 4 | Ferguson, L.H. | *Chem.Rev.* | **1946** | 38 | 244 |
| 5 | Mosettig, E. | *Org.React.* | **1954** | 8 | 232 |

## M E E R W E I N   Alkylating Reagent

$R_3O^+BF_4^-$ reagent for O-alkylation of amides (see 1st edition).

| 1 | Meerwein, H. | *J.Prakt.Chem.* | **1937** | 147 | 17 |
|---|---|---|---|---|---|
| 2 | Eschenmoser, A. | *Pure Appl.Chem.* | **1963** | 7 | 297 |
| 3 | Fujita, A. | *Chem.Pharm.Bull.* | **1965** | 13 | 1183 |
| 4 | Curphey, T.J. | *J.Org.Chem.* | **1966** | 31 | 1199 |
| 5 | Ayers, W.A. | *Can.J.Chem.* | **1967** | 45 | 451 |
| 6 | Potts, K.T. | *J.Chem.Soc.Chem.Commun.* | **1970** | | 1025 |
| 7 | McMinn, D.G. | *Synthesis* | **1976** | | 824 |

## M C M U R R Y Coupling

Formation of olefins by coupling or cross coupling of ketones, mediated by low valent titanium. Also coupling enol ethers of 3-dicarbonyl compounds or of aldehydes (see 1st edition)

(94%; mainly *E*)

**1**          **2 (80%)**

| | | | | | |
|---|---|---|---|---|---|
| 1 | McMurry, J.E. | *J. Am. Chem. Soc.* | **1974** | *96* | 1708 |
| 2 | Finocchiaro, P. | *La Chimia e L'industria* | **1982** | *64* | 644 |
| 3 | McMurry, J.E. | *Acc. Chem. Res.* | **1983** | *16* | 405 |
| 4 | Coe, P.L. | *J. Chem. Soc. Perkin I* | **1986** | | 475 |
| 5 | Breitmaier, E. | *Chem. Ber.* | **1986** | *119* | 1734 |
| 6 | Breitmaier, E. | *Synthesis* | **1987** | | 96 |
| 7 | McMurry, J.E. | *J. Org. Chem.* | **1989** | *54* | 3748 |
| 8 | Ephritikhine, M. | *J. Chem. Soc. Chem. Commun.* | **1998** | | 2549 |
| 9 | Hong, B.C. | *Synth. Commun.* | **1999** | *29* | 3097 |
| 10 | Gautier, S. | *Tetrahedron* | **2000** | *56* | 703 |
| 11 | McMurry, J.E. | *Chem. Rev.* | **1989** | *89* | 1513 |

**Cyclotetradecene 2.**[7] $TiCl_3(DME)_{1.5}$ (5.2 g; 17.8 mmol) and zinc-copper couple (3.8 g; 58.1 mmol) were added to a flask under Ar and were stirred while DME (150 mL) was added by syringe. After the mixture was heated at 80°C for 4 h to form the active titanium coupling reagent, tetradecanedial **1** (500 mg; 2.2 mmol) in DME (50 mL) was added via syringe pump over a period of 35 h. The reaction was heated an additional 6 h after addition was complete and then cooled at 20°C. The reaction mixture was diluted with pentane (100 mL) and the slurry was filtered through Florisil. After washing the filter with pentane, the filtrate was concentrated under vacuum (0°C) to give 340 mg of pure **2** (80%), as a colorless oil. The ratio *E*:*Z* = 9:1.

# M E E R W E I N - P O N N D O R F - V E R L E Y  Reduction

Reduction of carbonyl groups to alcohols by means of Al(iPrO)$_3$ and iPrOH or with lantanide alkoxides (see 1st edition).

| 1 | Meerwein, H. | *Liebigs Ann.* | **1925** | *444* | 221 |
|---|---|---|---|---|---|
| 2 | Verley, A. | *Bull.Soc.Chim.Fr.* | **1925** | *37* | 537 |
| 3 | Ponndorf, W. | *Angew.Chem.* | **1926** | *39* | 138 |
| 4 | Lund, H. | *Chem.Ber.* | **1937** | *70* | 1520 |
| 5 | Snyder, C.H. | *J.Org.Chem.* | **1970** | *35* | 264 |
| 6 | Merbach, A. | *Helv.Chim.Acta.* | **1972** | *55* | 44 |
| 7 | Kagan, H.B. | *Tetrahedron Lett.* | **1991** | *32* | 2355 |
| 8 | Huskens, J. | *Synthesis* | **1994** | | 1007 |
| 9 | Denno, N.C. | *Chem.Rev.* | **1960** | *60* | 7 |
| 10 | Wilds, A.L. | *Org.React.* | **1944** | *2* | 178 |

# M E I N W A L D  Rearrangement

Unusual course of the peracid oxidation of bicyclic olefins leading to a carboxyaldehyde rather than an epoxide.

| 1 | Meinwald, J. | *J.Am.Chem.Soc.* | **1958** | *80* | 6303 |
|---|---|---|---|---|---|
| 2 | Meinwald, J. | *J.Am.Chem.Soc.* | **1960** | *82* | 5235 |
| 3 | Meinwald, J. | *J.Am.Chem.Soc.* | **1963** | *85* | 582 |
| 4 | Meinwald, J. | *Tetrahedron Lett.* | **1965** | | 1789 |
| 5 | Kobayashi, S. | *Tetrahedron Lett.* | **1993** | *34* | 665 |

**M E I S E N H E I M E R**  N-Oxide Rearrangement

Rearrangement of tertiary amine oxides to trisubstituted hydroxylamines via a [2,3] sigmatropic shift. Also chlorination of pyridines via N-oxides (see 1st edition).

| 1 | Meisenheimer, J. | *Chem.Ber.* | **1926** | 59 | 1848 |
|---|---|---|---|---|---|
| 2 | Albert, A. | *J.Chem.Soc.* | **1960** | | 1790 |
| 3 | Brown, E.V. | *J.Org.Chem.* | **1967** | 32 | 241 |
| 4 | Pandler, W.L. | *J.Org.Chem.* | **1971** | 36 | 1720 |
| 5 | Thyagarajan, B.S. | *Tetrahedron Lett.* | **1974** | | 1999 |
| 6 | Majumdar, K.C. | *J.Chem.Soc.Perkin 1* | **1993** | | 715 |
| 7 | Majumdar, K.C. | *J.Org.Chem.* | **1997** | 62 | 1506 |

**2-(and-4-) Chloro-1,5-naphthyridine (3) and (4).[3]**   1,5-Naphthyridine **1** (4.5 g, 34 mmol) was treated with a mixture of AcOH (10 mL) and 40% peracetic acid (5 mL) for 3 h at 70°C. From the mixture of mono and di-N-oxides, the mono N-oxide **2** was obtained by recrystallization from methylcyclohexane. **2** (770 mg, 5 mmol) was heated in $POCl_3$ (30 mL) and $P_2O_5$ for 30 min. The product was collected and analyzed by GC (15% SE-30 on Chromosorb W 240°C He, 40 psi) to be a mixture of **3** (56.8%) and **4** (43.2%).

**O-{2-(2-Methylbut-3-enyl)}-N-methyl-N-phenylhydroxylamine (7).[7]**  mCPBA (3.44 g 50%, 10 mmol) in $CHCl_3$ (50 mL) was added to a solution of amine **5** (1.75 g, 10 mmol) in $CHCl_3$ (50 mL) at 0-5°C over a period of 20 min. After 10 h stirring the reaction mixture was washed with an aq soln of $K_2CO_3$, dried ($Na_2SO_4$), the solvent evaporated and the residue chromatographed (silica gel, petroleum ether) to afford 3.06 g of **7** (89%).

## M E I S E N H E I M E R - J A N O V S K Y  Complex

The adduct formed from a polynitroaromatic compound in alkaline solution with $RO^-$, $HO^-$ (Meisenheimer) or with acetone (Janovsky) (see 1st edition).

| 1 | Janovsky, I.V. | *Chem.Ber.* | **1886** | *19* | 2155 |
|---|---|---|---|---|---|
| 2 | Meisenheimer, J. | *Liebigs.Ann.* | **1902** | *323* | 205 |
| 3 | Fendler, J.H. | *J.Org.Chem.* | **1967** | *82* | 2507 |
| 4 | Jones, P.R. | *J.Org.Chem.* | **1986** | *51* | 3016 |
| 5 | Niclas, H.J. | *Synth.Commun.* | **1989** | *19* | 2789 |
| 6 | Kind, N. | *Synth.Commun.* | **1993** | *23* | 1569 |
| 7 | Terrier, F. | *J.Chem.Soc.Chem.Commun.* | **1997** | | 789 |
| 8 | Straw, M.J. | *Chem.Rev.* | **1970** | *70* | 667 |

## M E L D R U M ' S  Acid

A cyclic malonate derivative **3** (acidic methylene) used in place of malonate in alkylations, acylations, or reaction with aldehydes (see 1st edition).

| 1 | Meldrum, A.N. | *J.Chem.Soc.* | **1908** | *93* | 598 |
|---|---|---|---|---|---|
| 2 | Davidson, D. | *J.Am.Chem.Soc.* | **1948** | *70* | 3426 |
| 3 | Chau, C.C. | *Synthesis* | **1984** | | 224 |
| 4 | Ping, L. | *Org.Prep.Proceed.Intn.* | **1992** | *24* | 185 |
| 5 | M'Zia Ebrahimi | *Synthesis* | **1996** | | 215 |
| 6 | Yamamoto, Y. | *J.Chem.Soc.Chem.Commun.* | **1997** | | 359 |

## **M E N C K E - L A S Z L O**  Nitration of Phenols

Ortho nitration of phenols and nitration of others aryls by metal nitrates or alkyl nitrates catalyzed by bentonite clay (see also 1st edition).

| 1 | Mencke, J.B. | *Rec. Trav. Chim. Pays Bas* | **1925** | *44* | 141 |
|---|---|---|---|---|---|
| 2 | Laszlo, P. | *Tetrahedron Lett.* | **1982** | *23* | 5035 |
| 3 | Laszlo, P. | *Tetrahedron Lett.* | **1983** | *24* | 3101 |
| 4 | Laszlo, P. | *J. Org. Chem.* | **1983** | *48* | 4771 |
| 5 | Laszlo, P. | *Pure Appl. Chem.* | **1990** | | 2027 |
| 6 | Braibante, M.E.F. | *J. Org. Chem.* | **1994** | | 898 |
| 7 | Kwork, T.J. | *J. Org. Chem.* | **1994** | 59 | 4942 |

Clayfen preparation from K-10-bentonite clay and Fe(NO₃)₃ in acetone, ref. 4

## **M E N Z E R**  Benzopyran Synthesis

Benzopyranone synthesis from phenols and β-ketoesters or unsaturated acids (see 1st edition).

| 1 | Menzer, Ch. | *C.R.* | **1952** | 232 | 1488 |
|---|---|---|---|---|---|
| 2 | Lacey, R.N. | *J. Chem. Soc.* | **1954** | | 859 |
| 3 | Mercier, Ch. | *C.R. Serie C.* | **1973** | 273 | 1053 |

## M E Y E R S Asymmetric Synthesis

Chiral oxazoles in asymmetric synthesis of carboxylic acids, chiral naphthalenes (see 1st edition).

| 1 | Meyers, A.I. | J.Am.Chem.Soc. | **1974** | *96* | 268 |
|----|--------------|-----------------|----------|------|-----------|
| 2 | Meyers, A.I. | Acc.Chem.Res. | **1978** | *11* | 375 |
| 3 | Meyers, A.I. | J.Am.Chem.Soc. | **1976** | *98* | 567 |
| 4 | Meyers, A.I. | J.Org.Chem. | **1980** | *45* | 2785 |
| 5 | Meyers, A.I. | J.Org.Chem. | **1987** | *52* | 4592 |
| 6 | Meyers, A.I. | J.Am.Chem.Soc. | **1988** | *110* | 4611, 7854 |
| 7 | Meyers, A.I. | Tetrahedron | **1989** | *45* | 6949 |
| 8 | Meyers, A.I. | Tetrahedron | **1991** | *47* | 9503 |
| 9 | Meyers, A.I. | Tetrahedron | **1992** | *48* | 2589 |
| 10 | Meyers, A.I. | Tetrahedron Lett. | **1998** | | 5301 |

**(S)-(+)-2-methylhexanoic acid 3.**[3] (4S, 5S-1 (15.4 g, 70 mmol) in THF (160 mL) under N$_2$ at −78°C, was treated with LDA (from 9.8 mL of iPr$_2$NH and 2.2 M n-BuLi (33 mL)) in THF (75 mL) over 20 min. After 20 min the mixture was cooled to −98°C and BuI (14.7 g, 80 mmol) in THF (20 mL) was added over 20 min. After 2 h the mixture was warmed to 20°C, poured into brine and extracted with Et$_2$O. Bulb to bulb distillation afforded pure 2 [α]$_{589}^{24}$ = -32.2°. The crude oxazoline 2 (17.2 g) was refluxed for h in 4N H$_2$SO$_4$. Extraction with Et$_2$O (3.75 mL), washing with 5% K$_2$CO$_3$ (3x100 mL), acidification (pH = 1) of the aqueous extract with 12 M HCl and extraction with Et$_2$O, gave on distillation 5.8 g of 3 (66%), α]$_{589}^{24}$ = +14.5°.

## M E Y E R - S C H U S T E R  Rearrangement

Acid catalyzed rearrangement of acetylenic alcohols into $\alpha,\beta$-unsaturated carbonyl compounds, (see 1st edition).

| # | | | | | |
|---|---|---|---|---|---|
| 1 | Meyer, K.H.; Schuster, K. | *Chem.Ber.* | **1922** | *55* | 819 |
| 2 | McGregor, W.S. | *J.Am.Chem.Soc.* | **1948** | *72* | 183 |
| 3 | Swaminathan, S. | Chem.Rev. | **1971** | *71* | 429 |
| 4 | Huggil, H.P.W. | *J.Chem.Soc.* | **1950** | | 335 |
| 5 | Yoshimatsu, M. | *J.Org.Chem.* | **1995** | *60* | 4798 |

## M I C H A E L I S - B E C K E R - N Y L E N  Phosphonylation

Nucleophilic attack of lithium dialkylphosphonates on pyridium salts to produce 2-pyridine phosphonates, (see 1st edition).

| # | | | | | |
|---|---|---|---|---|---|
| 1 | Michaelis, A.  Becker | *Chem.Ber.* | **1897** | *30* | 1003 |
| 2 | Michaelis, A. | *Chem.Ber.* | **1898** | *31* | 1048 |
| 3 | Nylen, T. | *Chem.Ber.* | **1924** | *57* | 1023 |
| 4 | Gordon, M. | *J.Org.Chem.* | **1966** | *31* | 333 |
| 5 | Redmore, D. | *J.Org.Chem.* | **1970** | *35* | 4114 |
| 6 | Kemm, K.M. | *J.Org.Chem.* | **1981** | *46* | 5188 |

**Diethyl pyridine-2-phosphonate (3).**[4]  BuLi (23% in hexane) (63 ml, 0.15 mol) was added dropwise to diethyl phosphonate (25.0 g, 0.18 mol) at −20 to −30°C over 2 h. To the resulting 2 was added 1 (from pyridine N-oxide 14.3 g, 0.15 mol and dimethyl sulfate 18.9 g, 0.15 mol) in diethyl phosphonate (40 ml) over 1 h at −15°C. The mixture was stirred at rt overnight and 100 ml water was added. After extraction with CHCl₃ (3x75 ml), the organic layer was extracted with 4N HCl, basified and reextracted to yield 22.9 g of 3 (67%), bp 105-112°C (0.08 mm).

## **M I C H A E L** Addition

Base promoted 1,4-additions of nucleophiles (usually C) to $\alpha,\beta$-unsaturated esters, ketones, nitriles, sulfones, nitro compounds; often stereoselective addition (see 1st edition).

| 1 | Komnenos, A. | *Liebigs Ann.* | **1883** | *218* | 145 |
|----|----------------|------------------------|----------|------|--------|
| 2 | Michael, A | *J.prakt.Chem.* | **1887** | *35* | 348(2) |
| 3 | Piers, E. | *Can.J.Chem.* | **1969** | *47* | 137 |
| 4 | Yamaguchi, M. | *Tetrahedron Lett.* | **1984** | *25* | 5661 |
| 5 | Seebach, D. | *Helv.Chim.Acta* | **1985** | *68* | 1592 |
| 6 | Heathcock, C.H. | *Tetrahedron Lett.* | **1986** | *27* | 6169 |
| 7 | Enders, D. | *Tetrahedron* | **1986** | *42* | 2235 |
| 8 | Bunce, R.A. | *Org.Prep.Proced.Int.* | **1987** | *19* | 471 |
| 9 | Pfau, M. | *Tetrahedron Asymm.* | **1997** | *8* | 1101 |
| 10 | Macquarrie, D.J. | *Tetrahedron Lett.* | **1998** | *39* | 4125 |
| 11 | Bergman, E.D. | *Org.React.* | **1959** | *10* | 179 |

# MICHAELIS - ARBUZOV Phosphonate Synthesis

Ni catalyzed phosphonate synthesis from phosphites and aryl halides. Reaction of alkyl halides with phosphites proceeds without nickel salts (see 1st edition).

$$NiCl_2 + (EtO)_3P \xrightarrow{150^0C} [(EtO)_3P]_4Ni \xrightarrow[160^0]{PhI\ 3} Ph-P(OEt)_2$$

$$1 \qquad\qquad 2 \qquad\qquad \underset{O}{\overset{\|}{}}$$

$$4\ (94\%)$$

5 → (MeO)$_3$P, 1, 50$^0$ → 6 (64%)

| 1 | Michaelis, A. | *Chem.Ber.* | **1898** | *31* | 1048 |
|---|---|---|---|---|---|
| 2 | Arbuzov, A. | *J.Russ.Phys.Chem.Soc.* | **1906** | *38* | 687 |
| 3 | Balthazar, T.M. | *J.Org.Chem* | **1980** | *45* | 5425 |
| 4 | Montero, J.L | *Tetrahedron Lett.* | **1987** | *28* | 1163 |
| 5 | Kemm, M.K. | *J.Org.Chem.* | **1970** | *36* | 5118 |
| 6 | Redmore, D. | *J.Org.Chem.* | **1981** | *46* | 4114 |
| 7 | Coward, J.K. | *J.Org.Chem.* | **1994** | *59* | 7625 |
| 8 | Brill, Th.B. | *Chem.Rev.* | **1984** | *84* | 577 |
| 9 | Kosolapov, G.M. | *Org.React.* | **1951** | *6* | 276 |

**Tetrakis(triethylphosphite)nickel(0) 2.**[3]
**Diethyl phenylphosphonate 4**. To **2** (20 mg) in PhI (10 g, 49 mmol) at 160$^0$C was added slowly **1** (9.37 g, 56.4 mmol). The solution (red upon each addition of **1**) faded to yellow and EtI was distilled. Vacuum distillation afforded 9.88 g of **4** (94%), bp 94-101$^0$C/0.1 mm.

**Dimethyl((S)-3-(N-Benzyloxycarbonylamino)-4-carbethoxypropyl)phosphonate 6.**[7]
A solution of **5** (188 mg, 0.546 mmol) in (MeO)$_3$P (5mL, 42mmol) was heated to reflux. The reflux condenser was flushed with water at 50$^0$C and an Ar stream was maintained to remove MeBr. Concentration in vacuum, distillation and flash chromatography (CHCl$_3$ EtOAc 1:1) afforded 130 mg of **6** (64%).

## M I D L A N D Asymmetric Reduction

Asymmetric reduction of propargyl ketones with (R) or (S) Alpine borane
(B-isopinocamphenyl-9-borabicyclo, [3,3,1] nonane(A))

**2 (86%, 94%ee)**

**100%ee**

| 1 | Midland, M.M. | *J.Am.Chem.Soc.* | **1980** | *102* | 867 |
|---|---------------|------------------|----------|-------|------|
| 2 | Midland, M.M. | *Tetrahedron* | **1984** | *40* | 1371 |
| 3 | Shigemasa, Y. | *J.Org.Chem.* | **1991** | *56* | 910 |
| 4 | Midland, M.M. | *Chem.Rev.* | **1989** | *89* | 1553 |

**R-(+)-1-Octyn-3-ol 2.**[2] To Alpine borane (prepared from 9-BBN (9-bora-bicyclo [3,3,1] nonane)), 800 mL of 0.5M THF solution (0.4 mol) and (+)-(α)-pinene (61.3 g, 0.45 mol) was added. After 4 h reflux, excess α-pinene and THF were removed in vacuum (0.05 mm, 40°C). To the thick oil of A 1-octyn-3-one **1** (35.3 g, 0.285 mol) was added under ice cooling and N$_2$. The ice cooling was removed and the reaction mixture was allowed to warm to 20-25°C. After 8 h (GC monitoring) the excess of Alpine-borane was destroyed by addition of propionaldehyde (0.3 mol) and stirring for 1 h. α-pinene was removed in vacuum, then THF (200 mL) was added followed by 3M NaOH (150 mL) and 30% H$_2$O$_2$ (150 mL). After 3 h stirring at 40°C, the reaction mixture was extracted with Et$_2$O (3x50 mL). The ether extract after drying (MgSO$_4$) was evaporated and the residue distilled to afford 31 g of **2** (86%), bp 60-65°C/3 mm Hg, [α]$_D^{25}$ =7.5°, 94%ee.

## M I E S C H E R   Degradation

Three carbon degradation of a carboxylic acid side chain (see Barbier-Wieland) (see 1st edition).

| 1 | Miescher, K. | *Helv.Chim.Acta.* | **1944** | *27* | 1815 |
|---|---|---|---|---|---|
| 2 | Spring, F.S. | *J.Chem.Soc.* | **1950** |  | 3355 |
| 3 | Wettstein, A. | *Experientia* | **1954** |  | 407 |

## M I G I T A - S A N O   Quinodimethane Synthesis

Quinodimethane synthesis by proton induced 1,4-elimination of stannanes.

| 1 | Kauffmann, T. | *Angew.Chem.Int.Ed.Engl.* | **1982** | *21* | 410 |
|---|---|---|---|---|---|
| 2 | Migita, T.; Sano, H. | *J.Am.Chem.Soc.* | **1988** | *110* | 2014 |

**Anhydride (3).**[2] To a solution of **1** (500 mg, 1.22 mmol) and **2** (358 mg, 3.65 mmol) in $CH_2Cl_2$ (1 mL) was added TFA (0.19 mL, 2.43 mmol) at 20°C and the mixture was stirred for 1 h. The $CH_2Cl_2$, TFA and unreacted **2** were removed in vacuo and the residue was treated with n-heptane (5 mL). The precipitate was filtered to give 235 mg of **3** (96%).

**M I L A S**   Olefin Hydroxylation

Hydroxylation of a double bond to a 1,2-diol with hydrogen peroxide and $OsO_4$ as catalyst (see 1st edition).

OsO4
30% H2O2

**2a** (23%)
**3a** (4β, 5β) (28%)

| 1 | Milas, W.A. | *J.Am.Chem.Soc.* | **1936** | *58* | 1302 |
|---|---|---|---|---|---|
| 2 | Milas, W.A. | *J.Am.Chem.Soc.* | **1959** | *81* | 3114 |

**M I L L E R - S N Y D E R**   Aryl Cyanide Synthesis

Synthesis of benzonitriles from aldehydes via oxime ethers in the presence of p-nitrobenzonitrile. Formation of p-cyanophenol fron p-nitrobenzaldoxime and p-nitrobenzonitrile (used as a sometimes recyclable chain carrier) (see 1st edition).

DMSO, 114⁰
Na2CO3

**3** (83%)

+ **2**

NH2OH, 70⁰
KOH, **2**

+ **3**

| 1 | Miller, M.J.; Loudon, G.M. | *J.Org.Chem.* | **1975** | *40* | 126 |
|---|---|---|---|---|---|
| 2 | Snyder, M.R. | *J.Org.Chem.* | **1974** | *39* | 3343 |
| 3 | Snyder, M.R. | *J.Org.Chem.* | **1975** | *40* | 2879 |

## M I N I S C I  Radical Aromatic Substitution

Iron catalyzed free radical amination of aromatics or free radical carbamylation, alkylation of protonated heterocycles (see 1st edition).

| 1 | Minisci, F. | *Tetrahedron Lett.* | **1965** | | 433 |
|---|---|---|---|---|---|
| 2 | Minisci, F. | *Chem.Ind.Milano* | **1966** | *48* | 716 |
| 3 | Minisci, F. | *Tetrahedron Lett.* | **1970** | | 15 |
| 4 | Heinisch, G. | *Synthesis* | **1988** | | 119 |
| 5 | Bourguignon, J. | *Tetrahedron Lett.* | **1995** | | 7875 |
| 6 | Minisci, F. | *J.Org.Chem.* | **1987** | *52* | 730 |
| 7 | Minisci, F. | *Heterocycles* | **1989** | *28* | 489 |
| 8 | Minisci, F. | *J.Heterocyclic Chem.* | **1990** | *27* | 79 |

**N,N-Dimethylaniline 2.**[2] To N-chlorodimethylamine 1 (4.3 g, 54 mmol), HOAc (50 mL), PhH (30 mL) and $H_2SO_4$ (83 mL) was added with stirring $FeSO_4$. After 15 min the mixture was quenched with ice, basified (NaOH) and extracted (PhH). Distillation gave 5 g of 2 (76%), bp 193-194°C.

**Quinoxaline-2-carboxamide 4.**[3] 3 (13 g, 0.1 mol) and 98% $H_2SO_4$ (5.5 mL) in $HCONH_2$ (100 mL) was treated simultaneously with 34% $H_2O_2$ (15 mL, 0.15 mol) and $FeSO_4 \cdot 7H_2O$ (41.7 g, 0.15 mol) under efficient stirring. After 15 min at 10-15°C, $HCONH_2$ was distilled, the residue extracted ($CHCl_3$) and the solvent evaporated, to give 14.2 g of 4 (82%), mp 200°C.

**1-Dioxanoisoquinoline 7.**[5] A mixture of 5 (258 mg, 2 mmol), TFA (228 mg, 2mmol) and 60% $H_2O_2$ (6 mL) in $Me_2CO$ (5 mL) and dioxane 6 (5 mL) were refluxed for 10 h. The mixture was diluted with water (20 mL), basified ($NH_4OH$) and extracted with $CH_2Cl_2$. Evaporation of the solvent and chromatography (silica gel hexane:EtOAc) afforded 275 mg of 7 (64%).

## M I S L O W - B R A V E R M A N - E V A N S  Rearrangement

Reversible 2,3-sigmatropic rearrangement of allylic sulfoxides to allyl sulfenates which are cleaved by phosphites or thiols to allylic alcohols (see 1st edition).

| 1 | Mislow, K. | *J.Am.Chem.Soc.* | **1966** | *88* | 3138 |
| 2 | Braverman, S. | *J.Chem.Soc.Chem.Comunn.* | **1967** | | 270 |
| 3 | Evans, D.A. | *J.Am.Chem.Soc.* | **1971** | *93* | 4956 |
| 4 | Evans, D.A. | *Acc.Chem.Res.* | **1974** | *7* | 147 |
| 5 | Grieco, P.A. | *J.Chem.Soc.Chem.Comunn.* | **1972** | *38* | 2245 |
| 6 | Grieco, P.A. | *J.Org.Chem.* | **1975** | *38* | 2245 |
| 7 | Biellmann, J.F. | *J.Org.Chem.* | **1992** | *57* | 6301 |
| 8 | Ruano Garcia J.L. | *J.Org.Chem.* | **1994** | *59* | 3421 |

**(+)-(E)-Nuciferole (3).[6]** To 1 (195 mg, 1 mmol) in THF (10 mL) at –50°C under $N_2$ was added dropwise 1.66 M BuLi in hexane (0.65 mL, 1.08 mmol). 2 (548 mg, 15 mmol) in THF (1 mL) was added dropwise over 10 min. After 1 h stirring at –50°C and 2 h at 25°C the mixture was poured into brine and extracted with $Et_2O$:hexane (3:1). The residue obtained after evaporation of the solvent was dissolved in MeOH (1.5 mL) and treated with Ph-SH (660 mg, 5.4 mmol) in MeOH (20 mL). BuLi (0.78 mL) was added under $N_2$. Heating for 7 h at 65°C and preparative TLC ($Et_2O$:hexane) gave 127 mg of 3 (58%).

**Allyl alcohol (5).[7]** 4 (320 mg, 1.38 mmol) and m-CPBA (380 mg, 1.52 mmol) was stirred for 15 h at -78°C. Hydrolysis with aq.$NH_4Cl$, extraction with $CH_2Cl_2$ and evaporation of the solvent gave 346 mg of an oil. Reflux with MeOH (15 mL) and $Et_2NH$ (730 mg, 10 mmol) followed by work up and chromatography (silica gel, hexane:$Et_2O$ 1:1) afforded 165 mg of 5 (86%).

# M I T S U N O B U  Displacement

Inter and intramolecular nucleophilic displacement of alcohols with inversion by means of diethyl azodicarboxylate (DEAD)-triphenylphosphine and a nucleophile. Also dehydration, esterification of alcohols or alkylation of phenols and one step synthesis of nitriles from alcohols (see 1st edition).

| 1 | Mitsunobu, O. | *Bull. Chem. Soc. Jpn.* | **1967** | *40* | 2380 |
| 2 | Miller, M.J. | *J.Am.Chem.Soc* | **1980** | *102* | 7026 |
| 3 | Berchtold, G.A. | *J.Org.Chem.* | **1981** | *46* | 2381 |
| 4 | Mitsunobu, O. | *Synthesis* | **1981** | | 1 |
| 5 | Evans, S.A. | *J.Org.Chem.* | **1988** | *53* | 2300 |
| 6 | Crich, D. | *J.Org.Chem.* | **1988** | *54* | 257 |
| 7 | Hassner, A. | *J.Org.Chem.* | **1990** | *55* | 2243 |
| 8 | Wilk, B. | *Synth.Commun.* | **1993** | *23* | 2481 |
| 9 | Macor, J.E. | *Heterocycles* | **1993** | *35* | 349 |

| 10 | Szantay, C. | *Synth.Commun.* | **1995** | 25 | 1545 |
| 11 | Procopiou, P.A. | *J.Chem.Soc.Perkin 1* | **1996** | | 2249 |
| 12 | Hughes, D.L. | *Org.Prep.Proced.Intn.* | **1996** | 28 | 127 |
| 13 | Katritzky, A. | *Synth.Commun.* | **1997** | 27 | 1613 |

**(-)Methyl cis-3-hydroxy-4,5-epoxycyclohex-1-enecarboxylate (2).**[3] To (-) methyl shikimate **1** (220 mg, 106 mmol) and triphenylphosphine (557 mg, 2.12 mmol) in THF at $0^0$C, under $N_2$ was added with stirring (DEAD) (370 mg, 2.42 mmol). After 30 min at $0^0$C and 1h at $20^0$C, the product was vacuum distilled (kugelrohr) at 165 $^0$C (0.1 mm) and taken up in $Et_2O$. Cooling gave bis (carbethoxy) hydrazine (10 mg, mp $133^0$C). The filtrated was concentrated and chromatographed (preparative TLC, silica gel, $Et_2O$) to afford on standing 140 mg of **2** (77%); recrystallized from $Et_2O$-petroleum ether, mp 81-82$^0$C, $\alpha_D^{25}= 55.4^0$.

**1-Benzylbenzotriazole (5).**[13] To a solution of benzyl alcohol **4** (1.06 g, 10 mmol) and $Ph_3P$ (2.62 g, 10 mmol) in THF (8 mL) cooled at -18$^0$C under stirring, was added NBS (1.78 g, 10 mmol) over 2-4 min in portions. After 5 min benzotriazole **3** ( 2.86 g, 24 mmol) was added and stirring was continued until 20$^0$C was reached. Workup and chromatography afforded 1.77 g of **5** (85%), mp 115-116$^{0C}$ (from EtOH).

**(Z,Z)-Nona-3,6-dienenitrile (7).**[10] To a stirred solution of triphenylphosphine (1.0 g, 3.8 mmol) in $Et_2O$ (10 mL) was added dropwise diethyl azodicarboxylate (0.66 g, 3.8 mmol) at -20$^0$C under $N_2$. After 20 min stirring under cooling, octa-3,6-dienol **6** (315 mg, 2.5 mmol) was added dropwise at -20$^0$C. After another 20 min stirring at 20$^0$C, a solution of acetone cyanhydrin (320 mg, 3.75 mmol) in $Et_2O$ (1 mL) was added and the mixture was stirred for another 4 h at -20$^0$C. After warming to 20$^0$C, the mixture was stirred for 10 h, filtered and the filtrate concentrated in vacuum. The residue, after flash chromatography (hexane:$Me_2CO$ 10:0.5) afforded 236 mg of **7** (70%).

## M O O R E  Cyclobutenone Rearrangement

Thermal rearrangement of alkyl or alkenylcyclobutanones to benzofurans, quinones, phenols.

| 1 | Moore, H.W. | J.Am.Chem.Soc. | **1985** | *107* | 3392 |
| 2 | Moore, H.W. | J.Org.Chem. | **1986** | *51* | 3067 |
| 3 | Moore, H.W. | J.Org.Chem. | **1988** | *53* | 4166 |
| 4 | Moore, H.W. | J.Org.Chem. | **1991** | *56* | 6104 |
| 5 | Wulff, W.D. | J.Am.Chem.Soc. | **1996** | *118* | 1808 |

## M O R I N  Penicillin Rearrangement

Ring expansion of penams to cephems under acidic catalysis.

| 1 | Morin, R.B. | J.Am.Chem.Soc. | **1963** | 85 | 1896 |
| 2 | Morin, R.B. | J.Am.Chem.Soc. | **1969** | 91 | 1401 |
| 3 | Conway, T.T. | Can.J.Chem. | **1978** | 56 | 1335 |
| 4 | Cooper, L.E. | Chem.& Ind. | **1978** | | 794 |
| 5 | Farina, V. | Tetrahedron Lett. | **1992** | 33 | 3559 |

**Cephalosporin (2).**[1] Reflux of phenoxymethylpenicillin sulfoxide methyl ester **1**, with a trace of p-toluenesulfonic acid in xylene gave **2** (15%), mp 141-142°C, $\alpha_D$ + 94°.

## MORI–SHIBASAKI   Catalytic Nitrogenation

Introduction of nitrogen or N-heterocycles in organic molecules in the presence of a titanium-nitrogen catalyst.

$$TiCl_4 + THF, Mg/N_2 \text{ (1 atm)} \longrightarrow (THF.Mg_2Cl_2TiN) \longrightarrow (3THF, Mg_2Cl_2TiNCO)$$

**1**          **2**                                                    **3**

| 1 | Mori, M.; Shibasaki, M. | *Tetrahedron Lett.* | **1987** | *28* | 6187 |
|---|---|---|---|---|---|
| 2 | Mori, M.; Shibasaki, M. | *J.Am.Chem.Soc.* | **1989** | *111* | 3725 |
| 3 | Mori, M.; Shibasaki, M. | *J.Chem.Soc.Chem.Commun.* | **1991** | | 81 |
| 4 | Mori, M.; Shibasaki, M. | *J.Synth.Org.Chem.Jpn.* | **1991** | *49* | 937 |

**Titanium complex 3.[2]** To a mixture of Mg (7 g, 0.29 mmol) in THF (50 mL) was added TiCl₄ 1 (1.9 g, 10 mmol) at -78°C under Ar. After degassing, the mixture was stirred at 20°C under $N_2$ for 16 h with change of color and exothermicity. The unreacted Mg was filtered under $N_2$ and the solution was stirred for 1 h at 20°C under $CO_2$. Under cooling (ice) the black suspension was treated with hexane (1 mL) and the precipitate filtered and washed with Et₂O and dried in vacuum.

**3-Methyleneisoindolinone 5.** To a mixture of o-bromoacetophenone 4 (39.8 mg, 0.2 mmol), K₂CO₃ (55.2 mg, 0.4 mmol), Pd(Ph₃P)₄ (11.54 mg, 0.01 mmol) and 3 (264.8 mg, 0.6 mmol) in N-methylpyrrolidone (2 mL) after degassing, the mixture was heated to 100°C for 16 h under CO (1atm) (monitoring by TLC). The cooled mixture was diluted with EtOAc and stirred with water a few hours. Filtration through Celite and washing after evaporation and chromatography are obtained 13 mg of 5 (48%).

# M O R I T A - B A Y L I S - H I L L M A N   Vinyl Ketone Alkylation

Amine catalyzed conversion of acrylates to α-(hydroxyalkyl) acrylates or of vinyl ketones to α-(hydroxyalkyl) vinyl ketones, also with chiral induction (see 1st edition).

| 1 | Morita, K. | *Japan.Pat.* 6003,364(1967) C.A.**1969** | | 70 | 19613u |
|---|---|---|---|---|---|
| 2 | Morita, K. | *Bull.Chem.Soc.Jpn.* | **1968** | 41 | 2816 |
| 3 | Baylis, A.B.; Hillman, M.E.D. | *Ger.Pat.*2155113 | C.A.**1972** | 77 | 3417 |
| 4 | Basavaiah, D. | *Tetrahedron Lett.* | **1986** | 27 | 2031 |
| 5 | Perlmutter, P.T. | *J.Org.Chem.* | **1995** | 60 | 6515 |
| 6 | Scheeren, H.W. | *Tetrahedron* | **1996** | | 1253 |
| 7 | Leahy, J.W. | *Tetrahedron* | **1997** | 53 | 1642 |
| 8 | Ciganek, E. | *Org.React.* | **1997** | 51 | 201 |
| 9 | Shi, M. | *J. Org. Chem.* | **2001** | 66 | 406 |

**4-Hydroxy-3-methylenetridecan-2-one (3).[2]**   A solution of decanal **1** (3.12 g, 20 mmol), methyl vinyl ketone **2** (1.4 g, 20 mmol) and 1,4-diazabicyclooctane (DABCO) (0.33 g, 3 mmol) in THF (5 mL) was allowed to stand at 20°C for 10 days. The reaction mixture was taken up in Et$_2$O (25 mL), washed with 2N HCl, NaHCO$_3$ and the solution dried (MgSO$_4$). Purification by column chromatography (5% EtOAc in hexane) and distillation gave 2.95 g of **3** (65%), bp 117-120°C/0.5 mm.

**2(R),6(R)-2,6-Dimethyl-5-methylene-1,3-dioxan-4-one (6).[7]**   A stirred solution of chiral acrylamide **4** (1 g, 3.7 mmol) in CH$_2$Cl$_2$ (2 mL), cooled to 0°C, was treated with acetaldehyde **5** (2.38 g, 54 mmol), followed by DABCO (270 mg, 1.85 mmol). After 8 h stirring at 0°C, evaporation of the solvent and chromatography gave 448 mg of **6** (85%), 99% ee), α$_D$ = +73.4° c=1.8 CHCl$_3$.

## M O S H E R ' S   A C I D   for Chirality Determination

Synthesis and use of $\alpha$-methoxy-$\alpha$-(trifluoromethyl)phenylacetic acid (MTPA) **4**, a chiral reagent for determination of enantiomeric purity of alcohols or amines by NMR.

$$PhCOCF_3 \;+\; Cl_3C\text{---}COOSiMe_3 \xrightarrow[K_2CO_3,\ 150^0]{18\text{-}Crown\text{-}6} Ph\text{---}\underset{OSiMe_3}{\overset{CF_3}{\underset{|}{\overset{|}{C}}}}\text{---}CCl_3 \xrightarrow[60^0]{KOH/MeOH}$$

**1**            **2**                                              **3 (83%)**

$$Ph\text{---}\underset{OMe}{\overset{CF_3}{\underset{\vdots}{\overset{|}{C}}}}\text{---}COOH \xrightarrow{\quad HO\text{---}\underset{R''}{\overset{R'}{\underset{}{\overset{|}{CH}}}}\quad} \text{2 diastereomeric esters}$$

**4 (79%)**

| | | | | | |
|---|---|---|---|---|---|
| 1 | Mosher, H.S. | *J.Org.Chem.* | **1969** | *34* | 2543 |
| 2 | Alper,H. | *J.Org.Chem.* | **1992** | *57* | 3731 |
| 3 | Mosher, H.S. | *J.Am.Chem.Soc.* | **1973** | *95* | 512 |
| 4 | Mosher, H.S. | *J.Org.Chem* | **1973** | *38* | 2143 |
| 5 | Ugi, I. | *Tetrahedron* | **1986** | *42* | 547 |
| 6 | Villani, F.G. | *J.Org.Chem.* | **1986** | *51* | 3715 |
| 7 | Alexakis, A. | *J.Org.Chem.* | **1992** | *57* | 1224 |
| 8 | Snyder, J.K. | *J.Org.Chem.* | **1988** | *53* | 5335 |
| 9 | Ohtouri, J. | *J.Am.Chem.Soc.* | **1991** | *113* | 4092 |
| 10 | Oikawa, H. | *Tetrahedron* | **1994** | *50* | 13347 |

For the synthesis of **4** see ref. 6.

**Determination of enantiomeric purity of an amine or alcohol.**[3,4] To a dried (150°C) test tube fitted with a rubber septum, the reagents were injected via syringe in the following order: pyridine (0.3 mL, 300 mg), (+)-MTPA-chloride (4-chloride) (35 mg, 0.026 mL, 0.14 mmol), $CCl_4$ (0.3 mL) and the corresponding amine or alcohol (0.1 mmol). The reaction mixture was shaken and allowed to stand at 20°C until the reaction was complete. 3-Dimethylamino-1-propylamine (20 mg, 0.024 mL, 0.20 mmol) was added to convert unreacted MTPA-chloride (or anhydride) into a basic amide, which can be removed by washing. After dilution with ether, washing (dil. HCl, $Na_2CO_3$, aq. brine), drying, evaporation and passing through a short column of silica gel, the optical purity was determined by NMR integration.

# MOUSSERON - FRAISSE - MCCOY Cyclopropanation

Stereoselective synthesis of cyclopropane-1,2-dicarboxylic acids or 1,2-dicyano substituted cyclopropanes by Michael addition (see also Hassner – Ghera - Little).

$$H_2C=C-COOMe + H-C-COOMe \xrightarrow{NaH} $$

1                    2                                3 (74%)

$$H_2C=C-CN + H-C-CN \longrightarrow $$

| | | | | | |
|---|---|---|---|---|---|
| 1 | Fraisse, J. | *Bull.Soc.Chim.Fr.* | **1957** | | 986 |
| 2 | McCoy, L.L. | *J.Am.Chem.Soc.* | **1958** | 80 | 6568 |
| 3 | Mousseron, M. | *C.R.* | **1959** | 248 | 887;1465;2840 |
| 4 | Warner, D.T. | *J.Org.Chem.* | **1959** | 24 | 1536 |
| 5 | Wawzonek, S. | *J.Am.Chem.Soc.* | **1960** | 82 | 439 |
| 6 | McCoy, L.L. | *J.Org.Chem.* | **1960** | 25 | 2078 |

## M U K A I Y A M A   Aldolization

Stereoselective aldol condensation of aldehydes with silyl enol ethers catalyzed by Lewis acids (Ti (IV), Sn (II), Yb (OTf)$_3$, InCl$_3$, chiral Cu-oxazolines) (see 1st edition).

| 1 | Mukaiyama, T. | *J.Am.Chem.Soc.* | **1973** | *95* | 967 |
| 2 | Mukaiyama, T. | *Chem.Lett.* | **1982** | | 353 |
| 3 | Mukaiyama, T. | *Chem.Lett.* | **1986** | | 187 |
| 4 | Shibasaki, M. | *Tetrahedron Asymm.* | **1995** | *6* | 71 |
| 5 | Corey, E.J. | *Tetrahedron Lett.* | **1992** | *33* | 6907 |
| 6 | Loh, T.P. | *Tetrahedron Lett.* | **1997** | *38* | 3465 |
| 7 | Mukaiyama, T. | *Org.React.* | **1982** | *28* | 187 |
| 8 | Mukaiyama, T. | *Aldrichchim.Acta* | **1996** | *29* | 59 |
| 9 | Collins, S. | *J.Org.Chem.* | **1998** | *63* | 1885 |
| 10 | Evans, D.A. | *J.Am.Chem.Soc.* | **1996** | *118* | 5814 |

**Syn 3-(Ethylthiomethyl)-4-hydroxy-6-phenyl-2-hexanone (3) and (4).[3]** To ethane thiol (10 mg, 0.17 mmol) in THF (2 mL) was added 1.54 M n-butyl-lithium in hexane (0.11 mL) at $0^{\circ}C$ under Ar. Stannous triflate (69.0 mg, 0.17 mmol) was added and after 20 min the mixture was cooled to $45^{\circ}C$. Methyl vinyl ketone **1** (118 mg, 1.98 mmol) in THF (1.5 mL) was added followed by 3-phenylpropanal **3** (350 mg, 2.61 mmol) in THF (1.5 mL). After 12 h aq. citric acid was added and the organic material extracted with $CH_2Cl_2$. The residue after evaporation was dissolved in MeOH and treated with citric acid. After 30 min stirring, the mixture was quenched with pH 7 phosphate buffer, extracted with $CH_2Cl_2$, the solvent evaporated and the residue chromatographed to afford 336 mg of **3** and **4** (75%), syn:anti = 90:10.

**(R)-1-Hydroxy-1-phenyl-3-heptanone (8).[5]** To a solution of catalyst **7** (28.6 mg, 0.056 mmol) in EtCN (0.5 mL) cooled at $-78^{\circ}C$ was added **5** (0.028mL, 0.25 mmol) followed by 2-trimethylsiloxy-1-hexene **6** (0.08 mL, 0.41 mmol). After 14 h stirring at $-78^{\circ}C$, the mixture was quenched with saturated $NaHCO_3$ (10 mL). Usual workup and chromatography (silica gel, 5-20% EtOAc in hexane) gave 58 mg of **8** (100%, 90% ee).

**Ethyl 2,2-dimethyl-3-hydroxybutanoate (11) and (12).[4]** Bistriflamide of (1S,2S)-1,2-diphenylethylenediamine (0.06 mmol) was reacted with NaH (0.24 mmol) in THF (1.2 m) at $0^{\circ}C$ for 30 min and 1 mL of the supernatant solution was added to $Yb(OTf)_3$ (0.05 mmol) in THF (1 mL). The reaction mixture was stirred at $40^{\circ}C$ for 12 h and the solvent removed under reduced pressure. $CH_2Cl_2$ (1 mL) was added to the residue and the supernatant solution was used as catalyst solution. The catalyst solution was cooled at $-40^{\circ}C$ and **5** (0.25 mmol) was added followed by ketene silyl acetal (0.3 mmol) added over 6.5 h (syringe pump) and stirring continued for another 5.5 h. Workup and chromatography gave 43% of **11** (51% ee) and 43% of silylated **12** (48% ee). Total yield 84% with 49% ee.

**Anti + syn 2-hydroxybenzylcyclohexanone (14).[6]** To $InCl_3$ (22 mg, 0.1 mmol) was added **5** (51 µL, 0.5 mmol) and the mixture was prestirred for 30 min before addition of **13** (0.19 mL, 1 mmol) and water (5 mL). After 15 h stirring at $20^{\circ}C$, extraction with $CH_2Cl_2$ followed by chromatography gave 70.1 mg of **14** (69%), 61:39 = anti:syn.

## **N A Z A R O V** Cyclopentenone Synthesis
Acid catalyzed of dienones to cyclopentenones (see 1st edition).

**1**          **2 (46%)[5]**

**3**          **4 (92%)**

| 1 | Nazarov, J. N. | *Bull. Acad. Sci. (USSR)* | **1946** |    | 633  |
| 2 | Eaton, P. E.   | *J. Org. Chem.*           | **1973** | 38 | 4071 |
| 3 | Denmark, S. E. | *Tetrahedron*             | **1986** | 42 | 2821 |
| 4 | Peel, M.L.     | *Tetrahedron Lett.*       | **1986** | 27 | 5947 |
| 5 | Motoyoshiya, J.| *J. Org. Chem.*           | **1991** | 56 | 735  |
| 6 | Jchikaora, J.  | *J. Org. Chem.*           | **1995** | 60 | 2320 |
| 7 | Denmark, S. E. | *Org. React.*             | **1994** | 45 | 1    |
| 8 | Pridgen, L. N. | *Synlett*                 | **1999** |    | 1612 |

**Dihydrojasmone (4).[2]** γ-Methyl-γ -decanolactone **3** (4.91 g, 26.6 mmol) was added to rapidly stirred 1:10 $P_2O_5$:$MeSO_3H$ (410 g). The homogeneous reaction mixture was stirred for 33 h at 25 °C. After quenching ($H_2O$) extraction ($CHCl_3$), extract washing (aq. gave 4.08 g of **4** (92%), bp 90-91 °C/2 Torr, purity 97% (GC).

## **N E B E R** Rearrangement
Rearrangement of N, N-dimethylhydrazone or tosylate derivatives of oxime to azirines and from there to α-amino ketones (see 1st edition).

**1**          **2 (42%)[2]**

| 1 | Neber, P. W.  | *Liebigs Ann.*  | **1926** | 449 | 109  |
| 2 | Neber, P. W.  | *Liebigs Ann.*  | **1936** | 526 | 277  |
| 3 | Morow, D. H.  | *J. Org. Chem.* | **1965** | 30  | 579  |
| 4 | Hyatt, J. A.  | *J. Org. Chem.* | **1981** | 46  | 3953 |
| 5 | O'Brine, C.   | *Chem. Rev.*    | **1964** | 64  | 81   |
| 6 | Yamura, Y.    | *Synthesis*     | **1973** |     | 215  |

## NEBER–BOSSET Oxindole Cinnoline Synthesis

Synthesis of N-aminooxindoles or of cinnolines (see 1st edition).

| | | | | | |
|---|---|---|---|---|---|
| 1 | Neber, P.W. | *Liebigs, Ann.* | **1929** | *471* | 113 |
| 2 | Bosset, G. | *C. Z.(Ph. D. Thesis)* | **1920** | *II* | 3015 |
| 3 | Bruce, J. M. | *J. Chem. Soc.* | **1959** | | 2366 |
| 4 | Baumgartner, H. F. | *J. Am. Chem. Soc.* | **1960** | *82* | 3977 |
| 5 | Bruce, J. M. | *J. Chem. Soc.* | **1964** | | 4037 |

## NEF Reaction

Conversion of nitroalkanes to carbonyl compounds by acidification of nitronates, compare McMurry use of TiCl$_3$ (see 1st edition).

| | | | | | |
|---|---|---|---|---|---|
| 1 | Nef, J. U. | *Liebigs Ann.* | **1894** | *280* | 286 |
| 2 | Weinstein, B. | *J. Org. Chem.* | **1962** | *27* | 4049 |
| 3 | Langrene, M. | *C. R. (C)* | **1974** | *284* | 1533 |
| 4 | Seebach, D. | *Chimia,* | **1979** | *33* | 1 |
| 5 | Miyakoshy, T. | *Synthesis* | **1986** | | 766 |
| 6 | Hwu, J. R. | *J. Am. Chem. Soc.* | **1991** | *113* | 5917 |
| 7 | Noland, W. E. | *Chem. Rev.* | **1955** | *55* | 137 |
| 8 | McMurry, J. | *Acc. Chem. Res.* | **1974** | *7* | 281 |
| 9 | Pinnick, H. W. | *Org. React.* | **1990** | *38* | 655 |

## N E G I S H I Cross Coupling

Pd or Ni catalyzed cross coupling, hydrometallation-cross coupling, and carbometallation-cross coupling using organometals of intermediate electronegativity e.g Al, Zn.

| 1 | Negishi, E. | *J. Org. Chem.* | **1975** | *40* | 1676 |
|---|---|---|---|---|---|
| 2 | Negishi, E. | *J. Org. Chem.* | **1978** | *43* | 358 |
| 3 | Negishi, E. | *Tetrahedron Lett.* | **1990** | *31* | 4393 |
| 4 | Negishi, E. | *Tetrahedron Lett.* | **1997** | *38* | 525 |
| 5 | Negishi, E. | *Acc.Chem. Res.* | **1982** | *15* | 340 |
| 6 | Negishi, E. | *Chem. Rev.* | **1996** | *96* | 365 |

**p-Methoxyphenylethyne (4).[2]** To a saturated solution of acetylene in THF (50 mL) at −78 °C was added n-BuLi (50 mmol) diluted with THF (50 mL) followed by a solution of anh.ZnCl₂ (50 mmol) in THF. The mixture was warmed to 20 °C. To this solution of **2** was added p-iodoanisole **3** (4.68 g, 20 mmol) in THF (20 mL) and Pd (PPh₃)₄ (1.15 g, 1 mmol) in THF. Work up and distillation gave 1.48 g of **4** (56%).

**(Z)-1'-[1-Methyl-(E)-2'-heptenylidene] indane (7).[3]** To ZrCp(H)Cl (380 mg, 1.5 mmol) in PhH (3 mL) was added 1-hexyne (0.23 mL, 2 mmol) at 25 °C. After 3 h the volatiles were evaporated in vacuum and THF (2 mL) was added to the residue. This solution was added to **5** (260 mg, 0.95 mmol) and Pd (PPh₃)₄ (55 mg, 0.05 mmol) in THF (2 mL). After 5 h reflux, cooling, work up and chromatography (hexane) afforded 140 mg of **7** (70%) and <3% of **8**.

## N E N I T Z E S C U Indole Synthesis

Synthesis of indoles by reductive cyclisation of o, ω-dinitrostyrenes.

| | | | | | |
|---|---|---|---|---|---|
| 1 | Nenitzescu, C. D. | *Chem. Ber.* | **1925** | 58 | 1063 |
| 2 | Schroeder, D. C. | *J. Am. Chem. Soc.* | **1953** | 75 | 5887 |
| 3 | Schultz, T. W. | *J. Org. Chem.* | **1985** | 50 | 2790 |

## N E N I T Z E S C U 5-Hydroxyindole Synthesis

5-Hydroxyindole synthesis from quinones and β-aminocrotonates (see 1st edition).

| | | | | | |
|---|---|---|---|---|---|
| 1 | Nenitzescu, C.D. | *Chem. Ber.* | **1925** | 58 | 1063 |
| 2 | Nenitzescu, C.D. | *Bull. Soc. Chim. Rom.* | **1929** | 11 | 37 |
| 3 | Allen, G. R. | *J. Org. Chem.* | **1968** | 33 | 198 |
| 4 | Bernici, J. L. | *J. Org. Chem.* | **1981** | 46 | 4197 |
| 5 | Rapderey, T. | *Austr. J. Chem.* | **1984** | 37 | 1263 |
| 6 | Martinelli, J. A. | *J. Org. Chem.* | **1996** | 61 | 9058 |
| 7 | Allen, G. R. Jr. | *Org. React.* | **1973** | 20 | 337 |

**3-(Carbomethoxy)-2-ethyl-1-leuzyl-1H-5-hydroxyindole (5).**[6] Methyl propionyl acetate **2** (131 g, 1 mol) and benzylamine **1** (112 g, 1.05 mole) in PhMe (500 mL) was stirred with p-TsOH.H$_2$O (9.5 g, 50 mmol) under reflux for 4 h with a Dean-Stark trap to remove water (18.9 g, 1.05 mol). The mixture was cooled to 10 °C and filtered. Evaporation afforded a crude oil **3** (220 g). 1,4-Benzoquinone **4** (149 g, 1.38 mol) in Me-NO$_2$ (500 mL) was treated dropwise with **3** (220 g) in Me-NO$_2$ (250 mL) at 20 °C under N$_2$ over 30 min (endothermic reaction). After 48 h at 20 °C the mixture was cooled (ice/water), filtered and crude **5** was washed (Me-NO$_2$) and dried to give **5** 214 g (69%), mp194-195 °C.

## **NERDEL** Enol Ether Homologation

Homologation of enol ethers by dihalocarbenes.

$$H_2C=CH-OBu + CHFCl_2 + H_2C-CH_2 \xrightarrow[Et_4NBr]{150°, 5h} \left[ \begin{array}{c} H_2C-CH-OBu \\ C \\ F \quad Cl \end{array} \right] \longrightarrow$$

1                    2                    3 (O)                              4

$$\left[ \begin{array}{c} H_2C=C-CH-OBu \\ F \quad Cl \end{array} \right] \xrightarrow{3} H_2C=C-CH \begin{array}{c} OBu \\ OCH_2 \\ F \quad CH_2Cl \end{array} \xrightarrow[120°]{5\% H_2SO_4} H_2C=C \begin{array}{c} CHO \\ F \end{array}$$

5 (53%)                                    6 (41%)

| 1 | Nerdel, F. | *Tetrahedron Lett.* | **1965** |       | 3585 |
|---|-----------|---------------------|----------|-------|------|
| 2 | Nerdel, F. | *Tetrahedron Lett.* | **1966** |       | 5379, 5383 |
| 3 | Nerdel, F. | *Chem.Ber* | **1967** | *100* | 1858 |

**2-Fluroacrolein n-butyl (2-chloroethyl) acetal 5.**[2] A mixture of n-butyl vinyl ether **1** (60 g, 0.6 mol), CHFCl$_2$ **2** (62 g, 0.6 mol), ethylene oxide **3** (120 mL, 2.4 mol) and tetraethylammonium bromide (4.0 g, 19 mmol) was heated for 5 h at 150°C. Distillation gave ethylene chlorhydrin (24 g, bp 35 C/ 13 mm) and 67 g of **5** (53%), bp 97°C / 13 mm.

**2-Fluoroacrolein 6.** Acetal **5** (21 g, 109 mmol) was added slowly under stirring to a 5% solution of H$_2$SO$_4$, followed by heating to 120°C. Separation, drying (CaCl$_2$) and distillation afforded 3 g of **6** (41%), bp 71° C; 2,4-DNPH, mp 200°C dec.

## N E S M E J A N O W  Aromatic Mercuric Halides

Preparation of aromatic mercuric halides from aromatic amines via diazonium salts:

| | | | | | |
|---|---|---|---|---|---|
| 1 | Nesmejanow, A. N. | *Chem. Ber.* | **1929** | 62 | 1010 |
| 2 | Nesmejanow, A. N. | *Chem. Ber.* | **1929** | 62 | 1018 |
| 3 | McClure, R. E. | *J. Am. Chem. Soc.* | **1931** | 53 | 319 |
| 4 | Larock, R. C. | *Tetrahedron.* | **1982** | 38 | 1713 |

α-**Naphthylmercuric chloride (3).**[1] To dilute (1:1) HCl (100 mL), was added α-naphthylamine **1** (14.3 g, 0.1 mol) under stirring. Under cooling (5 °C), NaNO$_2$ (6.9 g, 0.1 mol) was added (starch-iodine paper). With cooling and stirring a solution of HgCl$_2$ (27.1 g, 0.1 mol) in 36% HCl (30 mL) was added. After 30 min the mercury complex **2** is filtered and washed with water (2 × 40 mL) and Me$_2$CO (2 × 15 mL) to give 38 g of **2** (82%) (handle with case). **2** (4.6 g, 10 mmol) and copper powder (1.26 g) in Me$_2$CO (25 mL) was stirred at 20 °C for 1 h and after 18 h, the solid is filtered and extracted with xylene under reflux. On cooling **3** crystallized, 1.75 g (40%), mp 266-267 °C.

## N I C K L  Benzofurans Synthesis

Synthesis of benzofurans from phenols and allyl halides.

| | | | | | |
|---|---|---|---|---|---|
| 1 | Nickl, J. | *Chem. Ber.* | **1958** | 91 | 553 |
| 2 | Casiraghi, G. | *Angew. Chem. Int. Ed.* | **1978** | 17 | 684 |
| 3 | Casiraghi, G. | *J. Org. Chem.* | **1979** | 44 | 803 |
| 4 | Kawase, J. | *Chem. Lett.* | **1979** | | 253 |
| 5 | Casiraghi, G. | *Tetrahedron* | **1983** | 39 | 169 |

## N I C O L A O U Oxidations

One step oxidation of alcohols or ketones to enones (see also Saegusa); selective oxidation of benzylic groups (methyl to aldehydes) by o-iodoxybenzoic acid (IBX).

| 1 | Nicolaou, K. C. | *J. Am. Chem. Soc.* | **2000** | *122* | 7596 |
| 2 | Nicolaou, K. C. | *J. Am. Chem. Soc.* | **2001** | *123* | in press |

**Oxidation of alcohols or carbonyl compounds (general method). Synthesis of 2-cyclooctenone.**[1] To a solution of cyclooctanol 1 (1 mmol) in fluorobenzene:DMSO (2:1, 0.1 M) was added 2.2 equiv of IBX and the solution was heated to 55-65 °C (or to 85 °C for synthesis of 3). The reaction was monitored by TLC. Dilution with Et₂O and usual work up followed by flash chromatography afforded 2-cyclooctenone 2 in 77% yield.

**Benzylic Oxidation. Synthesis of p-t-butylbenzaldehyde (5).**[2] To a solution of p-t-butyltoluene 4 (148 mg, 1 mmol) in a mixture of fluorobenzene:DMSO (2:1) (7.5 mL) was added IBX (840 mg, 3 mmol) and the mixture was heated to 85 °C for 12 h. The mixture was cooled, diluted with Et₂O, washed (5% NaHCO₃, water, brine) and dried (MgSO₄). Chromatography (silica gel, hexane :Et₂O 10:1 to 5:1 ) afforded 138 mg of p-tert.butylbenzaldehyde 5 (85%) and 15 mg of unreacted 4(10%).

## N I E M E N T O W S K I Quinazolone Synthesis

Synthesis of quinazolone from anthranilic acid and amides or isatoic anhydride and amides (see 1st edition).

| 1 | Niementowsky, V.S | *J. Prakt. Chem.* | **1895** | 51 | 564 |
|---|---|---|---|---|---|
| 2 | Meyer, V. E.; Bellmann, Th. | *J. Prakt. Chem.* | **1886** | 33 | 18(2) |
| 3 | Endicot, M. M. | *J. Am. Chem. Soc.* | **1946** | 68 | 1300 |
| 4 | Pater, R. | *J. Heterocyclic. Chem.* | **1970** | 7 | 1113 |
| 5 | Pater, R. | *J. Heterocyclic. Chem.* | **1971** | 8 | 699 |

## O L E K S Y S Z Y N α-Aminophosphonic Acid Synthesis

Synthesis of 1-aminoalkanephosphonic and 1-aminoalkanephosphinic acid from ketones or aldehydes, chlorophosphines and carbamates (see 1st edition).

| 1 | Oleksyszyn, J. | *Synthesis* | **1978** | 479 |
|---|---|---|---|---|
| 2 | Soroka, M. | *Liebigs Ann.* | **1990** | 331 |

**1-Aminocyclohexylphosphonic acid (4).**[1] Cyclohexanone 1 (7.35 g, 75 mmol) was added at 20 °C to a stirred mixture of benzyl carbamates 3 (7.55 g, 50 mmol) and $PCl_3$ 2 (6.87 g, 50 mmol) in AcOH(10 mL). The mixture was refluxed for 40 min, treated with 4 M HCl (50 mL) and again refluxed for 0.5 h. After cooling, the organic layer was removed and the aqueous solution was refluxed with charcoal. After filtration and evaporation in vacuum, the residue was dissolved in MeOH (25-40 mL). The filtration and evaporation in vacuum, the residue was dissolved in MeOH (25-40 mL). The methanolic solution was treated with propene oxide until pH 6-7 is reached. The precipitates was filtered, washed with $Me_2CO$. and recrystallized from MeOH-water to give 7.74 g of 4 (58%), mp 264-265 °C.

## NISHIMURA–CRISTESCU N-Glycosidation

N-Glycosidation of disilyl uracyl derivatives by fusion with acylated α-halo sugars (Nishimura) or by condesation in the presence of $Hg(OAc)_2$ (Cristescu) (see also Vorbruggen).

| 1 | Handschumacher, R.T. | *J. Biol. Chem.* | **1960** | *236* | 764 |
|---|---|---|---|---|---|
| 2 | Nishimura, T | *Chem. Pharm. Bull. (Jap)* | **1963** | *11* | 1470 |
| 3 | Nishimura, T. | *Chem. Pharm. Bull.* | **1964** | *12* | 352; 1471 |
| 4 | Cristescu, C. | *Rev. Roum. Chim.* | **1968** | *13* | 365 |

**6-Azauridine 4.**[4] 6-Azauracil **1** (2.26 g; 20 mmol) in PhH (200 mL), after drying (azeotropic distillation) was treated with $Me_3SiCl$ (4.34 g; 40 mmol) and $Et_3N$ (4.04 g; 40 mmol). After 8 h reflux and usual work up there were obtained 4.7 g of crude **2** (90%) (sensitive to atmospheric moisture). A suspension of **2** (4.7 g; 23.4 mmol), **3** (from 1-O-acetyl-2,3,5-tri-O-benzoylribofuranose (10.35 g; 21.5 mmol) and HCl in ether) and $Hg(OAc)_2$ (6.48 g; 20.4 mmol) in PhMe (100 mL) was stirred at 20°C for 48 h, and refluxed 90 min. Work up and recrystallization (PhH) afforded 5.3 g (60%) of **4**, mp 189-190°C.

## N O Y O R I Chiral Homogeneous Hydrogenation

Homogeneous chiral hydrogenation of unsaturated alcohols, or carboxylic acids, enamides, ketones in the presence of a BINAP Ru or Rh complex as catalyst (see 1st edition).

| | | | | | |
|---|---|---|---|---|---|
| 1 | Noyori, R. | *J. Am. Chem. Soc.* | **1980** | *102* | 7932 |
| 2 | Noyori, R. | *J. Org. Chem.* | **1986** | *51* | 629 |
| 3 | Noyori, R. | *J. Am. Chem. Soc.* | **1986** | *108* | 7117 |
| 4 | Noyori, R. | *J. Am. Chem. Soc.* | **1987** | *109* | 9134 |
| 5 | Noyori, R. | *J. Am. Chem. Soc.* | **1989** | *111* | 9134 |
| 6 | Smrcina, M. | *Synlett* | **1991** | | 231 |
| 7 | King, A. S. | *J. Org. Chem.* | **1992** | *57* | 6689 |
| 8 | Otsuka, S. | *Synthesis* | **1991** | | 668 |
| 9 | Noyori, R. | *Acc. Chem. Res.* | **1990** | *23* | 345 |
| 10 | Noyori, R. | *Chem. Soc. Rev.* | **1989** | *18* | 187 |
| 11 | Noyori, R. | *Angew. Chem. Int. Ed.* | **1991** | *30* | 49 |
| 12 | Noyori, R. | *Acta. Chim. Scand.* | **1996** | *50* | 390 |

**(R)-(+)-2,2'-Bis (diphenylphosphino)-1,1'-binaphthyl (BINAP) (7).[2]** To Mg (2.62 g, 0.108 g-at) under $N_2$ was added $I_2$ (50 mg), THF (40 mL), 1,2-dibromoethane (0.51 mL). 2.2'-Dibromo-1,1'-dinaphthyl **1** (20 g, 46.4 mmol) in PhMe (360 mL) was added dropwise over a period of 4 h at 50-75 °C. After 2 h stirring at 75 °C the mixture was cooled to 0 °C and diphenylphosphinyl chloride **2** (23.2 g, 98 mmol) in PhMe (23 mL) was added over 30 min. The mixture was heated to 60 °C for 3 h, cooled , quenched with water (60 mL), stirred at 60 °C for 10 min and the organic layer concentrated to 60 mL. After 24 h at 20 °C, the product was filtered, stirred with heptane (45 mL) and PhMe (5 mL), filtered and dried to afford 27.5 g of (±) **3** (91%) mp 295-298 °C (pure 304-305 °C). (±) **3** (65.4 g, 0.1 mol), (1S)-(-)-camphorsulfonic acid monohydrate **4** (25 g, 0.1 mol) and EtOAc (270 mL) were heated to reflux and HOAc (90 mL) was added to get a clear solution. Gradual cooling to 2-3 °C, filtration and washing (EtOAc) gave 35.3 g of 1:1:1 complex of 3:4:AcOH. The complex was suspended in PhMe (390 mL), treated with water (30 mL) at 60 °C and cooled. The organic layer was concentrated to 50 mL and treated with hexane (50 mL). Filtration and drying gave 22.2 g of (R)-(+) **5** (68%);mp 262-263 °C, $\alpha_D^{24}$=399° (c 0.5 PhH).(R) **5** (50 g, 76.4 mmol) xylene (500 mL), $Et_3N$ (32.4 g, 320 mmol) and $Et_3SiH$ (41.4 g, 304 mmol) under Ar were heated 1 h at 100 °C, 1 h at 120 °C and 5 h at reflux, 30 % NaOH (135 mL) was added under stirring at 60 °C, the organic layer was concentrated and the residue treated with MeOH (200 mL) to give 47.5 g of (R)-BINAP **7** (95%), mp 241-242 °C, $\alpha_D^{24}$=-228° (c 0.679 PhH). $RuCl_2(BINAP)_2NEt_2$[7] To (1,5-Cyclooctadiene) ruthenium dichloride **8** (214 mg, 0.76 mmol) and **7** (500 mg, 0.8 mmol) under $N_2$, was added PhMe (17 mL) and $Et_3N$(1.7 mL). The mixture was heated to 140 °C, for 4 h, and after cooling was filtered under $N_2$ and dried in vacuum to give 760 mg of **9** (75%).

**t-Butyl 3(R)-hydroxybutyrate (11).** [7] t-Butyl acetoacetate **10** (14.5 g, 90 mmol) and MeOH (30 mL) after deoxygenation with $N_2$ was treated with **9** (36 mg, 0.041 mmol) and HCl (2 N, 0.041 mL). The mixture was hydrogenated in a Paar bottle under 50 psi $H_2$ at 40 °C. After 8 h the reaction was complete, the mixture was treated with hexane (30 mL) to remove **9** and the filtrated was concentrated to give 14.5 g of **11** (97 %).

## NUGENT–RAJANBABU Epoxide Homolysis

Selective generation of free radicals from epoxides promoted by (cyclopentadienyl) titanium (III) chloride, followed by trapping, usually with olefin.

| 1 | Nugent, W.A.; Rajanbabu, T.V. | *J.Am.Chem.Soc.* | **1988** | *110* | 8561 |
|---|---|---|---|---|---|
| 2 | Rajanbabu, T.V.; Nugent, W.A. | *J.Am.Chem.Soc.* | **1989** | *111* | 4525 |
| 3 | Rajanbabu, T.V.; Nugent, W.A. | *J.Am.Chem.Soc.* | **1990** | *112* | 6408 |
| 4 | Rajanbabu, T.V.; Nugent, W.A. | *J.Am.Chem.Soc.* | **1994** | *116* | 986 |
| 5 | Matty, G.; Roy, S.C. | *J.Chem.Soc.Perkin 1* | **1996** | | 403 |
| 6 | Gold, H.J. | *Synlett* | **1999** | | 159 |

**Bicyclic Tetrahydrofuran 8.[4]** To a solution of epoxide **7** (250 mg, 1 mmol) in THF (25 mL) was added dropwise a solution of $Cp_2TiCl$ (430 mg, 2 mmol) in THF (25 mL). After 10 min iodine (250 mg, 1 mmol) was added and the reaction mixture was stirred for 1 h. Quenching with saturated $NH_4Cl$ solution (50 mL) and extraction with $Et_2O$ followed by washing the organic phase with aqueous NaHS solution afforded after evaporation of the solvent a crude residue. Chromatography (silica gel, hexane:ethyl acetate 60:40) yielded 130 mg of **8** (52%) as a colorless liquid.

## N Y S T E D - T A K A I  Olefination

Organozinc reagent for olefination (alkylidenation) of aldehydes, ketones, enolizable ketones, esters, in the presence of a Lewis acid.

$$Zn + CH_2X_2 \xrightarrow{THF} \mathbf{1} \xrightarrow[BF_3, Et_2O, 0°]{Ph.CHO} Ph-CH=CH_2 \quad 3\ (96\%)$$

$$X = Cl, Br, I$$

$$Ph-CH=CH-CHO + CH_2I_2 \xrightarrow[0°-25°]{Zn, Me_3Al} Ph-CH=CH-CH=CH_2$$

$$PhH-C \underset{SMe}{\overset{O}{<}} \xrightarrow[\text{TMEDA, THF}]{MeCHBr_2, Zn, TiCl_4} \underset{MeS}{\overset{Ph}{>}}C=CH \overset{Me}{\underset{}{}} \quad \mathbf{5}\ (77\%)^5$$

**4**

$$\xrightarrow[THF, CH_2Cl_2]{CH_2Br_2, Zn, TiCl_4}$$

**6**          **7**

| 1 | Nysted, L.N. | *U.S.Pat. 3 865 848* | **C.A. 1975** | **83** | 10406q |
|---|---|---|---|---|---|
| 2 | Nysted, L.N. | *U.S.Pat. 3. 960 904* | **C.A. 1976** | **85** | 94618n |
| 3 | Matsubara, S. | *Synlett* | **1998** | | 313 |
| 4 | Oshima, H. | *Tetrahedron Lett.* | **1978** | | 2417 |
| 5 | Takai, K. | *Tetrahedron Lett.* | **1989** | **30** | 211 |
| 6 | Takai, K. | *J.Org.Chem.* | **1994** | **59** | 2668 |
| 7 | Lombardo, L. | *Org.Synth.* | **1987** | **65** | 81 |
| 8 | Pine, G.H. | *Org.React.* | **1993** | **43** | 1 |
| 9 | Breit, B. | *Angew.Chem.Int.Ed.* | **1998** | **37** | 453 |

**Styrene (3).[3]** Under Ar Nysted reagent **1** (20% suspension, 2.3 g, 1 mmol) and THF (3 mL) was cooled to 0°C. BF$_3$.OEt$_2$ (0.14 g, 0.1 mmol) in THF (2 mL) was added and the mixture was stirred for 5 min at 0°C. Benzaldehyde **2** (110 mg, 1 mmol) in THF was added at 0°C. benzaldehyde **2** (110 mg, 1 mmol) in THF was added at 0°C and the mixture was stirred for 2 h at 18°C. Quenching with 1M HCl and usual work up gave 99.8 mg of **3** (96%).

**1-Phenyl-1-methylthio-1-propene (5).[5]** To TiCl$_4$ (1 M, 4 mmol) in CH$_2$Cl$_2$ and THF (10mL) under Ar at 0° was added TMEDA (1.2 mL, 8 mmol) and all was stirred for 10 min at 25°C. Zn (0.59 g, 9 mmol) was added and the mixture was stirred for 30 min. A mixture of **4** (152 mg, 1 mmol) and 1,1-dibromoethane (414 mg, 2.2 mmol) in THF was added and stirring was continued for 15 min. Et$_2$O (10 mL) was added, the mixture was filtered through silica gel and the filtrate evaporated. Chromatography afforded 127 mg of **5** (77%).

## O' D O N N E L L Amino Acid Synthesis

Synthesis of amino acids from a Schiff base substrate of glycine, enantioselective alkylation by phase transfer catalysis (PTC) (see 1st edition).

$$Ph_2C=NH \quad + \quad H_2N-CH_2-CO_2Et \quad \xrightarrow[20^\circ C]{CH_2Cl_2} \quad Ph_2C=N-CH_2-CO_2Et$$

HCl

| **1** | **2** | **3 (97%)** |

Ph$_2$C=N—CO$_2$tBu $\xrightarrow[\text{PTC, Cinconidine}]{p-ClC_6H_4CH_2Br}$

H$_2$N    CO$_2$tBu          100% e,e
   \C/
   CH$_2$-C$_6$H$_4$-Cl        50% yield

$$3 \quad + \quad \xrightarrow[\text{DMF}]{\text{NBS / NaOAc}} \quad Ph_2C=N-CH-CO_2Et \xrightarrow{(2\text{-Thioph.})_2Cu(CN)Li_2} \quad Ph_2C=N-CH-CO_2Et$$

OAc.                                      2–Thioph.

**4 (71%)**

| 1 | O' Donnell, M. J. | *J. Org. Chem.* | **1982** | *47* | 2663 |
|---|---|---|---|---|---|
| 2 | O' Donnell, M. J. | *Tetrahedron Lett.* | **1985** | *26* | 695; 699 |
| 3 | O' Donnell, M. J. | *J. Am. Chem. Soc.* | **1989** | *111* | 2325 |
| 4 | O' Donnell, M. J. | *Tetrahedron* | **1994** | *50* | 4507 |
| 5 | de Meijere, A. | *Synlett.* | **1995** | | 226 |

**Ethyl N-(diphenylmethylene) glycinate (3).**[1] Benzophenone imine 1 (25 g, 0.138 mol) and ethyl glycinate HCl 2 (14.21 g, 0.138 mol) finely ground were stirred in CH$_2$Cl$_2$ (500 mL) at 20 °C for 24 h. Removal by filtration of NH$_4$Cl and evaporation of the solvent gave crude 3 . The residue was taken up in Et$_2$O, washed with water, dried (MgSO$_4$) to the solvent evaporated and the residue recrystallised from Et$_2$O/hexane to afford 32 g of 3 (97%), mp 51-55 °C.

**Ethyl N-(diphenylmethylene)-2-acetoxyglycinate (4).**[2] A solution on NBS (13.9 g, 78 mmol) in THF (40 mL) was added under stirring at 20 °C in 3 h to a solution of 3 (16.05 g, 60 mmol) and anh. NaOAc (16.5 g, 201 mmol) in DMF (60 mL). After overnight stirring at 20 °C, the mixture was poured into water and extracted with Et$_2$O. Normal work up afforded after recrystallisation from Et$_2$O/ligroin 13.7 g of 4 (71%), mp 62-65 °C.

## **OHSHIRO** Bromoalkene Reduction

Reduction of gem-dibromoalkenes to monobromoalkenes with diethyl phosphite and triethylamine.

| 1 | Ohshiro, Y. | *J.Org.Chem.* | **1981** | *46* | 3745 |
| 2 | Ohshiro, Y. | *Bull.Chem.Soc.Jpn.* | **1982** | *55* | 909 |
| 3 | Hayes, C.J. | *Tetrahedron Lett.* | **2000** | *41* | 3215 |

**β-Bromostyrene (2).**[1] To a solution of β,β-dibromoallene **1** (1.05 g, 4.0 mmol) and diethyl phosphite (2.21 g, 16 mmol) was added triethyl amine (0.81 g, 8 mmol) and the mixture was stirred for 5 h at 90°C. Et$_2$O (50 mL) was added, and then Et$_3$N.HBr was removed by filtration. After evaporation of the filtrate, the residue was chromatographed on a silica gel column (n-hexane) to afford 702 mg of **2** (96%), E:Z ratio 94:6.

**1-Bromo-2-phenylcyclopropane (4).** From 1,1-dibromo-2-phenylcyclopropane **3** (6.10 g, 4 mmol), diethyl phosphite (2.21 g, 16 mmol) and triethyl amine (0.81 g, 8 mmol) are obtained by the same procedure 730 mg of **4** (93% yield, ratio E:Z 75:25).

## O L O F S O N  Reagent

The use of vinyl chloroformate **2** for N-dealkylation of tertiary amines, protection of amino groups, protection of hydroxyl groups formation of 2-ketoimidazoles. Synthesis of vinyl carbonates by means of fluoro or chloroformates (see 1st edition).

**3 (90%)**

**5 (89%)**

| 1 | Olofson, R. A. | *Tetrahedron Lett.* | **1977** | | 1567 |
|---|---|---|---|---|---|
| 2 | Olofson, R. A. | *Tetrahedron Lett.* | **1977** | | 1570 |
| 3 | Pratt, P. F. | *Tetrahedron Lett.* | **1981** | 22 | 2431 |
| 4 | Cooley, J. H. | *Synthesis* | **1989** | | 1 |
| 5 | Olofson, R. A. | *J. Org. Chem.* | **1990** | 55 | 1 |
| 6 | Olofson, R. A. | *Pure Appl. Chem.* | **1988** | 60 | 1715 |

## O P P E N A U E R  Oxidations

A mild oxidation of alcohols to ketones using metal alkoxides (Al, K) and a ketone or with lantanide catalyst, zirconium or hafnium complexes (see 1st edition).

**2 (58%)**

| 1 | Oppenauer, R. V. | *Rec. Trav. Chim.* | **1937** | 56 | 137 |
|---|---|---|---|---|---|
| 2 | Woodward, R. B. | *J. Am. Chem. Soc.* | **1945** | 67 | 1425 |
| 3 | Kagan, H. B. | *J. Org. Chem.* | **1984** | 49 | 2045 |
| 4 | Ogawa, M. | *J. Org. Chem.* | **1986** | 51 | 240 |
| 5 | Djerassi, C. | *Org. React.* | **1951** | 6 | 207 |
| 6 | Huskens, J. | *Synthesis* | **1994** | | 1007 |

For catalyst preparation see Meerwein-Ponndorf-Verley.
**Acetophenone (2)**[3] A mixture of 0.023 M La(iPrO)$_3$ in PhMe (17 mL, 0.4 mmol), 1-phenylethanol **1** (490 mg, 4 mmol) and 2-butanone (290 mg, 4 mmol) was stirred at 20 °C for 24 h to afford **2** in 58% yield.

**O P P O L Z E R** Asymmetric Allyl Alcohol Synthesis

Asymmetric synthesis of secondary (E)-allyl alcohols from acetylenes and aldehydes, catalyzed by a chiral catalyst

| 1 | Oppolzer, W. | *Tetrahedron Lett.* | **1988** | *29* | 5645 |
| 2 | Oppolzer, W. | *Tetrahedron Lett.* | **1991** | *32* | 5777 |
| 3 | Oppolzer, W. | *Helv.Chim.Acta* | **1992** | *75* | 173 |

**Allyl alcohol 6.**[3] Under Ar, to a cooled (0°C) and stirred solution of borane-methyl sulfide complex (1M, 1.0 mL, 1mmol), was added cyclohexene (2.05 mL, 2 mmol) in hexane 1 mL. After 3 h at 0°C, oct-1-yne **1** (1.50 mL, 1 mmol) was added, and the mixture was stirred at 20°C for 1 h. Then the solution was cooled to -78°C and a hexane solution of $Et_2Zn$ (1M, 1.05 mL, 1.05 mmol) was added over 10 min and was followed by addtion of DAIB (-)-3-exo-(dimethylamino isoborneol) **4** (2 mg, 0.01 mmol). The mixture was cooled to 0°C and a solution of propionaldehyde **5** (0.072 mL, 1mmol) in hexane (4 mL) was added during 20 min. The reaction mixture was stirred for 1 h at 0°C, quenched with sat. aq. $NH_4Cl$ and chromatographed (silica gel, hexane:$Et_2O$) to afford 155 mg of **6** (91%), 84% ee.

## OPPOLZER Cyclopentenone Synthesis

Pd catalyzed cyclization of 1,6-dienes or 6-en-1-ynes to mono- or bicyclic cyclopentenones with CO insertion.

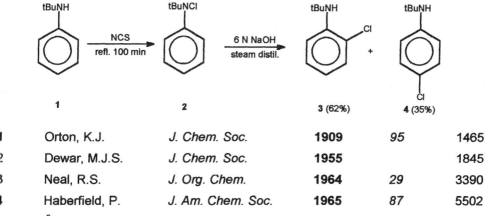

| | | | | | |
|---|---|---|---|---|---|
| 1 | Oppolzer, W. | *Tetrahedron Lett.* | **1989** | 30 | 5883 |
| 2 | Oppolzer, W. | *Pure Appl. Chem.* | **1990** | 62 | 1941 |
| 3 | Oppolzer, W. | *Helv. Chim. Acta* | **1991** | 74 | 465 |
| 4 | Heathcook, C. H. | *J. Org. Chem.* | **1993** | 58 | 560 |

**Ketone (2).**[4] To Pd$_2$(dba)$_3$.CHCl$_3$ (35 mg, 34 μmol), Ph$_3$P (17.1 mg, 65 μmol), LiCl (65 mg, 1.53 mmol) and THF (5 mL) was added degassed water (2.5 mL), and 1 (389 mg, 1.38 mmol) in THF (5 mL). CO was bubbled through for 2 min. The yellow solution was heated at 70 °C under CO for 24 h poured into 1% HCl and extracted with Et$_2$O. Work up and chromatography afforded 147.6 mg of 2 (43%).

## ORTON Haloaniline Rearrangement

Rearrangement of N-haloanilides or anilines to o- or p-haloaniline derivatives (see 1st edition).

| | | | | | |
|---|---|---|---|---|---|
| 1 | Orton, K.J. | *J. Chem. Soc.* | **1909** | 95 | 1465 |
| 2 | Dewar, M.J.S. | *J. Chem. Soc.* | **1955** | | 1845 |
| 3 | Neal, R.S. | *J. Org. Chem.* | **1964** | 29 | 3390 |
| 4 | Haberfield, P. | *J. Am. Chem. Soc.* | **1965** | 87 | 5502 |

**3 and 4.**[3] 1 (4.91 g; 33 mmol) and NCS (3.88 g; 30 mmol) were heated in PhH (100 mL) to reflux for 100 min. Filtration, evaporation and steam distillation from 6N NaOH afforded 62% of 3 and 35% of 4.

## O V E R M A N  Pyrrolidine Synthesis

Consecutive Aza-Cope-Mannich reactions for formation of pyrrolidines with stereo-control (see 1st edition).

**2 (87%)**

3                                    4 (92%)

| | | | | | |
|---|---|---|---|---|---|
| 1 | Overman, L.E. | *J. Am. Chem. Soc.* | **1979** | *101* | 1310 |
| 2 | Overman, L.E. | *Tetrahedron Lett.* | **1979** | | 4041 |
| 3 | Overman, L.E. | *J. Am. Chem. Soc.* | **1983** | *105* | 6629 |
| 4 | Padwa, A. | *J. Org. Chem.* | **1990** | *55* | 4801 |
| 5 | Kakimura, K. | *Tetrahedron* | **1993** | *49* | 4527 |
| 6 | Overman, L.E. | *J. Am. Chem. Soc.* | **1995** | *117* | 5776 |
| 7 | Overman, L.E. | *Isr. J. Chem.* | **1997** | *37* | 23 |
| 8 | Overman, L.E. | *Aldrichimica Acta* | **1995** | *28* | 107 |

**3-Acetyl-5-phenyl-1-propylpyrrolidine (2).**[3] A mixture of tetrafluoroborate salt of amino ether **1** (735 mg; 3 mmol), benzaldehyde (350 mg; 3.3 mmol) in PhH (5 mL) was heated to reflux for 5 h. The cooled mixture was treated with 1N NaOH (3 mL), extracted with $Et_2O$, the organic layer was dried ($MgSO_4$) and the solvent evaporated. Bulb-to-bulb distillation (oven temperature 95°C, 0.01 mm) afforded 599 mg of **2** (87%).

**(2S,3aS,7aR)-Octahydro-1,2-dimethyl-3a-phenyl-4H-indol-4-one (4).**[7] A solution of oxazolidine **3** (63 mg; 0.26 mmol), (±)-10-camphorsulfonic acid (CSA) (54 mg; 0.23 mmol), and MeCN (7.4 mL) was maintained at 60°C for 24 h. After cooling to 20°C, $CH_2Cl_2$ and 1M NaOH (20 mL each) were added and the layers separated. The aqueous layer was extracted with $CH_2Cl_2$ (3x20 mL). The organic layers were dried, the solvent evaporated and the residue chromatographed (hexane:EtOAc:$Et_3N$ 9:1:0.1) to give 58 mg of **4** (92%) as a colorless oil.

## P A A L – K N O R R  Pyrrole Synthesis

Pyrrole synthesis from 1,4-butanediones and amines (see 1st edition)

| 1 | Paal, C. | *Chem.Ber.* | **1885** | *18* | 367 |
|---|----------|-------------|----------|------|-----|
| 2 | Knorr, L. | *Chem.Ber.* | **1885** | *18* | 299 |
| 3 | Buu-Hoi, Ng.P. | *J.Org.Chem.* | **1955** | *20* | 639 |
| 4 | Wasserman, H.H. | *Tetrahedron* | **1976** | *32* | 1863 |
| 5 | Gossauer, A. | *Synthesis* | **1996** | | 1336 |
| 6 | Ogura, K. | *Tetrahedron Lett.* | **1999** | *40* | 8887 |

## P A D W A  Pyrroline Synthesis

Pyrrolines and pyrroles by (4+1) annulation of 2,3-bis(phenylsulfonyl)-1,3-butadiene and amines (see 1 st edition).

| 1 | Padwa, A. | *Tetrahedron Lett.* | **1988** | | 2417 |
|---|-----------|---------------------|----------|------|------|
| 2 | Padwa, A. | *Tetrahedron Lett.* | **1989** | | 3259 |
| 3 | Padwa, A. | *J.Org.Chem.* | **1989** | *54* | 810, 2862 |
| 4 | Padwa, A. | *J.Org.Chem.* | **1990** | *55* | 4801 |
| 5 | Padwa, A. | *Org.Prep.Proc.Int* | **1991** | *23* | 545 |

## P A R H A M Cyclization

Benzoheterocycle synthesis by lithiation (see 1st edition)

Tol: $pCH_3C_6H_4$

| 1 | Parham, W.E. | *J.Org.Chem.* | **1975** | *40* | 2394 |
|---|---|---|---|---|---|
| 2 | Parham, W.E. | *J.Org.Chem.* | **1976** | *41* | 1184 |
| 3 | Brewer, P.D. | *Tetrahedron Lett.* | **1977** | | 4573 |
| 4 | Bradsher, C.K. | *J.Org.Chem.* | **1978** | *43* | 3800 |
| 5 | Bradsher, C.K. | *J.Org.Chem.* | **1981** | *46* | 1384, 4600 |
| 6 | Sudani, M. | *Tetrahedron Lett.* | **1981** | *22* | 4253 |
| 7 | Bracher, F. | *Synlett* | **1991** | | 95 |
| 8 | Larsen, S.D. | *Synlett* | **1997** | | 1013 |
| 9 | Parham, W.E. | *Acc.Chem.Res.* | **1982** | *15* | 300 |

## P A R N E S Geminal dimethylation

Gem dimethylation of cyclohexane derivatives from vicinal dihalocyclohexanes or methylcyclohexane with tetramethylsilane (TMS) and $AlX_3$ (see 1st edition)

| 1 | Parnes, Z.N. | *Chem.Commun.* | **1980** | *16* | 748 |
|---|---|---|---|---|---|
| 2 | Parnes, Z.N. | *Zh.Org.Khim.* | **1981** | *17* | 1357 |
| 3 | Parnes, Z.N. | *J.Org.Chem.USSR(Engl.)* | **1988** | *24* | 291 |
| 4 | Parnes, Z.N. | *Dokl.Akad.Nauk.SSSR* | **1991** | *317* | 405 |

## P A S S E R I N I Condensation

Synthesis of α-hydroxycarboxamides by acid catalyzed reaction of an isocyanide with an aldehyde or ketone (see 1st edition)

| 1 | Passerini, M. | *Gazz.Chim.Ital.* | **1921** | *51* | 126 |
|---|---------------|-------------------|----------|------|------|
| 2 | Passerini, M. | *Gazz.Chim.Ital.* | **1945** | *55* | 726 |
| 3 | Baecker, J. | *J.Am.Chem.Soc.* | **1948** | *70* | 3712 |
| 4 | Uggi, J. | *Angew.Chem.* | **1962** | *74* | 9 |
| 5 | Eckert, H. | *Synthesis* | **1977** | | 332 |
| 6 | Kaiser, C. | *J.Med.Chem.* | **1977** | *20* | 1258 |
| 7 | Lumna, W.C. | *J.Org.Chem.* | **1981** | *46* | 3668 |

## P A S T O – M A T T E S O N Rearrangement

Rearrangement of α-bromoorganoboranes by C migration

| 1 | Pasto, D.J. | *J.Am.Chem.Soc.* | **1962** | *84* | 4991 |
|---|-------------|------------------|----------|------|------|
| 2 | Matteson, D.S. | *J.Am.Chem.Soc.* | **1963** | *85* | 2595 |
| 3 | Pasto, D.J. | *J.Am.Chem.Soc.* | **1963** | *85* | 2118 |
| 4 | Mikhailov, B.M. | *J.Organomett.Chem.* | **1982** | *226* | 115 |

**Borane 3.**[4] **1** (8.0 g, 45 mmol) in PhH (35 mL) was treated with $Br_2$ (8.0 g, 50 mmol) in PhH (10 mL) for 30 min at 3-5°C, then stirred for 15 min under vacuum (100 torr) and a slow stream of Ar. Evaporation and distillation of the residue afforded 10.1 g of **3** (87%), bp 86-87°C/2 torr.

## PATERNO–BÜCHI  2+2 Cycloaddition

Photochemical 2+2 cycloaddition of carbonyls and olefins to oxetanes (see 1st edition).

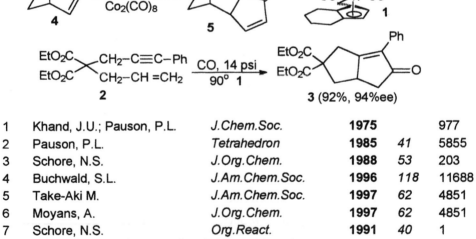

| 1 | Paterno, E. | *Gazz.Chim.Ital.* | **1909** | *39* | 341 |
|---|---|---|---|---|---|
| 2 | Büchi, G. | *J.Am.Chem.Soc.* | **1954** | *76* | 4327 |
| 3 | Lange, G.C. | *Tetrahedron lett.* | **1971** | *12* | 715 |
| 4 | Carless, H.A.J. | *Tetrahedron lett.* | **1987** | *28* | 5933 |
| 5 | Bach, Th. | *Chem.Ber.* | **1995** | *126* | 2457 |
| 6 | Neckers, D.C. | *J.Org.Chem.* | **1997** | *62* | 564 |

## PAUSON–KHAND  Cyclopentenone Annulation

Cyclopentenone synthesis from carbon monoxide, an acetylene and an olefin catalyzed by cobalt carbonyl or $Cp_2Ti(CO)_2$ (see 1 st edition).

| 1 | Khand, J.U.; Pauson, P.L. | *J.Chem.Soc.* | **1975** |  | 977 |
|---|---|---|---|---|---|
| 2 | Pauson, P.L. | *Tetrahedron* | **1985** | *41* | 5855 |
| 3 | Schore, N.S. | *J.Org.Chem.* | **1988** | *53* | 203 |
| 4 | Buchwald, S.L. | *J.Am.Chem.Soc.* | **1996** | *118* | 11688 |
| 5 | Take-Aki M. | *J.Am.Chem.Soc.* | **1997** | *62* | 4851 |
| 6 | Moyans, A. | *J.Org.Chem.* | **1997** | *62* | 4851 |
| 7 | Schore, N.S. | *Org.React.* | **1991** | *40* | 1 |

**Bicyclic cyclopentenone 3.[4]** Under Ar a Schlenk flask was charged with (s,s)-(EBTHI)TiMe$_2$ 1 (8 mg, 0.025 mmol), PhMe (2 mL) and 1,6-enyne 2 (157 mg, 0.5 mmol). Under CO pressure of 14 psi, the mixture was heated to 90°C for 12-16 h. After releasing CO, the reaction mixture was filtered (silica gel, Et$_2$O). Flash chromatography afforded 146 mg of 3 (92%), 94%ee.

## P A Y N E Rearrangement

Stereoselective base catalyzed rearrangement of epoxy alcohols; also of aziridinyl alcohols (see 1st edition)

| 1 | Payne, G.B. | *J.Org.Chem.* | **1962** | 27 | 3818 |
|---|---|---|---|---|---|
| 2 | Swindell, C.S. | *J.Org.Chem.* | **1990** | 55 | 3 |
| 3 | Bullman Page, P.C. | *J.Chem.Soc.Perkin 1* | **1990** | | 1375 |
| 4 | Fujii, N.; Ibuca, T. | *J.Org.Chem.* | **1995** | 60 | 2045 |
| 5 | Ibuka, T. | *Chem.Soc.Rev.* | **1998** | 27 | 145 |

**(2S)-2-(Aminomethyl)-N-(p-toluenesulfonyl)-1-oxaspiro** [3,6] **octane4.**[4]
N-Tosylaziridinylcyclohexanol 3 (59 mg, 0.2 mmol) was treated with 0.36 N NaOH (2.8 mL) in t-BuOH-water (2:5) at 0°C for 18 h. Preparative TLC afforded 1.2 mg of 3 (2%) and 57 mg of epoxide 4 (98%) as colorless crystals from hexane:Et$_2$O, mp 80°C, $[\alpha]_D^{20}$ = -50 (c 0.50, CHCl$_3$).

## P E A R L M A N Hydrogenolysis Catalyst

A neutral and non pyrophoric Pd catalyst active in hydrogenolysis of benzyl-N or O-benzyl bonds using Pd(OH)$_2$ on carbon.

| 1 | Pearlman, W.M. | *Tetrahedron Lett* | **1967** | | 1663 |
|---|---|---|---|---|---|
| 2 | Glaudemans, C.P.J. | *J.Org.Chem.* | **1963** | 28 | 3004 |
| 3 | Hanessian, S. | *Synthesis* | **1981** | | 396 |

**α-Methyl glucoside 2.**[3] Tetrabenzyl ether 1 (406 mg, 1 mmol) in EtOH (8 mL) was treated with cyclohexene (4 mL) and Pd(OH)$_2$ catalyst (40.6 mg) prepared by heating PdCl$_2$ charcoal and LiOH [1] and stirred and refluxed for 2 h (TLC monitoring). The catalyst was filtered and the filtrate was evaporated to afford 190 mg of 2 (98%), mp 168-170°C.

## v o n  P E C H M A N Diazo-olefin Cycloaddition

A (3+2) dipolar cycloaddition usually regioselective of diazo compounds to olefine leading to pyrazolines (see 1st edition)

| 1 | Pechman, von H. | *Chem.Ber.* | **1898** | *31* | 2950 |
|---|---|---|---|---|---|
| 2 | Sheenan, V. | *J.Am.Chem.Soc.* | **1949** | *71* | 4059 |
| 3 | Matteson, D.S. | *J.Org.Chem.* | **1962** | *27* | 4293 |
| 4 | Shioiri, T. | *Tetrahedron Lett* | **1984** | *25* | 433 |
| 5 | Aoyama, T. | *Heterocycles* | **1988** | *27* | 343 |
| 6 | Huisgen, R. | *Angew.Chem.* | **1964** | *75* | 616 |

**4-Phenyl-3-trimethylsilylpyrazole 9.**[4] A solution of **8** prepared from TMSCHN$_2$ and BuLi was treated with cinnamonitrile **7** (129 mg, 1 mmol) in THF (2 mL). After 0.5 h stirring at -78 C and 1.5 h at -45 C, the reaction mixture was quenched with aq. NH$_4$Cl. Usual work up and chromatography (silica gel, CHCl$_3$:Et$_2$O 20:1) gave 190 mg of **9** (88%), mp 117-118.5 C

## v o n  P E C H M A N – D U I S B E RG Coumarin Synthesis

Synthesis of coumarins from phenols and β-oxo esters catalyzed by homogeneous acids, Lewis acids or clays (montmorillonite)

| 1 | v.Pechman, M.; Duisberg, C. | *Chem.Ber.* | **1883** | *16* | 2119 |
|---|---|---|---|---|---|
| 2 | Israelstam, J. | *J.Org.Chem.* | **1961** | *26* | 240 |
| 3 | Kaufmann, K.D. | *J.Org.Chem.* | **1967** | *32* | 504 |
| 4 | Miyano, M. | *J.Org.Chem.* | **1972** | *37* | 259 |
| 5 | Hvao Bekkum. | *Chem.Commun.* | **1995** | | 225 |
| 6 | Li, T.S. | *J.Chem.Research(S)* | **1998** | | 38 |
| 7 | Sethna, S. | *Chem.Rev.* | **1945** | *36* | 10 |
| 8 | Sethna, S. | *Org.React.* | **1953** | *7* | 2 |

## PEDERSEN Crown Ethers

Crown ether formation and its use in substitutions, oxidations (see 1st edition).

$C_6H_{13}$

**2 (62%)**

15-crown-5

**5 (80%)**

| 1 | Lutringhaus, A. | *Liebigs Ann* | **1937** | *528* | 155 |
|---|---|---|---|---|---|
| 2 | Pedersen, C.J. | *J.Am.Chem.Soc.* | **1967** | *89* | 2495 |
| 3 | Sam, D.D. | *J.Am.Chem.Soc.* | **1972** | *94* | 4024 |
| 4 | Mitsuo, O. | *J.Org.Chem.* | **1980** | *45* | 5855 |
| 5 | Manning, M. | *J.Org.Chem.* | **1981** | *46* | 1944 |
| 6 | Palomo, C. | *Synthesis* | **1986** | | 52 |
| 7 | Gokel, W.G. | *Synthesis* | **1976** | | 168 |
| 8 | Krakowiak, K.E. | *Chem.Rev.* | **1989** | *89* | 929 |

**4-Nitrobenzyl 6,6-dibromopenicillinate 5.**[6] To sodium 6,6-dibromopenicillinate **3** (11.4 g, 30 mmol) and 15-crown-5 **2**[4] (1.5 mL) in MeCN (60mL) was added 4-nitrobenzyl bromide **4** (6.05 g, 28 mmol) and stirring at 20°C was continued for 24 h. After addition of $CH_2Cl_2$ (50 mL) and washing with water (3X30 mL), the organic solution was dried and evaporated in vacuum to give **5**, recrystallized from EtOH, 11.5 g of **5** (80%), mp 122-124°C.

## P E D E R S E N Niobium Coupling Reagents

NbCl$_3$·(DME) and NbCl$_4$·(THF)$_2$ catalysts in the synthesis of Vic diamines, 2-aminoalcohols or 2,3-disubstituted-1-naphthols by coupling of imines, imines with ketones or dialdehydes acetylenes

| 1 | Pedersen, S.F. | *J.Am.Chem.Soc.* | **1987** | *109* | 3152, 6551 |
| 2 | Manzer, L.E. | *Inorg.Chem.* | **1977** | *16* | 525 |
| 3 | Pedersen, S.F. | *Organometallics* | **1990** | *9* | 1414 |

**1,2-Diamino-1,2-diphenylethane 3.**[1] To an orange solution of NbCl$_4$·(THF)$_2$ 1 (10 g, 26.4 mmol) in DME (300 mL) was added a solution of N-trimethylsilylbenzylideneimine 2 (4.68 g, 26.4 mmol) in DME. The mixture was stirred for 4 h, the color changed to green and a precipitate was formed. After removing the solvent in vacuum, the residue was stirred with 10% w/v of KOH (125 mL), the mixture was extracted with Et$_2$O, the combined extract dried (MgSO$_4$) and the solvent evaporated. The residue after recrystallization from hexane/Et$_2$O gave 1.93 g of 3 (69%), ratio dl:meso 19:1.

## P E R K I N Carboxylic Acid (Ester) Synthesis

Synthesis of cycloalkanecarboxylic acids from α,ω–dihaloalkanes and diethyl sodiummalonate (see 1st edition)

$BrCH_2-(CH_2)_3-CH_2Br$ **2** $+$ $H_2C$ $CO_2Et$ / $CO_2Et$ **1** $\xrightarrow{NaOEt}$ [cyclohexane] $CO_2Et$ / $CO_2Et$ **3 (34%)** $\xrightarrow{H^+}$ [cyclohexane]$-CO_2H$

| | | | | | |
|---|---|---|---|---|---|
| 1 | Perkin, W.H. | *Chem.Ber.* | **1883** | 16 | 1793 |
| 2 | Perkin, W.H. | *J.Chem.Soc.* | **1888** | 53 | 202 |
| 3 | Dox, A.W. | *J.Am.Chem.Soc.* | **1921** | 43 | 1366 |
| 4 | Heyningen | *J.Am.Chem.Soc.* | **1954** | 76 | 2241 |
| 5 | Rice, L.M. | *J.Org.Chem.* | **1961** | 26 | 54 |

## P E R K I N Coumarin Rearrangement

Rearrangement of coumarins to benzofurans (see 1st edition)

$\xrightarrow[95^\circ C, \ 1 \ h]{5\% \ KOH}$

**1** **2 (78%)**

| | | | | | |
|---|---|---|---|---|---|
| 1 | Perkin, W.H. | J.Chem.Soc. | **1867** | 20 | 568 |
| 2 | Perkin, W.H. | J.Chem.Soc. | **1871** | 24 | 37 |
| 3 | Holton, G.W. | J.Chem.Soc. | **1949** | | 2049 |
| 4 | Johnson, I.R. | Org.React. | **1942** | 1 | 210 |
| 5 | | Org.Synth.Coll.Vol. | **III.-165** | | 209 |

**6-Hydroxybenzofuran-2,3-dicarboxylic acid 2.**[3] A solution of coumarin **1** (1.5 g, 6.25 mmol) in 5% KOH was heated on a water bath for 1 h. After cooling the mixture was acidified with 32% HCl and the product filtered off, to afford 1.1 g of **2** (78%), mp 227°C(dec), from dil.HCl or EtOAc petroleum ether.

## P E R K O W  Vinyl Phosphate Synthesis

Reaction of α-haloketones with trialkylphosphite to give ketophosphonate or vinylphosphate (see 1st edition)

| 1 | Perkow, W. | *Naturwissenschaften* | **1952** | *39* | 353 |
|---|---|---|---|---|---|
| 2 | Perkow, W. | *Chem.Ber.* | **1954** | *87* | 755 |
| 3 | Borowitz, I.J. | *J.Org.Chem.* | **1971** | *36* | 3282 |
| 4 | Hennig, M.L. | *J.Org.Chem.* | **1973** | *38* | 3434 |
| 5 | Mitsonobu, S. | *J.Org.Chem.* | **1981** | *46* | 4030 |
| 6 | Lichtenthaler, F.W. | *Chem.Rev.* | **1961** | *61* | 607 |

# PETERSON Olefination

Synthesis of alkenes from α–silyl carbanions and carbonyl compounds. In cases where separation of β–silyl alcohol diastereomers (e.g. **6**) can be achieved, pure Z or E olefins can be isolated (see 1st edition).

**5** $(46\%)^4$ E:Z 1:1

98%

**7** (86%, 92% Z)$^7$

| | | | | | |
|---|---|---|---|---|---|
| 1 | Peterson, D. J. | *J. Organomet. Chem.* | **1968** | *33* | 780 |
| 2 | Peterson, D. J. | *J. Org. Chem.* | **1967** | *32* | 1717 |
| 3 | Peterson, D. J. | *J. Am. Chem. Soc.* | **1975** | *97* | 1464 |
| 4 | Chan, T. H. | *J. Org. Chem.* | **1974** | *39* | 3264 |
| 5 | Mikami, K. | *Tetrahedron Lett.* | **1986** | *27* | 4198 |
| 6 | Taylor R. T. | *Synthesis* | **1982** | | 672 |
| 7 | Emslie, N. D. | *Tetrahedron* | **1998** | *54* | 3255 |
| 8 | Pulido, F. J. | *Synthesis* | **2000** | | 1223 |
| 9 | Ager, D. J. | *Org. React.* | **1990** | *38* | 1 |

**1-Phenylheptene (5).$^4$** To stirred n-BuLi in Et$_2$O (2.2 mL, 5 mmol), was added dropwise triphenylvinylsilane **2** (1.43 g, 5 mmol) in Et$_2$O (50 mL). Benzaldehyde **4** (530 mg, 5 mmol) was added over 5 min, the mixture was refluxed for 3 h and then poured into 10% NH$_4$Cl (50 mL). Extraction (Et$_2$O) evaporation of the solvent and distillation afforded 400 mg of **5** (46%), bp 46 °C/0.01 mm, mixture of Z:E=1:1.

**(Z)-1-Phenylprop-1-ene(7).$^7$** KH (103 mg, of a 50% slurry in oil 1.25 mmol) was stirred with hexane (3×4 mL) and the supernatant layer was removed with a-syringe. To the residue was added THF (5 mL) and a solution of β–hydroxy-silane **6** (141.6 mg, 0.4 mmol) in THF (2 mL). After 2 h stirring at 20 °C the mixture was added to 10% NH$_4$Cl and Et$_2$O. Work up gave 40.5 mg of **7** (86%, Z:E=92:8).

## P F A U – P L A T T N E R Cyclopropane Synthesis

Diazoalkane insertion into olefins with formation of cyclopropanes or ring enlargement of aromatics to cycloheptatrienes; see also formation of pyrazolines (von Pechman) (see 1st edition)

| 1 | Pfau, A.S.; Plattner P.A. | *Helv.Chim.Acta* | **1939** | *22* | 202 |
|---|---|---|---|---|---|
| 2 | Pfau, A.S.; Plattner P.A. | *Helv.Chim.Acta* | **1942** | *25* | 590 |
| 3 | Huisgen, R. | *Angew.Chem.* | **1964** | *75* | 616 |
| 4 | Seyferth, D. | *J.Organomet.Chem.* | **1972** | *44* | 279 |
| 5 | Gordon, M. | *Chem.Rev.* | **1952** | *50* | 141 |
| 6 | Hafner, K. | *Angew.Chem.* | **1958** | *70* | 419 |

**Anti and syn 7-trimethylsilylnorcarane 4.[4]** To CuCl (500 mg, 5.05 mmol) in cyclohexene **3** (3.82 mL) under $N_2$ was added a benzene solution of trimethylsilyldiazomethane (6.12 mmol) under stirring and occasional cooling in an ice bath. After 1 h stirring vacuum distillation and chromatography afforded anti **4** (65%) and syn **4** (7%).

## P F I T Z I N G E R Quinolin Synthesis

Quinoline-4-carboxylic acids from isatin and α–methylene carbonyl compounds (see 1st edition)

| 1 | Pfitzinger, W. | *J.Prakt.Chem.* | **1886** | *33* | 100(2) |
|---|---|---|---|---|---|
| 2 | Pfitzinger, W. | *J.Prakt.Chem.* | **1888** | *38* | 582(2) |
| 3 | Borsche, D. | *Liebigs Ann.* | **1910** | *377* | 70 |
| 4 | Henze, H.R. | *J.Am.Chem.Soc.* | **1948** | *70* | 2622 |
| 5 | Buu Hoi, N.P. | *J.Chem.Soc.* | **1949** | | 2882 |
| 6 | Buu Hoi, N.P. | *Bull.Soc.Chim.Fr.* | **1966** | | 2765 |

## P I C T E T – H U B E R T – G A M S  Isoquinoline Synthesis

Isoquinolines from phenethylamides, phenanthridine from o-acylamino biaryls with
$POCl_3$ or $POCl_3$-$SnCl_4$ (see 1st edition).

| 1 | Pictet, A.; Hubert, A. | *Chem.Ber.* | **1896** | 29 | 1182 |
|---|---|---|---|---|---|
| 2 | Pictet, A.; Gams, A. | *Chem.Ber.* | **1909** | 42 | 2943 |
| 3 | Falk, J.R. | *J.Org.Chem.* | **1981** | 46 | 3742 |
| 4 | Boyer, J.H. | *Synthesis* | **1978** | | 205 |
| 5 | Whaley, M.W. | *Org.React.* | **1951** | 6 | 151 |

## P I C T E T – S P E N G L E R  Isoquinoline Synthesis

Isoquinoline synthesis of phenethylamine and pyruvic acid derivatives (see 1st edition).

| 1 | Pictet, A.; Spengler, F. | *Chem.Ber.* | **1911** | 44 | 2030 |
|---|---|---|---|---|---|
| 2 | Valentine, D. | *Synthesis* | **1978** | | 329 |
| 3 | Hudlicki, T. | *J.Org.Chem.* | **1981** | 46 | 1738 |
| 4 | Bates, H.A. | *J.Org.Chem.* | **1986** | 51 | 3061 |
| 5 | Goel, Q.P. | *Synth.Commun.* | **1995** | 25 | 49 |
| 6 | Nakagawa, M. | *Synlett.* | **1997** | | 761 |
| 7 | Govindachari, T.R. | *Org.React.* | **1951** | 6 | 151 |

**5.**[6] A solution of $(PhO)_3B$ (232.4 mg, 0.8 mmol) in $CH_2Cl_2$ (5 mL) was added to a stirred
suspension of (S)-2,2'-dihydroxy-1,1'-binaphthyl (458 mg, 1.6 mmol) and powdered
molecular sieves (2 g) at 20°C under Ar. A solution of **4** (104.3 mg, 0.39 mmol) in
$CH_2Cl_2$ (10 mL) was added and after 48 h stirring at 20°C, the mixture was filtered
(Celite), washed ($NaHCO_3$) and evaporated. Chromatography (silica gel, EtOAc:
hexane 1:6   1:10) gave 85.7 mg of **5** (82%), 78%ee.

## PINNER Imino Ether Synthesis

Synthesis of imino ethers, amidines and orto esters from nitriles (see 1st edition)

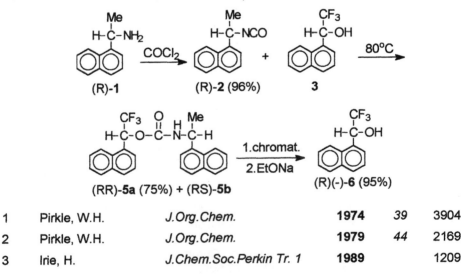

| 1 | Pinner, A. | *Chem.Ber.* | **1877** | *10* | 1889 |
| 2 | Rapoport, H. | *J.Org.Chem.* | **1981** | *46* | 2465 |
| 3 | Cheng,C.C. | *Org.Prep.Proc.Int.* | **1990** | *22* | 643 |
| 4 | Roger, R. | *Chem.Rev.* | **1961** | *61* | 179 |

**2,4-Dimethoxybenzamidinium HCl 6.³ 4** (22 g, 0.44 mol) in CHCl₃ (400 mL) was treated with saturated HCl in MeOH (10 g, 0.31 mol). After 4 h at 0°C the mixture was heated slowly to 20°C and maintained for 24 h. After evaporation in vacuum, the residue was treated with a MeOH solution saturated with NH₃ (800 mL) and maintained for 4 days at 20°C. Evaporation to dryness, extraction with Et₂O (5X100 mL) and recrystallization from n-BuOH gave 18.4 g of 6 (63%), mp 238-238°C.

## PIRKLE Resolution

1-(1-Naphthyl)ethyl isocyanate **2** for chromatographic resolution of alcohols, hydroxy esters thiols via diastereomeric derivatives (see 1st edition)

| 1 | Pirkle, W.H. | *J.Org.Chem.* | **1974** | *39* | 3904 |
| 2 | Pirkle, W.H. | *J.Org.Chem.* | **1979** | *44* | 2169 |
| 3 | Irie, H. | *J.Chem.Soc.Perkin Tr. 1* | **1989** | | 1209 |

## POLONOVSKY N-Oxide Rearrangement

Rearrangement of heterocyclic N-oxide to α–acetoxyheterocycles and elimination or fragmentation of acylated N-oxide (see 1st edition).

| 1 | Polonosvsky, M. & M. | *Bull.Soc.Chim.Fr.* | **1927** | *41* | 1190 |
|---|---|---|---|---|---|
| 2 | Bell, C.C. | *J.Org.Chem.* | **1962** | *27* | 1601 |
| 3 | Huisgen, R. | *Chem.Ber.* | **1959** | *92* | 3223 |
| 4 | Sternbach, L.M. | *J.Org.Chem.* | **1965** | *30* | 3576 |
| 5 | Ahond, A. | *Bull.Soc.Chim.Fr.* | **1970** | | 2707 |
| 6 | Lalonde, R.T. | *J.Am.Chem.Soc.* | **1971** | *93* | 2501 |
| 7 | Lewin, G. | *Tetrahedron* | **1990** | *46* | 7775 |
| 8 | Kende, A.S. | *J.Am.Chem.Soc.* | **1995** | *117* | 10597 |
| 9 | Volz, H. | *Kontakte (Darmstadt)* | **1984** | *3* | 14 |

# POMERANZ–FRITSCH–SCHLITTER–MULLER

## Isoquinoline Synthesis

Isoquinoline synthesis from aromatic aldehydes or benzyl halides and aminoacetal (Pomeranz-Fritsch) or from benzylamines and glyoxal acetal (Schlitter-Muller) (see 1st edition).

$$(CH_3O)_2CH\text{-}CH_2NH_2 \;+\; TsCl \longrightarrow (CH_3O)_2CH\text{-}CH_2NHTs$$
$$\textbf{1} \hspace{6cm} \textbf{2}$$

**3**                           **4 (93%)**                           **5 (81%)**

| | | | | | |
|---|---|---|---|---|---|
| 1 | Pomeranz, C. | *Monatsh.* | **1893** | *14* | 116 |
| 2 | Fritsch, P. | *Chem. Ber.* | **1893** | *26* | 419 |
| 3 | Schlitter, E.; Muller, I. | *Helv. Chim. Acta* | **1948** | *31* | 914; 1119 |
| 4 | White, J.D. | *J. Org. Chem.* | **1967** | *32* | 2689 |
| 5 | Jackson, A.H. | *J. Chem. Soc. Perkin I* | **1974** | | 3185 |
| 6 | Boger, D.A. | *Tetrahedron* | **1981** | *37* | 3977 |
| 7 | Popp, F.D. | *Chem. Rev.* | **1958** | *58* | 328 |
| 8 | Gensler, W.J. | *Org. React.* | **1951** | *6* | 192 |

**6,7-Methylenedioxyisoquinoline 5.**[6] To a supension of NaH (0.5 g; 12.5 mmol) in THF was added **2** (3.24 g; 12.5 mmol) in THF (20 mL). After evolution of $H_2$ ceased, piperonyl bromide **3** (2.56 g; 11.9 mmol) in THF (20 mL) was added under stirring. After 2 h at 20°C, usual work up and chromatography afforded 4.33 g of **4** (93%).
A solution of **4** (2.0 g; 5.1 mmol) in dioxane (48 mL) and 6N HCl (3.7 mL) was refluxed for 24 h. Work up and chromatography ($Et_2O$) afforded 720 mg of **5** (81%).

## **P O S N E R** Trioxane Synthesis

Reaction of triethylsilyl hydrotrioxide with electron-rich olefins to give dioxetanes that react intramolecularly with a keto group in the presence of t-butyldimethyl silyl triflateto afford 1,2,4-trioxanes; also oxydative cleavage of alkenes. Also used in cleavage of olefins (see 1st edition)

| 1 | Corey, E.J. | J.Am.Chem.Soc. | **1986** | *108* | 2472 |
|---|---|---|---|---|---|
| 2 | Posner, G.H. | J.Am.Chem.Soc. | **1987** | *109* | 278 |
| 3 | Posner, G.H. | J.Org.Chem. | **1989** | *54* | 3252 |
| 4 | Posner, G.H. | Tetrahedron Lett. | **1991** | *32* | 4235 |

## P R E V O S T – W O O D W A R D   Olefin Hydroxylation

Difunctionalization of alkenes with iodine and silver (or sodium) carboxylates (see 1st edition).

| 1 | Prevost, C. | C.R. | **1933** | *196* | 1129 |
| 2 | Smissman, E.E. | J. Org. Chem. | **1970** | *35* | 3532 |
| 3 | Johnson, R.G. | Chem. Rev. | **1956** | *56* | 261 |
| 4 | Wilson, C.V. | Org. React. | **1957** | *9* | 350 |
| 5 | Woodward, R.B. | J. Am. Chem. Soc. | **1958** | *80* | 209 |
| 6 | Lwowski, W. | Angew. Chem. | **1958** | *70* | 490 |
| 7 | Brimble, M.A | J. Org. Chem. | **1996** | *61* | 4801 |
| 8 | Welzel, P. | Tetrahedron | **2000** | *56* | 1345 |

**4(a)-Isopropyl-6-benzoyloxymethyl-5H,6H-furo[2,3-d]-Δ$^{1,7}$-2,4-(3H)pyrimidinedione (2).**[2] A supension of silver benzoate (11.50 g; 50 mmol) in PhH (200 mL) was treated with I$_2$ (5.35 g; 25 mmol) in PhH (100 mL). After 15 min stirring, 5-isopropyl-5-allylbarbituric acid **1** (5.25 g; 25 mmol) in hot PhH (100 mL) was added and the mixture was refluxed for 2 h. Cooling, filtration, concentration in vacuum and chromatography of the residue on silica gel (CHCl$_3$) afforded 2.28 g of **2** (28%), mp 170-172°C (Me$_2$CO-petroleum ether).

**Iodoacetate 6b.** To a stirred suspension of **5**[7] (94 mg; 0.56 mmol), AgAcO (280 mg; 1.68 mmol) and H$_2$O (0.11 mL; 6 mmol) in AcOH (10 mL) was added iodine (171 mg; 0.76 mmol) portion wise over 10 min. After 18 h stirring, the insoluble matter was removed by filtration, the filtrate diluted with Et$_2$O and washed (water, aq. NaHCO$_3$). Evaporation of the solvent and chromatography of the residue (hexane:EtOAc 95:5 → 90:10) give 32 mg of **6a** (16%), 85 mg of **6b** (43%) and 24 mg of **6c** (12%).

## PRINS – KRIEWITZ Hydroxymethylation

Acid catalyzed hydroxymethylation of alkenes (see 1st edition).

| 1 | Kriewitz, O. | *Chem.Ber.* | **1899** | *32* | 57 |
|---|---|---|---|---|---|
| 2 | Prins, H. | *Chem.Weeckblad* | **1919** | *16* | 1072 |
| 3 | Dolby, L.J. | *J.Org.Chem.* | **1963** | *28* | 1456 |
| 4 | Adam, D.R. | *Synthesis* | **1977** | | 661 |
| 5 | Andersen, N.A. | *J.Org.Chem.* | **1985** | *50* | 4144 |
| 6 | Rychnovsky, S.D. | *Tetrahedron Lett.* | **1996** | *37* | 8679 |
| 7 | Arundale, R. | *Chem.Rev.* | **1952** | *51* | 505 |
| 8 | Willis, C.L. | *Chem. Commun.* | **2001** | | 832 |

**Tetrahydropyran 6.[6]** 4-Allyl-1,3-dioxane **5** (200 mg, 0.88 mmol), AcOTMS (265 L, 1.76 mmol) and AcOH (506 L, 880 mmol) in cyclohexane (13 mL) under $N_2$ at 20°C was treated dropwise with $BF_3 \cdot Et_2O$ (435L, 3.53 mmol). After 4 h, the mixture was quenched with aq. $NaHCO_3$ extracted with $CH_2Cl_2$. The solvent evaporated and the residue treated with $Ac_2O$ and $Et_3N$ (and a catalytic amount of DMAP) in $CH_2Cl_2$. Aqueous work up and chromatography gave 245 mg of **6** (84%), 94:4 diastereoisomers.

## P S C H O R R Arylation

Formation of polycyclics from a diazonium salt. Intramolecular Cu catalyzed arylation of diazonium salts (see Gomberg-Bachmann) (see 1st edition).

| 1 | Pschorr, R. | *Chem.Ber.* | **1900** | *33* | 1810 |
|---|---|---|---|---|---|
| 2 | Hey, D.H. | *J.Chem.Soc.* | **1949** | | 3162 |
| 3 | Kupchan, S.N. | *J.Org.Chem.* | **1973** | *56* | 405 |
| 4 | Gokel, G.W. | *Tetrahedron Lett* | **1980** | *21* | 4141 |
| 5 | Le Maire, J. | *Tetrahedron* | **1996** | *52* | 3953 |
| 6 | Leake, P.N. | *Chem.Rev.* | **1956** | *56* | 27 |
| 7 | De Tar, D.L.F. | *Org.React.* | **1957** | *9* | 409 |

**Phenanthren-9-carboxylic acid 2.**[3] A solution of 1 (1.45 g, 6 mmol) in HCl (3.3 mL) and water (100 mL) was diazotized with $NaNO_2$ (0.7 g, 10 mmol) in water (40 mL). To the diazonium salt was added copper bronze (1 g), the mixture was heated on water bath to complete the reaction. Usual work up and crystallization (AcOH) gave 0.5 g of 2 (40%), mp 250-252°C.

**Dimethyldibenzothiophene 5.**[5] 3 (12.3 g, 54 mmol) in 30% $H_2SO_4$ (175 mL) was treated with 40% $NaNO_2$ (75 mmol), the with a solution of $NaBF_4$ (106 mmol). The precipitate after filtration and drying was used in the next step. 4 (17.6 g) was added to a suspension of copper (10.5 g) in DMSO (800 mL). After 2h stirring water (2000 mL) was added, the precipitate filtered and chromatographed (silica gel, cyclohexane). There were obtained 4.5 g of crude 5. Recrystallization (cyclohexane:iPrOH 25:80) afforded 3 g of pure 5 (26%), mp 152°C.

## **P U D O V I K** Reaction

Base catalyzed synthesis of α-hydroxyphosphonates from aromatic aldehydes and diethyl phosphite

| 1 | Pudovik, A.N. | *Synthesis* | **1979** | | 79 |
|---|---|---|---|---|---|
| 2 | Sum, V. | *J.Chem.Soc.Perkin 1* | **1993** | | 2071 |
| 3 | Shibuya, S. | *Tetrahedron Asymm.* | **1995** | *4* | 1779 |
| 4 | Sasai, H. | *J.Org.Chem.* | **1996** | *61* | 2926 |
| 5 | Spiling, C.D. | *J.Org.Chem.* | **1995** | *60* | 931 |
| 6 | Shibuya, S. | *J.Chem.Soc.Perkin 1* | **1997** | | 1527 |

**(S)-Diethyl hydroxy (4-methylphenyl) methylphosphonate 3.**[6] A stock solution of LLB (La-Li-(S)-BINOL) (100 mL) was prepared from $LaCl_3 \cdot 7H_2O$ (1.85 g, 5 mmol), dilithium (S)-binaphthoxide (5 mmol), NaOBu (496 mg, 5 mmol) and water ($3.6 \times 10^{-1}$ mL, 20 mmol) (see Shibasaki *Tetrahedron Lett*, **1993**, *34*, 855). To a stirred solution of p-tolualdehyde **1** (240 mg, 2 mmol) and diethyl phosphite **2** (331 mg, 2.4 mmol) in THF (4.5 mL) was added the THF solution of LLB (8mL) over 5 min at -40°C. After being stirred for 15 h the reaction mixture was quenched with 1 M HCl and extracted with $Et_2O$. Flash chromatography ($SiO_2$, hexane:EtOAc 1:1 to 1:20) afforded 513 mg of **3** (95%), 82%ee.

## **P U M M E R E R** Sulfoxide Rearrangement

Rearrangement of a sulfoxide to an α-acetoxysulfide (see 1st edition)

| 1 | Pummerer, R. | *Chem, Ber.* | **1991** | *43* | 1401 |
|---|---|---|---|---|---|
| 2 | Johnson, C.R. | *J.Am.Chem.Soc.* | **1969** | *91* | 682 |
| 3 | Dinizo, St.D. | *Synthesis* | **1977** | | 181 |
| 4 | Ishibashi, H. | *J.Chem.Res.* | **1987** | | 296 |
| 5 | Sugehara, T. | *Synthesis* | **1978** | | 881 |
| 6 | Takahashi, T. | *J.Chem.Soc.Perkin 1* | **1991** | | 1667 |
| 7 | Padwa, A. | *J.Org.Chem.* | **1997** | *62* | 774 |
| 8 | De Lucchi, O. | *Org.React.* | **1991** | *40* | 157 |

## R A M B E R G – B A C K L U N D - P A Q U E T T E  Olefin synthesis

Conversion of dialkyl sulfones to alkenes either by rearrangement of α-halosulfones with base (via SO$_2$ elimination from thiaranedioxides) (Ramberg-Backlund) or by desulfonation of sulfones with BuLi-LAH(Paquette) (see 1st edition).

| 1 | Ramberg, L.; Backlund, B. | *Ark. Kem. Mineral Geol.* | **1940** | *13A* | 50 |
|---|---|---|---|---|---|
| 2 | Paquette, L. | *J. Org. Chem.* | **1981** | *46* | 4021 |
| 3 | Opitz, G. | *Angew. Chem.* | **1963** | *77* | 411 |
| 4 | Paquette, L. | *Org. React.* | **1977** | *25* | 1 |
| 5 | Taylor, R. J. | *J. Chem. Soc. Perkin 1* | **1993** | | 2317 |
| 6 | Taylor,.J.K. K. | *Tetrahedron Lett.* | **2001** | *42* | 1197 |

**1,4,9,10-Tetrahydro-5,6-benzo-4a, 10a-ethenophenantrene (3).[3]** To 1,4,9,10 - Tetrahydro -5,6-benzo-4a, 10a-mathanothiomethanophenanthrene **1** (14.6 g, 50 mmol) were added N-chlorosuccinimide (NCS) (6.72 g, 50.5 mmol) and dry CCl$_4$. The mixture was refluxed under N$_2$ for 29 h, cooled, filtered, and evaporated to give a mixture of isomeric α–chlorosulfides. To this product in CHCl$_3$ (200 mL) at –23 °C was added dropwise 0.624 N ethereal monoperphthalic acid (163 mL). After 10 h at 20 °C, work up gave an isomeric mixture of α–chloro sulfone **2**.
Sulfone **2** dissolved in dioxane (250 mL) was treated with t-BuOK (35.1 g, 0.313 mol) under N$_2$ at 0 °C, then heated to reflux for 20 h. Dilution with water and chromatography on silca gel (hexane) gave 5.13 g of **3** (40%) as a yellow oil.

**4,4-Dibutoxycyclopentene (5).[5]** 4, 4-Dibutoxy-2-iodothiane-1,1-dioxide **4** (800 mg, 1.98 mmol) in THF (20 mL) at 20 °C under N$_2$ was treated with K-OtBu (0.67 g, 5.97 mmol) in THF (5 mL). Standard work up and distillation (kugelrohr) gave 390 mg, of **5** (93%), bp 80 °C/0.5 mm.

## RAPP–STOERMER Benzofuran Synthesis

Benzofuran synthesis from salicylaldehydes and $\alpha$-haloketones (see 1st edition).

| | | | | | |
|---|---|---|---|---|---|
| 1 | Rapp, E. | *Gazz. Chim. Ital.* | **1895** | *25II* | 285 |
| 2 | Stoermer, R. | *Liebigs Ann.* | **1900** | *312* | 331 |
| 3 | Buu Hoi, Ng. P. | *J. Chem. Soc.* | **1957** | | 2593 |

**Benzofuran (4).** [3] A mixture of **1** (15.0 g, 0.096 mol), **2** (22.0 g, 0.096 mol) and KOH (5.3 g, 0.096 mol) in EtOH (150 mL) was heated to reflux to give 10 g of crude **3** (35%). This was heated in pyridine hydrochloride for 30 min, cooled stirred with water, filtered, purified via its sodium salt and recrystallized from aq. EtOH to afford 6 g of **4** (67%), mp 238 °C.

## RATHKE $\beta$-Keto Ester Synthesis

Synthesis of $\beta$–keto esters by condensation of acyl chlorides with malonates in the presence of $MgCl_2$ and $Et_3N$.

| | | | | | |
|---|---|---|---|---|---|
| 1 | Rathke, M. W. | *J. Org. Chem.* | **1985** | *50* | 2622 |
| 2 | Wemple, J. | *Synthesis* | **1993** | | 290 |
| 3 | Krysan, D. J. | *Tetrahedron Lett.* | **1996** | *37* | 3303 |

**Ketoester (3).** [3] Ice cooled **1** (510 mg, 2.1 mmol) in THF (10 mL) was treated with $Et_3N$ (430 mg, 4.3 mmol) then $MgCl_2$ (230 mg, 2.3 mmol). The slurry was stirred at 0 °C for 2.5 h, then **2** (170 mg, 1 mmol) in THF (5 mL) was added. After 5 min, the mixture was stirred for 12 h at 20 °C, quenched with citric acid and extracted with EtOAc. Flash chromatography gave 280 mg of **3** (82%).

## **REETZ** Titanium Alkylation Reagent

Alkyl titanium reagents in stereoselective addition to aldehydes and dialkylamino titanium compounds as protecting groups of aldehydes in the presence of ketones (see 1st edition).

| # | | | | | |
|---|---|---|---|---|---|
| 1 | Reetz, M.T. | *Angew.Chem.Int.Ed.* | **1980** | *19* | 1011 |
| 2 | Reetz, M.T. | *Chem.Ind.* | **1981** | | 541 |
| 3 | Reetz, M.T. | *Tetrahedron Lett.* | **1981** | | 4691 |
| 4 | Reetz, M.T. | *Angew.Chem.Int.Ed.* | **1982** | *21* | 96 |
| 5 | Reetz, M.T. | *Top.Curr.Chem.* | **1982** | *106* | 1 |
| 6 | Reetz, M.T. | *J.Chem.Soc.Chem.Commun.* | **1983** | | 406 |
| 7 | Posner, G.H. | *Tetrahedron* | **1984** | *40* | 1401 |
| 8 | Reetz, M.T. | *Angew.Chem.Int.Ed.* | **1984** | *23* | 566 |
| 9 | Schollkopf, V. | *Synthesis* | **1985** | | 55 |

**Threo and erythro 2-phenylbutan-3-ol 2 and 3.**[1] A solution of **1** (2.7 g, 20 mmol) in $CH_2Cl_2$ (50 mL) cooled at -50°C was treated with $Me_2TiCl_2$ (2.98 g, 20 mmol) in $CH_2Cl_2$ (100 mL). After 1 h the mixture was poured into water, the organic phase separated and the solvent evaporated. Kugelrohr distillation afforded 2.45 g of **2** an **3** (82%) bp 80°C/1 torr.

**3-Phenyl-3-hydroxy-7-formyl-heptanoic acid ethyl ester (7).**[6] To **4** (190 mg, 1 mmol) in THF at -78°C was added $Ti(NEt_2)_4$ (336 mg, 1 mmol) under stirring. The mixture was allowed to warm to -50°C during 1 h and the Li enolate **6** (94 mg, 1 mmol) was added at the same temperature. Aqueous work up afforded 263 mg of **7** (95%).

## REFORMATSKY–BLAISE Zinc Alkylation

Synthesis of β-hydroxyesters from carbonyl derivatives and α-halo esters via a zinc reagent (Reformatsky). Synthesis of β-ketoesters from nitriles and α-halo esters via a zinc reagent (Blaise) (see 1st edition).

| | | | | | |
|---|---|---|---|---|---|
| 1 | Reformatsky, S.N. | *Chem. Ber.* | **1887** | 20 | 1210 |
| 2 | Rathke, M.W. | *Org. React.* | **1975** | 22 | 423 |
| 3 | Blaise, E. | *C.R.* | **1901** | 132 | 478 |
| 4 | Kitazume, T. | *Synthesis* | **1986** | | 855 |
| 5 | Toda, F. | *J. Org. Chem.* | **1991** | 56 | 4333 |
| 6 | Fuerster, A. | *Synthesis* | **1989** | | 571 |
| 7 | Meyers, A.I. | *J. Org. Chem.* | **1991** | 56 | 2091 |
| 8 | Ioshida, M. | *Synth. Commun.* | **1996** | 26 | 2523 |
| 9 | Lee, A.S.Y. | *Tetrahedron Lett.* | **1997** | 38 | 448 |
| 10 | Weisjohan, L. | *J. Org. Chem.* | **1997** | 62 | 3858 |
| 11 | Chattopathyay, A. | *Synthesis* | **2000** | | 561 |

**Hydroxy ester 3.**[5] Piperonal **1** (765 mg; 5.1 mmol), ethyl bromoacetate **2** (2.56 g; 15.3 mmol), zinc powder (5 g; 77 matg) and NH$_4$Cl (2 g), were thoroughly ground in an agate mortar and pestle, and the mixture was kept at 20°C for 2-3 h. After mixing with aqueous NH$_4$Cl and extraction with Et$_2$O, the organic layer was washed with water, dried (MgSO$_4$) and the volatiles evaporated to afford 1.14 g of **3** (94%).

**R E G I T Z**  Diazo transfer

Synthesis of diazo compounds from active methylenes with tosyl azide (diazo transfer) (see 1st edition).

$$(R\text{-}SO_2)_2CH_2 \xrightarrow{p\text{-}TsN_3} (R\text{-}SO_2)_2C^-\text{-}N_2^+$$

$$NC\text{-}CH_2\text{-}CN \xrightarrow{p\text{-}TsN_3} \underset{N_2^+}{NC\text{-}\overset{|}{C}\text{-}CN} \xrightarrow{Et_3N} \underset{N=N\text{-}N^+Et_3}{NC\text{-}\overset{-}{C}\text{-}CN}$$

| | | | | | |
|---|---|---|---|---|---|
| 1 | Regitz, M. | *Angew. Chem. Int. Ed.* | **1967** | *6* | 733 |
| 2 | Regitz, M. | *Chem. Ber.* | **1964** | *97* | 1482 |
| 3 | Ledon, H. | *Synthesis* | **1972** | | 351 |
| 4 | Ledon, H. | *Synthesis* | **1974** | | 347 |
| 5 | Koteswar, Rao Y. | *Indian J. Chem.* | **1986** | *25b* | 735 |

## REICH–KRIEF  Olefination

Synthesis of olefins by stereospecific reductive elimination of β-hydroxyalkyl selenides (a variant of the Peterson olefination) by means of $MeSO_2Cl$, $HClO_4$ or $P_2I_4$ (see 1st edition).

$$CH_3(CH_2)_5\text{-}\underset{\underset{OH}{|}}{CH}\text{-}\underset{\underset{SePh}{|}}{CH}\text{-}(CH_2)_5CH_3 \xrightarrow[Et_2O;\ 25°]{HClO_4} CH_3(CH_2)_5\text{-}CH=CH\text{-}(CH_2)_5CH_3$$

**3**                                                                 **4** (85%)

| | | | | | |
|---|---|---|---|---|---|
| 1 | Reich, J.R. | *J. Am. Chem. Soc.* | **1973** | 95 | 5813 |
| 2 | Krief, A. | *Angew. Chem. Int. Ed.* | **1974** | 13 | 804 |
| 3 | Reich, J.R. | *J. Chem. Soc. Chem. Commun.* | **1975** | | 790 |
| 4 | Krief, A. | *Tetrahedron Lett.* | **1976** | | 1385 |
| 5 | Reich, J.R. | *J. Am. Chem. Soc.* | **1979** | 101 | 6638 |
| 6 | Krief, A. | *Bull. Soc. Chim. Fr.* | **1990** | | 681 |
| 7 | Comins, D. | *Org. Lett.* | **1999** | | 1031 |
| 8 | Reich, J.R. | *Org. React.* | **1993** | 44 | 1 |

**1,4-Diphenyl-3-butene-1-one (2).**[1] To LDA in THF (3 mmol) under $N_2$ were added 1,4-diphenyl-1-butanone **1** (560 mg; 2.5 mmol) at –78°C. After 10 min stirring phenyl selenyl bromide (3 mmol) was added. The mixture was heated to 0°C, AcOH (0.3 mL) and $H_2O_2$ (1.4 g) were added. The temperature was raised to 25°C and gas evolved. Quenching with $NaHCO_3$, extraction, and separation of **1** by TLC followed by sublimation afforded 470 mg of **2** (85%), mp 40-41°C.

## REIMER–TIEMANN   Phenol Formylation

Synthesis of aromatic aldehydes by formylation of phenols, pyrroles with $CHCl_3$-base (dichlorocarbene) (see 1st edition).

| 1 | Reimer, K; Tiemann, F. | *Chem. Ber.* | **1876** | 9 | 1285 |
| 2 | Newmann, R. | *Synthesis* | **1986** | | 569 |
| 3 | De Angelis, F. | *J. Org. Chem.* | **1995** | 60 | 445 |
| 4 | Jung, M.E. | *J. Org. Chem.* | **1997** | 62 | 1553 |
| 5 | Wynberg, H. | *Chem. Rev.* | **1960** | 60 | 169 |
| 6 | Wynberg, H. | *Org. React.* | **1982** | 27 | 1 |

**Aldehyde 2.**[4] Powdered NaOH (1.71 g; 42.72 mmol) was added to a suspension of N-Boc-tyrosine **1** (2 g; 7.12 mmol), water (0.652 mL; 14.13 mmol) and $CHCl_3$ (30 mL). After 4 h reflux, a second portion of NaOH was added to give after usual work up 0.72 g of **2** (33%). Recovered **1** 0.62 g, (31%).

## RICHTER–WIDMAN–STOERMER   Cinnoline Synthesis

Synthesis of cinnolines from substituted anilines via diazonium salts (see 1st edition).

| 1 | Richter, v.W. | *Chem. Ber.* | **1883** | 16 | 677 |
| 2 | Scofield, K. | *J. Chem. Soc.* | **1949** | | 2393 |
| 3 | Leonard, N.I. | *Chem. Rev.* | **1945** | 37 | 270 |
| 4 | Widman, O. | *Chem. Ber.* | **1884** | 17 | 722 |
| 5 | Stoermer, R. | *Chem. Ber.* | **1909** | 42 | 3115 |
| 6 | Simpson, J.C.E. | *J. Chem. Soc.* | **1947** | | 808 |

# REISSERT–GROSHEINTZ–FISCHER Cyanoamine Reaction

Synthesis of aldehydes or alkaloids from acid chlorides via 1-cyano-2-acylisoquinoline or 2-cyanoquinoline intermediates.

| 1 | Reissert, A. | *Chem. Ber.* | **1905** | *38* | 1608; 3415 |
|---|---|---|---|---|---|
| 2 | Grosheintz, J.M.; Fischer, H.O. | *J. Am. Chem. Soc.* | **1941** | *63* | 2021 |
| 3 | Popp, F.D. | *Synthesis* | **1970** | | 591 |
| 4 | Koizumi, T. | *Synthesis* | **1977** | | 497 |
| 5 | Popp, F.D. | *Bull. Soc. Chim. Belge* | **1981** | *90* | 609 |
| 6 | McEwen, W.E. | *J. Org. Chem.* | **1981** | *46* | 2476 |
| 7 | Mosettig, E. | *Chem. Rev.* | **1955** | *55* | 511 |

**Reissert compound 3.**[4] A mixture of 4-bromoisoquinoline 1 (832 mg; 4 mmol) in CH$_2$Cl$_2$ (10 mL) and KCN (0.8 g; 12 mmol), benzyl triethylammonium chloride (4.3 g; 50 mmol) and water (10 mL) is stirred for 30 min at 20°C. Benzoyl chloride 2 (560 mg; 4 mmol) in CH$_2$Cl$_2$ (4 mL) is added over a period of 2 h under stirring. After 2 h additional stirring, quenching with water, work up and crystallization from EtOH gave 1.11 g of 3 (82%), mp 171-174°C.

## REMFRY–HULL Pyrimidine Synthesis

Synthesis of pyrimidines by condensation of malon diamides with carboxylic esters.

| 1 | Remfry, F.G.P. | *J. Chem. Soc.* | **1911** | | 610 |
|---|---|---|---|---|---|
| 2 | Hull, R. | *J. Chem. Soc.* | **1951** | | 2214 |
| 3 | Brown, D.J. | *J. Chem. Soc.* | **1964** | | 3204;1956;2312 |
| 4 | Budesinsky, Z. | *Coll. Czech. Chem. Commun.* | **1965** | *30* | 3730 |

**4,6-Dihydroxypyrimidine (3).**[2] To a solution of sodium ethoxide (from 4.6 g; 0.2 at g Na in 150 mL EtOH) was added malonamide **1** (10.2 g; 0.1 mol), followed by ethyl formate **2** (11.0 g; 0.14 mol). The mixture was refluxed for 2 h and after 24 h at 20°C the crystalline product was filtered off and washed with EtOH. The product was dissolved in water (50 mL) and acidified with 5N HCl. After filtration there are obtained 4.5 g of **3**·HCl (40%), mp > 300°C.

## REPPE Acetylene Reactions

Ni or Ti catalyzed tetramerization or trimerization of acetylene and reactions with alcohols, amines, carboxylic acids, thiols (see 1st edition).

| 1 | Reppe, W. | *Liebigs Ann.* | **1948** | 560 | 1-104 |
|---|---|---|---|---|---|
| 2 | Reppe, W. | *Experientia* | **1949** | 5 | 93-108 |
| 3 | Reppe, W. | *Liebigs Ann.* | **1953** | 582 | 1-133 |
| 4 | Reppe, W. | *Liebigs Ann.* | **1955** | 596 | 11-20 |
| 5 | Lutz, E.F. | *J. Am. Chem. Soc.* | **1961** | 83 | 2552 |
| 6 | Reppe, W. | *Angew. Chem. Int. Ed.* | **1969** | 8 | 727 |

**Cyclooctatetraene (2).**[2] A cooled (0-10°C) solution of $NiCl_2$ in EtOH was treated with 10% ethanolic HCN. After 12 h at 0°C the $Ni(CN)_2$ catalyst was filtered and washed. To $Ni(CN)_2$ (20 g) and calcium carbide (50 g) in THF (2000 mL) under $N_2$ at 5 atm acetylene was introduced at 15-20 atm and the mixture was heated to 30-60°C while acetylene was introduced from time to time. After removal of the catalyst, distillation afforded 320-400 g of **2**, bp 141-142°C.

## R I E C H E  Formylation

Ti mediated formylation or dichloromethylation of sterically hindered aromatics (compare with Reimer-Tiemann).

| 1 | Rieche, A. | *Chem. Ber.* | **1960** | *91* | 88 |
| 2 | Gross, H. | *Z. Chem.* | **1978** | *18* | 201 |
| 3 | Belen'kii, L.I. | *Tetrahedron* | **1993** | *49* | 3397 |

**2,4,6-Trimethylbenzaldehyde (3)**.[3] A solution of dichloromethyl methyl ether **2** (30 mL; 0.33 mol) and mesitylene **1** (23 mL; 0.17 mol) in $CH_2Cl_2$ (100 mL) was added at 25°C for 5 min to a solution of $TiCl_4$ (73 mL; 0.67 mol) in $CH_2Cl_2$ (150 mL). After 15 min stirring ice (500 g) was added. Extraction with $CH_2Cl_2$, washing and distillation afforded 20.6 g of **3** (84%), bp 108-111°C/10 mm.

## v o n   R I C H T E R  Aromatic Carboxylation

Reaction of *m*- and *p*-nitrohalobenzenes with CN⁻ leading to *o*- and *m*-halobenzoic acids with loss of the $NO_2$ group (see 1st edition).

| 1 | Richter, v.W. | *Chem. Ber.* | **1871** | *4* | 21 |
| 2 | Richter, v.W. | *Chem. Ber.* | **1875** | *8* | 1418 |
| 3 | Bunnett, J.E. | *J. Org. Chem.* | **1950** | *15* | 481 |
| 4 | Bunnett, J.E. | *J. Org. Chem.* | **1956** | *21* | 944 |
| 5 | Ibne Rasa, K.M. | *J. Org. Chem.* | **1963** | *28* | 3240 |
| 6 | Huisgen, R. | *Angew. Chem.* | **1960** | *72* | 314 |

### **R I L E Y**  Selenium Dioxide Oxidation

Oxidation of aldehydes or ketones to 1,2-dicarbonyl compounds with $SeO_2$ (sometimes oxidation to $\alpha,\beta$-unsaturated ketones) (see 1st edition).

| | | | | | |
|---|---|---|---|---|---|
| 1 | Riley, H.L. | *J. Chem. Soc.* | **1932** | | 1875 |
| 2 | Schaefer, J.P. | *J. Am. Chem. Soc.* | **1933** | 66 | 1668 |
| 3 | Waitkins, G.R. | *Chem. Rev.* | **1945** | 36 | 235 |
| 4 | Rabjohn, N. | *Org. React.* | **1976** | 24 | 263 |
| 5 | Sharples, K.B. | *J. Am. Chem. Soc.* | **1976** | 98 | 300 |

### **R I T T E R**  Amidation

Acid catalyzed reaction of nitriles with alkenes or alcohols via nitrilium ions to afford amides (see 1st edition).

| | | | | | |
|---|---|---|---|---|---|
| 1 | Ritter, J.J. | *J. Am. Chem. Soc.* | **1948** | 70 | 4045 |
| 2 | Ritter, J.J. | *J. Am. Chem. Soc.* | **1952** | 74 | 763 |
| 3 | Balaban, A. | *J. Org. Chem.* | **1965** | 30 | 879 |
| 4 | Wohl, R.A. | *J. Org. Chem.* | **1973** | 38 | 3099 |
| 5 | Ibatulin, V.G. | *Bull. Acad. Sci. USSR* | **1986** | 35 | 356 |
| 6 | Meyers, A.I. | *J. Org. Chem.* | **1973** | 38 | 36 |
| 7 | Ioachims, J.C. | *Tetrahedron* | **1992** | 48 | 8271 |
| 8 | Senanayaka, C.H. | *Tetrahedron Lett.* | **1995** | 26 | 3993 |
| 9 | Krimel, L.I. | *Org. React.* | **1969** | 17 | 218 |

## R O B I N S O N  Annulation

Synthesis of fused cyclohexenones by reaction of cyclanones with vinyl ketones (base or acid catalyzed), a tandem Michael addition-aldol condensation (see 1st edition).

| 1 | Robinson, R. | J. Chem. Soc. | 1937 | | 53 |
|---|---|---|---|---|---|
| 2 | Gawley, R.E. | Synthesis | 1976 | | 777 |
| 3 | House, H.O. | J. Org. Chem. | 1965 | 30 | 2513 |
| 4 | Zoretic, P.A. | J. Org. Chem. | 1976 | 41 | 3767 |
| 5 | Huffman, J.W. | J. Org. Chem. | 1985 | 50 | 4255 |
| 6 | Brewster, J.C. | Org. React. | 1953 | 7 | 113 |

## R O B I N S O N - A L L A N - K O S T A N E C K I  Chromone Synthesis

Synthesis of chromones or coumarins from o-acyloxy aromatic ketones (see 1st edition).

| 1 | Kostanecki, S. | Chem. Ber. | 1901 | 34 | 102 |
|---|---|---|---|---|---|
| 2 | Robinson, R.; Allan, J. | J. Chem. Soc. | 1924 | 125 | 2192 |
| 3 | Szell, Th. | J. Chem. Soc. (C) Org. | 1967 | | 2041 |
| 4 | Ziegler, F.E. | J. Org. Chem. | 1983 | 48 | 3349 |
| 5 | Sethna, S.M. | Chem. Rev. | 1945 | 36 | 8 |
| 6 | Hauser, C.R. | Org. React. | 1955 | 8 | 59 |

## R O B I N S O N - G A B R I E L  Oxazole Synthesis

Synthesis of oxazoles from amides of α-aminoketones (see 1st edition).

**1**                                    **2 (72%)**

**3**                                    **4**[3]

| 1 | Robinson, R. | *J. Chem. Soc.* | **1909** | *95* | 2165 |
|---|---|---|---|---|---|
| 2 | Gabriel, S. | *Chem. Ber.* | **1910** | *43* | 1283 |
| 3 | Balaban, A.T. | *Tetrahedron* | **1963** | *19* | 2199; 169 |
| 4 | Wasserman, H.H. | *J. Org. Chem.* | **1973** | *38* | 2407 |
| 5 | Krasowtsky, B.M. | *Chem. Heter. Compds.* | **1986** | *22* | 2291 |

## R O E L E N  Olefin Carbonylation

Synthesis of aldehydes or alcohols by cobalt catalyzed addition of CO-$H_2$ to olefins (see 1st edition).

**1**                                    **2 (70%)**

$$Me_3C\text{-}OH + 2H_2 + CO \longrightarrow Me_2CH\text{-}CH_2\text{-}CH_2\text{-}OH$$
$$(63\%)$$

| 1 | Roelen, O. | U.S. Pat. 2,327,006; 1943 | | | |
|---|---|---|---|---|---|
| 2 | Roelen, O. | *Angew. Chem.* | **1948** | *60* | 62 |
| 3 | Adkins, H. | *J. Am. Chem. Soc.* | **1948** | *70* | 383 |
| 4 | Keulemans, A.I.M. | *Rec. Trav. Chim.* | **1948** | *67* | 298 |
| 5 | Kropf, H. | *Angew. Chem. Int. Ed.* | **1966** | *5* | 648 |

**γ-Acetoxybutyraldehyde (2).**[3] A steel reaction vessel was filled with allyl acetate **1** (50.0 g; 0.5 mol) in $Et_2O$ (50 mL), $[Co(CO)_4]_2$ (2.2 g; 6.4 mmol) in $Et_2O$ (40 mL), followed by CO at 3200 psi and hydrogen at 4800 psi. The mixture was shaken and heated to 115°C (5050 psi) then slowly to 125°C (pressure 4000 psi). On cooling the pressure dropped to 2000 psi. Work up and distillation afforded 46 g of **2** (70%), bp 59-60°C/1 mm.

## R O S E N M U N D  Arsonylation

Cu catalyzed arsonylation by substitution of aromatic halides; see also Bart-Scheller (see 1st edition).

| 1 | Rosenmund, K.W. | *Chem. Ber.* | **1921** | 54 | 438 |
| 2 | Balaban, N.S. | *J. Chem. Soc.* | **1926** | | 569 |
| 3 | Hamilton, C.S. | *J. Am. Chem. Soc.* | **1930** | 52 | 3284 |
| 4 | Hamilton, C.S. | *Org. React.* | **1944** | 2 | 415 |

## R O S E N M U N D - B R A U N  Aromatic Cyanation

Cu catalyzed nucleophilic substitution of aromatic halogen by cyanide (see also Ullman-Goldberg) (see 1st edition).

| 1 | Rosenmund, K.W. | *Chem. Ber.* | **1916** | 52 | 1749 |
| 2 | Braun, J.v. | *Liebigs Ann.* | **1931** | 488 | 111 |
| 3 | Allen, R.E. | *J. Am. Chem. Soc.* | **1958** | 80 | 591 |
| 4 | Freedman, L. | *J. Org. Chem.* | **1961** | 26 | 2522 |
| 5 | Newmann, M.S. | *J. Org. Chem.* | **1961** | 26 | 2525 |
| 6 | Bunnett, J.F. | *Chem. Rev.* | **1951** | 49 | 392 |

**1-Naphthonitrile 2.**[4] A mixture of 1-bromonaphthalin **1** (207 g; 1 mol) and CuCN (103 g; 1.15 mol) in DMF (150 mL) was refluxed for 4 h. Work up afforded 114 g of **2** (94%), bp 160-161°C.

## R O S E N M U N D - S A I T Z E W   Reduction to Aldehydes

Hydrogenation of acyl chlorides to aldehydes in the presence of poisoned Pd catalyst (see 1st edition).

$$p\text{-Cl-C}_6\text{H}_4\text{-COCl} + \text{H}_2 \xrightarrow[\substack{2,4\text{-Dimethyl-}\\ \text{pyridine}}]{5\% \text{ Pd/BaSO}_4} p\text{-Cl-C}_6\text{H}_4\text{-CH=O}$$

**1**    **2** (77%)

(99%)    (95%)

| 1 | Saitzew, N. | *J. Prakt. Chem.* | **1873** | *114* | 1301 |
|---|---|---|---|---|---|
| 2 | Rosenmund, K.W. | *Chem. Ber.* | **1918** | *51* | 585 |
| 3 | Brown, H.C. | *J. Am. Chem. Soc.* | **1958** | *80* | 5372 |
| 4 | Burgsthaler, W. | *Synthesis* | **1976** | | 767 |
| 5 | Sonntag, A.D. | *Chem. Rev.* | **1953** | *52* | 245 |

## R O T H E M U N D – L I N D S E Y   Porphine Synthesis

Porphine synthesis from pyrrole and aldehydes modified by Lindsey (see 1st edition).

**1**    **2**    **3** (29%)

| 1 | Rothemund, P. | *J. Am. Chem. Soc.* | **1935** | *57* | 2010 |
|---|---|---|---|---|---|
| 2 | Lindsey, J.S. | *Tetrahedron Lett.* | **1986** | *27* | 4969 |
| 3 | Lindsey, J.S. | *J. Org. Chem.* | **1987** | *52* | 3069 |
| 4 | Lindsey, J.S. | *J. Org. Chem.* | **1989** | *54* | 828 |
| 5 | Collman, J.P. | *J. Org. Chem.* | **1995** | *60* | 1926 |

**meso-Tetramesitylporphyrin (3).**[4] To **1** (1.475 mL; 10 mmol) and **2** (694 µL; 10 mmol) in CHCl$_3$ (1000 mL), under N$_2$ was added 2.5 M BF$_3$·Et$_2$O (1.32 mL; 3.3 mmol). After 1 h stirring at 20°C, p-chloranil (1.844 g; 7.5 mmol) was added and the mixture was refluxed for 1 h. The cooled solution was treated with Et$_3$N (460 µL; 3.3 mmol) and the solvent evaporated. The residue was washed with MeOH (75 mL) to remove polypyrrolemethenes and quinone compounds, to afford 576 mg of **3** (29%), 95% purity.

## ROSINI–BARTOLI Reductive Nitroarene Alkylation

Synthesis of ortho alkyl anilines (Rosini) by reductive C-alkylation of nitroarenes. Also synthesis of indoles (Bartoli) by reaction of 2-substituted nitroarenes with vinyl Grignard reagents.

| 1 | Rosini, G. | *Synthesis* | **1976** | | 270 |
|---|------------|-------------|----------|----|------|
| 2 | Rosini, G. | *J. Chem. Soc. Perkin I* | **1977** | | 884 |
| 3 | Rosini, G. | *J. Chem. Soc. Perkin I* | **1978** | | 692 |
| 4 | Rosini, G. | *Synthesis* | **1978** | | 437 |
| 5 | Bartoli, G. | *Tetrahedron Lett.* | **1989** | *30* | 2129 |
| 6 | Bartoli, G. | *J. Chem. Soc. Perkin II* | **1991** | | 657 |
| 7 | Gilmore, J. | *Synlett.* | **1992** | | 79 |
| 8 | Bobbs, A.P. | *Synlett.* | **1999** | | 1954 |

**6-Amino-7-n-butylbenzothiazole (2).**[4] A solution of **1** (1.8 g; 10 mmol) in THF (10 mL) was added under $N_2$ at –10°C to a stirred solution of n-BuMgBr (50 mmol) in $Et_2O$ containing CuI (0.3 g; 1.5 mmol) . After 6 h stirring at 20°C, the mixture was quenched with 32% HCl, basified with $NH_4OH$ (pH=10), extracted with $CH_2Cl_2$ and chromatographed (silica gel, PhH:EtOAc 8:2) to give 1.13 g of **2** (55%), mp (HBr salt) 168-171°C.

**7-Formylindole (4).**[7] To a solution of **3** (70 g; 0.46 mol) in THF (2000 mL) at –65°C was added a solution of vinylmagnesium bromide in THF (1400 mL). After 15 min stirring a second portion (200 mL) was added and stirring was continued for another 30 min. Usual work up and chromatography (silica gel, EtOAc:PhH 2:8) followed by recrystallization afforded 45.5 g of **4** (68%), mp 86-87°C.

## R O Z E N  Hypofluorite Reagents

Acetyl hypofluorite (AcOF) and methyl hypofluorite (MeOF) as fluorinating agents of olefins and aromatics[3]; HOF·MeCN an oxygen transfer agent in epoxidation of electron poor olefins, in Baeyer-Villiger reaction, in oxidation of $\alpha$-amino acids to $\alpha$-nitro acids.

$$Ph_2C=CH_2 \xrightarrow[\text{MeCN; -40°}]{\text{MeOF}} Ph_2\overset{F}{\underset{}{C}}-\overset{OMe}{\underset{}{CH_2}}$$

$$Ph\text{-}CH=CH\text{-}CO_2Et \xrightarrow{\text{AcOF}} Ph\text{-}\overset{OAc}{\underset{}{CH}}-\overset{F}{\underset{}{CH}}\text{-}CO_2Et$$

**1** + AcOF $\xrightarrow{-75°}$ **2 (61%)** + **3 (8%)**

HOF·MeCN **6** →

**4**    **6** →    **5**

| 1 | Rozen, S. | *J. Chem. Soc. Chem. Commun.* | **1981** | | 443 |
|----|-----------|-------------------------------|----------|----|------|
| 2 | Rozen, S. | *J. Org. Chem.* | **1984** | *49* | 806 |
| 3 | Rozen, S. | *J. Am. Chem. Soc.* | **1991** | *113* | 2648 |
| 4 | Rozen, S. | *Acc. Chem. Soc.* | **1988** | *21* | 307 |
| 5 | Rozen, S. | *Chem. Rev.* | **1996** | *96* | 1717 |
| 6 | Rozen, S. | *Pure Appl. Chem.* | **1999** | *71* | 481 |
| 7 | Rozen, S. | *J. Org. Chem.* | **1990** | *55* | 5155 |
| 8 | Rozen, S. | *J. Chem. Soc. Chem. Commun.* | **1996** | | 627 |
| 9 | Rozen, S. | *Angew. Chem. Int. Ed.* | **1999** | *38* | 3471 |
| 10 | Rozen, S. | *Tetrahedron* | **1999** | *55* | 3457 |

**1-Methoxy-2-fluoronaphthalene (2).**[2] To **1** (3.16 g; 20 mmol) in $CH_2Cl_2$-$CFCl_3$ was added a solution of AcOF (10-50% excess). Quenching (water 500 mL), usual work up and chromatography afforded 2.18 g of **2** (61%) and 0.28 g of **3** (8%).

**HOF·MeCN 6.**[9] A mixture of 10-15% $F_2$ with $N_2$ was passed (ca. 400 ml/min) through a cold (-15°C) mixture of MeCN (400 mL) and $H_2O$ (40 mL). The product was monitored with KI/thiosulfate. Typical conc: 0.2-0.3 M. This solution was used without further purification or isolation of reagent.

**1,10-Phenanthroline-N,N-dioxide 5.**[9] A solution of 1,10-phenanthroline **4** (500 mg; 2.7 mmol) in $CHCl_3$ (20 mL) at 0°C was added to 2.2 equiv. of **6**. After 5 min the mixture was neutralized with $NaHCO_3$, extracted ($CHCl_3$), dried ($MgSO_4$) and the solvent evaporated. The crude product was recrystallized from EtOH/$H_2O$ (1/3) to give **5**, mp 200°C.

## RUFF–FENTON Degradation

Oxidative degradation of aldoses via α-hydroxy acids to lower chain aldoses (see 1st edition).

**1**

**2 (50%)**

| 1 | Ruff, O. | *Chem. Ber.* | **1898** | *31* | 1573 |
| 2 | Fenton, O. | *Proc. Chem. Soc.* | **1893** | *9* | 113 |
| 3 | Fletcher, H.G. | *J. Am. Chem. Soc.* | **1950** | *72* | 4546 |

**D-Arabinose (2).**[3] A mixture of **1** (200 g; 0.43 mol) from D-glucose, Ba(OAc)$_2$ (20 g; 0.08 mol) and Fe$_2$(SO$_4$)$_3$ (10 g; 0.025 mol) was stirred in water (2000 mL) until a precipitate appeared. The suspension was filtered and the brown solution was treated with 30% H$_2$O$_2$ (129 mL) at 35°C. A second portion of 30% H$_2$O$_2$ (120 mL) was added and the temperature was raised to 40°C. After filtration on Norrite and concentration under vacuum, MeOH was added and the precipitate filtered and recrystallized to give 55-60 g of **2** (50%), mp 162-164°C, [α]$_D^{20}$ = =103°.

## RUPE Rearrangement

Acid catalyzed isomerisation of ethynyl carbinols to unsaturated carbonyl compounds (see 1st edition).

**1**

**2 (49%)**

| 1 | Rupe, H. | *Helv. Chim. Acta* | **1920** | *9* | 672 |
| 2 | Rupe, H. | *Helv. Chim. Acta* | **1928** | *11* | 449; 656; 965 |
| 3 | Rupe, H. | *Helv. Chim. Acta* | **1938** | *17* | 238 |
| 4 | Chandey, J.D. | *J. Am. Chem. Soc.* | **1948** | *70* | 246 |
| 5 | Hurd, J. | *J. Am. Chem. Soc.* | **1948** | *70* | 118 |
| 6 | Olah, G.A. | *Synthesis* | **1981** | | 973 |

**1-Acetyl-1-cyclohexene (2).** **1** (65.0 g; 0.5 mol) in 90% HCOOH (400 mL) was refluxed for 45 min. The cooled mixture was poured into water (2000 mL) and extracted with petroleum ether. The organic layer was washed with 10% NaOH, the solvent evaporated and the residue was carefully fractionated. One obtains 32 g of **2** (49%), bp 111°C (49 mm).

## R U P P E R T   Perfluoroalkylation

Trifluoromethylation (perfluoroalkylation) by reaction of carbonyl compounds with (trifluoromethyl)trimethylsilane or (perfluoroalkyl)trimethylsilane.

$$Me_3SiCl \ + \ BrCF_3 \ \xrightarrow[\text{PhCN; -60°}]{\text{HMPT}} \ Me_3Si\text{-}CF_3$$

$$\textbf{1} \qquad\qquad \textbf{2} \qquad\qquad\qquad \textbf{3}\ (65\%)$$

$$\textbf{4} \qquad\qquad\qquad\qquad \textbf{5}\ (77\%)$$

| | | | | | |
|---|---|---|---|---|---|
| 1 | Ruppert, I. | *Tetrahedron Lett.* | **1984** | *25* | 2195 |
| 2 | Crindeman, R.S. | *J. Org. Chem.* | **1989** | *54* | 661 |
| 3 | Yagupolski, Yu.L. | *Synthesis* | **1990** | | 1151 |
| 4 | Surya Prakash, G.K. | *J. Org. Chem.* | **1991** | *56* | 984 |
| 5 | Nedelec, J.Y. | *Tetrahedron Lett.* | **1994** | *35* | 8623 |
| 6 | Iseko, K. | *Tetrahedron Lett.* | **1994** | *35* | 3137 |
| 7 | Bosmans, J.P. | *Jansen Chim. Acta* | **1992** | *24* | 23 |

**(Trifluoromethyl)trimethylsilane (3).**[3] To a solution of **1** (87.3 g; 0.83 mol) in PhCN (100 mL) cooled at -30°C was added **2** (261 g; 1.75 mol). The mixture was cooled progressively to -60°C and HMPT (216 g; 1.75 mol) in PhCN (175 mL) was added in 2 h. Stirring at -60°C was followed by slow warming to 20°C and stirring for 18 h at 20°C. The mixture was gently distilled (45°C, 20 mm Hg) and the distillate was collected into a trap (dry $CO_2$-$Me_2CO$). Usual work up and distillation afforded 77.1 g of **3** (65%), bp 55-55.5°C.

**1-(Trifluoromethyl)-1-cyclohexanol (5).**[4] A cooled (0°C) mixture of **4** (980 mg; 10 mmol) and **3** (1.704 g; 12 mmol) in THF (10 mL) was treated with tetrabutylammonium fluoride (TBAF) (20 mg). Under stirring and slow warming to 20°C the mixture was analyzed periodically by GC. Hydrolysis with aq. HCl, extraction ($Et_2O$) and distillation afforded 1.3 g of **5** (77%), bp 72-73°C/40 mm, mp 59-61°C.

**R U S S I G – L A A T S C H** Hydroquinone Monoether Formation

Regioselective monoalkyl ether formation from naphthalene-1, 4-diols using alcohols (prim. or sec.) containing HCl.

95%          90%

1          2          3 (89%)

| 1 | Russig, F. | *J. Prakt. Chem.* | **1900** | 62(2) | 30 |
|---|---|---|---|---|---|
| 2 | Laatsch, H. | *Liebigs Ann.* | **1980** | | 140,1321 |
| 3 | Laatsch, H. | *Liebigs Ann.* | **1991** | | 385 |
| 4 | Katz, T. J. | *J. Org. Chem.* | **1997** | 62 | 1274 |
| 5 | Katz, T. J. | *J. Org. Chem.* | **2000** | 65 | 806,815 |

**Binaphthol (2).[5]** To a solution of helicenebisquinone **1** in a mixture of 3:1 EtOAc-CH$_2$Cl$_2$ (0.07 M) was added twice the volume of water followed by Na$_2$S$_2$O$_4$ (25 mmol/mol). The mixture was shaken by means of a mechanical shaker until it was yellow (approx 1 h). The aqueous layer was removed and the organic layer was washed with brine and dried. Evaporation of the solvent afforded **2** (moderately air-sensitive).

**Methyl ether (3).** To a solution of **2** in 1,2-dichloroethane was added a saturated solution of HCl in MeOH and the reaction mixture was stirred under N$_2$ at 60 °C for 2 h. Dilution with EtOAc, washing with water drying (MgSO$_4$) and evaporation of the solvent afforded **3** in 89% yield. In the same manner from **3** can be obtained its monoethyl ether in 93% yield by reaction with HCl in EtOH.

## SAEGUSA Enone Synthesis

Conversion of silyl enol ethers of ketones to $\alpha,\beta$-unsaturated ketones or coupling to 1,4-diketones by means of $Ag_2O$ or $Pd(II)$; for a one pot conversion of ketones, aldehydes or alcohols to $\alpha,\beta$-unsaturated ketones (aldehydes) with iodoxybenzoic acid see Nicolaou.

**2 (81%)**

**3**                               **4 (85%)**

| | | | | | |
|---|---|---|---|---|---|
| 1 | Saegusa, T. | *J.Am.Chem.Soc.* | **1975** | *97* | 649 |
| 2 | Saegusa, T. | *J.Org.Chem.* | **1978** | *43* | 1011 |
| 3 | Boeckman, R.K. | *J.Am.Chem.Soc.* | **1989** | *111* | 2537 |
| 4 | Nicolaou, K.C. | *J.Am.Chem.Soc.* | **2000** | *122* | 7596 |

**2-Cyclohexenone 4.[2]** To a solution of $Pd(OAC)_2$ (112 mg, 0.5 mmol) and benzoquinone (54 mg, 0.5 mmol) in MeCN (4 mL) was added under stirring 1-trimethylsilyloxy-1-cyclohexene **3** (170 mg, 1 mmol). After 3 h stirring at 20°C under $N_2$, **4** was isolated in 85% yield after chromatography.

## SAKURAI–HOSOMI Allylation

Lewis acid (e.g Ti) mediated inter or intramolecular addition of allylic silanes to aldehydes, ketones or 1,4-addition to $\alpha,\beta$-unsaturated ketones (see 1st edition).

$$Ph-(CH_2)_2-CHO + Me_3Si-CH_2-CH=CH_2 \xrightarrow{TiCl_4} Ph-(CH_2)_2-\underset{\underset{OH}{|}}{\overset{\overset{H}{|}}{C}}H-CH_2-CH=CH_2$$

**1**  **2**  **3 (96%)**

| 1 | Sakurai, H.; Hosomi, A. | *J.Org.Chem.* | **1969** | *34* | 1764 |
|---|---|---|---|---|---|
| 2 | Hosomi, A.; Sakurai, H. | *Tetrahedron Lett.* | **1976,.** | | 1295 |
| 3 | Sakurai, H.; Hosomi, A. | *J.Am.Chem.Soc.* | **1977** | *99* | 1673 |
| 4 | Seebach, D. | *Angew.Chem.Int.Ed.* | **1985** | *24* | 765 |
| 5 | Magnus, P. | *Acta Chem.Scand.* | **1993** | *47* | 157 |
| 6 | Mikami, K. | *Tetrahedron Lett.* | **1994** | *35* | 3133 |
| 7 | Shiro, T.K.M | *J.Org.Chem.* | **1997** | *62* | 1230 |
| 8 | Sakurai, H. | *Pure Appl.Chem.* | **1983** | *54* | 1 |

**4-Hydroxy-6-6phenyl hexane-1 3.[2]** To a solution of 2-phenylpropanal **1** (268 mg, 2 mmol) in $CH_2Cl_2$ (3 mL) under $N_2$ at 20°C was added $TiCl_4$ (190 mg, 1 mmol) dropwise. After 5 min stirring allyl trimethyl silane **2** (228 mg, 2 mmol) was added at the same temperature. The mixture was stirred for 1 min, quenched with water , extracted (Et₂O), the organic phase dryed ($Mg_2SO_4$) and the solvent evaporated. Chromatography of the residue afforded 338 mg of **3** (96%).

**9-Allyl-2-decalone 6.[3]** To $\Delta^{1,9}$ 2-octalone **4** (300 mg, 2 mmol) $TiCl_4$ (380 mg, 2 mmol) in $CH_2Cl_2$ (5 mL) at -78°C was added a solution of trimethylsallyl silane **5** (159 mg, 2.8 mmol) in $CH_2Cl_2$ (3 mL); the reaction is exothermic. After stirring for 18 h at -78°C and 5 h at -30°C, work up and distillation afforded 353 mg of **6** (85%), bp 120 °C/5 mm.

## S A N D M E Y E R   Isatin Synthesis

Isatin synthesis from anilines (see 1st edition).

| 1 |              | Sandmeyer, T. | *Helv. Chim. Acta* | **1919** | *2*  | 234 |
| 2 |              | Sheilley, F.E. | *J. Org. Chem.*   | **1956** | *21* | 171 |

## S A N D M E Y E R - G A T T E R M A N N   Aromatic Substitution

Substitution of an amine group, via its diazonium salt, by nucleophiles such as Cl⁻, Br⁻, I⁻, CN⁻, R-S⁻, HO⁻, some by cuprous salt catalysis (see 1st edition).

| 1 | Sandmeyer, T.  | *Chem. Ber.*                | **1884** |     | 1633 |
| 2 | Gattermann, L. | *Chem. Ber.*                | **1890** | *23* | 1218 |
| 3 | Hodgson, H.H.  | *J. Chem. Soc.*             | **1944** |     | 22   |
| 4 | Suzuki, N.     | *J. Chem. Soc. Perkin Trans I* | **1987** |     | 645  |
| 5 | Condret, C.    | *Synth. Commun.*            | **1996** | *26* | 3143 |
| 6 | Hodgson, H.H.  | *Chem. Rev.*                | **1947** | *40* | 251  |
| 7 | Pfeill, E.     | *Angew. Chem.*              | **1953** | *65* | 155  |

**4-Iodopyridine (4).**[5] To a cooled (-10°C) **3** (6 g; 63.8 mmol) in 48% HBF₄ (50 mL) was added under stirring NaNO₂ (4.8 g; 69.5 mmol) at such rate that no nitric oxide evolution was detected. After 30 min the diazonium salt was filtered off and added to a solution of KI (17 g; 102.4 mmol) in 100 mL of Me₂CO:H₂O (40:60). The mixture was decolorized with Na₂S₂O₃, neutralized with Na₂CO₃ and extracted with Et₂O. Evaporation afforded 9.2 g of **4** (70%).

## S C H E I N E R  Aziridine Synthesis

Synthesis of triazolines or aziridines from azides by photodecomposition or flash vacuum pyrolysis of 1,2,3-triazolines.

| | | | | | |
|---|---|---|---|---|---|
| 1 | Wolff, A. | *Liebigs Ann.* | **1912** | 394 | 30 |
| 2 | Alder, K. | *Liebigs Ann.* | **1931** | 485 | 211 |
| 3 | Scheiner, P. | *J. Org. Chem.* | **1961** | 26 | 1923 |
| 4 | Scheiner, P. | *J. Org. Chem.* | **1965** | 30 | 7 |
| 5 | Scheiner, P. | *Tetrahedron* | **1967** | 24 | 349 |
| 6 | Hassner, A. | *Tetrahedron Lett.* | **1981** | 22 | 1863 |
| 7 | Hassner, A. | *J. Org. Chem.* | **1988** | 53 | 27 |
| 8 | Heine, P. | *Angew. Chem. Int. Ed.* | **1962** | 1 | 528 |
| 9 | Hassner, A. | *Acc. Chem. Res.* | **1971** | 9 | 1 |

**Triazoline 3**.[4] A solution of norbornene **1** (2.9 g; 31 mmol) and ethyl azidoformate **2** (3.6 g; 31 mmol) in pentane (10 mL) was maintained at 20°C for 7 days. Evaporation of the solvent afforded 6.1 g of **3** (94%).

**Aziridine 4**. A solution of **3** (1 g; 4.8 mmol) in $Me_2CO$ (25 mL) was irradiated with a General Electric sun lamp until gas evolution ceased. Evaporation of the solvent afforded 830 mg of **4** (95%), bp 99-100°C/2.4 mm.

**Vinylaziridine 7**.[6] Vinyl azide **5** (20 mmol) and trimethylsulfoxonium ylide (2 equiv) in DMSO was stirred at 20°C for 12 h. The reaction mixture wad diluted with $Et_2O$ (100 mL) and washed with water (5x100 mL). Evaporation of the solvent afforded triazoline **6** in 95% yield. Pyrolysis of **6** in refluxing PhMe (3 h) gave **7** in 65% yield; alternatively flash vacuum pyrolysis (FVP) afforded **7** in 93% yield.

## **SCHENCK** Allylic Oxidation

Ene reaction of alkenes and oxygen (with double bond migration) to form allyl
Hydroperoxides with double bond migration and derived allyl alcohols.

| | | | | | | | |
|---|---|---|---|---|---|---|---|
| 1 | Schenck, L.D. | *Ger.Pat.* | 1943 | | 933.925 |
| 2 | Schenck, L.D. | *Naturwissensch.* | 1945 | *32* | 157 |
| 3 | Schenck, L.D. | *Liebigs Ann.* | **1953** | *584* | 117 |
| 4 | Schenck, L.D. | *Liebigs Ann.* | **1958** | *618* | 185 |
| 5 | Adam, W. | *J.Org.Chem.* | **1994** | *59* | 3335 |
| 6 | Adam, W. | *J.Org.Chem.* | **1994** | *59* | 3341 |
| 7 | Adam, W. | *Synthesis* | **1994** | | 567 |
| 8 | Adam, W. | *J.Am.Chem.Soc.* | **1996** | *118* | 1899 |
| 9 | Stephenson, L.M. | *Acc.Chem.Res.* | **1980** | *13* | 419 |

**(E/Z)-1-Methyl-2-(dimethylphenylsilyl)-2-butenyl hydroperoxide 2.[5]** A solution of
(E/Z)-(1-Ethyl-1-propenyl) dimethylsilane **1** (200 mg, 0,98 mmol) and tetraphenyl
porphyrine (TPP) (0,3 mg) in CDCl$_3$ (1 mL) was photooxygenated at -5 to -10° C
by passing a slow stream of dry O$_2$ under continuous irradiation with two 150-W
sodium lamps for 2.5 h. After column chromatography (silica gel petroleum ether :
Et$_2$O 5:1 ) there were obtained 48 mg of pure E-**2** and 57 mg of an E/Z mixture of **2**.
Total yield 45% (1α ,2β ,6β )-2-Methyl-1-(trimethylsilyl)-7-oxabicyclo[4.1.0]–
**heptan - 2-ol 4.[6]** Vinylsilne **3** (336 mg, 2 mmol) was photooxygenated in the
presence of Ti (0-iPr)$_4$. Crystallization of the residue from pentane gave 172 mg of **4**
(43%), mp 56-57°C.

## S C H I E M A N N Aromatic Fluorination

Substitution of an aromatic amino group by fluorine via a diazonium salt using fluoroborates (compare Sandmeyer – Gattermann) (see 1st edition).

| 1 | Schiemann, G. | Chem. Ber. | **1927** | 60 | 1186 |
|---|---|---|---|---|---|
| 2 | Finger, G. C. | J. Org. Chem. | **1962** | 27 | 3965 |
| 3 | Roe, A. | Org. React. | **1949** | 5 | 194 |
| 4 | Kornblum, N. | Org. Synth. Coll. Vol. | | | II-188,295,299 |

## S C H M I D T Rearrangment

Conversion by means of $NH_3$ of carboxylic acids to amides, of aldehydes into nitriles or of ketones into tetrazoles or amides (see 1st edition).

| 1 | Schmidt, K. F. | Z. Angew. Chem. | **1923** | 36 | 511 |
|---|---|---|---|---|---|
| 2 | Greco, C.V. | Tetrahedron | **1970** | 26 | 4329 |
| 3 | Bach, R. D. | J. Org. Chem. | **1982** | 47 | 239 |
| 4 | Pavlov, P. A. | Chem. Heter. Compd. | **1986** | 22 | 140 |
| 5 | Hassner, A. | J. Org. Chem. | **1988** | 53 | 22 |
| 6 | Aube, J. | Tetrahedron | **1996** | 52 | 3403 |
| 7 | Applequist, D. E. | Chem. Rev. | **1954** | 54 | 1084 |
| 8 | Wolff, H. | Org. React. | **1964** | 3 | 307 |

**Benzoylpyrrolidine (4).**[6] To **3** (230 mg, 0.92 mmol) in $CH_2Cl_2$ (1 mL) at 0 °C was added TFA (1 mL); (vigorous gas evolution). The mixture was stirred for 16 h at 20 °C, the solvent removed in vacuo and replaced with a solution of NaI (276 mg, 1.87 mmol) in anh.$Me_2CO$ (2 mL). After 4 h at 70 °C work up gave 137 mg of **4** (79%).

**S C H M I T Z** Diaziridine Synthesis

Diaziridine synthesis from chloramine, ammonia and (excess) aldehyde. In the presence of excess aldehyde formation of bicyclic triazolidines takes place (see 1st edition).

| 1 | Schmitz, E. | *Angew. Chem.* | **1959** | *71* | 127 |
|---|---|---|---|---|---|
| 2 | Schmitz, E. | *Chem. Ber.* | **1962** | *95* | 680 |
| 3 | Nilsen, A.T. | *J. Org. Chem.* | **1976** | *41* | 3221 |
| 4 | Schmitz, E. | *Chem. Ber.* | **1967** | *100* | 142 |
| 5 | Brinker, U.H. | *Tetrahedron Lett.* | **2001** | *42* | 9161 |

**cis (trans)-2,4,6-tri-(n-pentyl)-1,3,5-triazabicyclo[3.1.0]hexane (3).**[3] A solution of t-butyl hypochlorite (2.71 g, 26 mmol) in t-BuOH (3 mL) was added at −35°C over 5 min to a stirred 10 N methanolic ammonia solution (25 mL), followed by hexanal 1 (5.0 g, 50 mmol). The mixture was stirred for 2.5 h at 20°C, the solvent was removed in vacuum and the residue was extracted with boiling hexane to afford 4.25 (87%) of a mixture of cis and trans 3 in a ratio of 3.3:6.7 (exo). The less soluble fraction from hexane gave 0.67 g of 3 trans-exo (13%), mp. 51-52°C; 3 cis-exo, mp. 50-54°C.

**S C H O L L** Polyaromatic Synthesis

Preparation of condensed polynuclear aromatics by Friedel-Crafts catalysts (see 1st edition).

| 1 | Scholl, R. | *Chem. Ber.* | **1910** | *43* | 2201 |
|---|---|---|---|---|---|
| 2 | Scholl, R. | *Monatsh.* | **1912** | *33* | 1 |
| 3 | Nenitzescu, C.D. | *Chem. Ber.* | **1958** | *91* | 2109 |
| 4 | Vingiello, F.A. | *J. Org. Chem.* | **1971** | *36* | 2053 |
| 5 | Allen, C.F.H. | *Chem. Rev.* | **1959** | *59* | 987 |

## S C H Ö L L K O P F Amino Acid Synthesis

Asymmetric synthesis of amino acids from dihydropyrazines(see 1st edition).

**3** X = Br
**4** X = S-tBu (93%)

**5** (72%, 95% ee)

| 1 | Schöllkopf, U. | *Synthesis* | **1981** | *969* |
| 2 | Schöllkopf, U. | *Liebig's Ann.* | **1982** | *1925* |
| 3 | Schöllkopf, U. | *Synthesis* | **1983** | *37* |
| 4 | Schöllkopf, U. | *Synthesis* | **1985** | *1052* |

**(3S, 6R) Pyrazine (4).**[3] To 1 (2.77 g, 14 mmol) in THF (25 mL) at –70 °C was added 1.8 N BuLi in hexane (8.3 mL, 15 mmol) followed after 15 min by $CH_2Br_2$ (26.1 g, 0.15 mol) in THF (15 mL). After stirring 30 h at –70 °C, work up afforded 3.2 g of 3 (79%), bp 760-80 °C (0.1 torr). Reaction of 3 will t-butylmercaptan in DMSO and KOBu for 5 h at 70 °C gave after work up and distillation 0.837 g of 4 (93%), bp 80-90 °C (0.1 torr).

## S C H O L T Z Indolizine Synthesis

Indolizine synthesis by reaction of pyridinyl ketones with aldehydes in the presence of ammonium acetate.(see 1st edition).

**3** (39%)

| 1 | Scholtz, M. | *Chem. Ber.* | **1912** | 45 | 734 |
| 2 | Barow, E. T. | *Chem. Ber.* | **1948** | 42 | 638 |
| 3 | Krohnke, F. | *Chem. Ber.* | **1971** | 104 | 1624 |
| 4 | Uchida, T. | *Synthesis* | **1976** | | 209 |

## S C H Ö L L K O P F- B A R T O N – Z A R D Pyrrole Synthesis

Synthesis of pyrroles form nitroolefins or β-acetoxy nitro compounds with α-isocyano esters in the presence of an organic base.

| 1 | Schöllkopf, U. | *J. Chem. Soc. Chem. Commun.* | **1985** | | 1098 |
|---|---|---|---|---|---|
| 2 | Barton, D. H. R.; Zard, S. Z. | *Tetrahedron* | **1990** | *46* | 7587 |
| 3 | Lash, T. D. | *Tetrahedron Lett.* | **1994** | *35* | 2494 |
| 4 | Lash, T. D. | *J. Heterocyclic Chem.* | **1991** | *28* | 1671 |
| 5 | Gribble, G. W. | *J. Chem. Soc. Chem. Commun.* | **1997** | | 1873 |
| 6 | Barton, D. H. R. | *Pure. Appl. Chem.* | **1994** | *66* | 1943 |
| 7 | Tardieux, C. | *Synthesis* | **1998** | | 267 |
| 8 | Lash, T. D. | *Synlett.* | **2000** | | 213 |

**t-Butyl 3-(p-methoxyphenyl)-4-methylpyrrole-2-carboxylate (3).** [2] To a solution of nitroolefin **1** (200 mg, 1 mmol) and isocyanide **2** (169 mg, 1.2 mmol) in a 1:1 mixture of THF and iPrOH (5 mL) was added tetramethyl-t-butylguanidine (TMBG) (180 mg, 1.05 mmol). After 3 h heating to 50 °C the mixture was diluted with water and extracted with $CH_2Cl_2$. The organic layer after drying ($MgSO_4$), was filtered through a short column of silica gel (eluent $CH_2Cl_2$). Evaporation of the solvent in vacuum afforded 272 mg of **3** (90%), mp 142-144 °C.

## S C H W A R T Z  Hydrozirconation

Hydrozirconation with Cp$_2$Zr(Cl)H; can be followed by Michael addition, or by reaction with O-mesitylsulfonyl hydroxylamine (MSH) to prepare amines (see 1st edition).

| 1 | Wailes, C.P. | *J. Organomet. Chem.* | **1970** | *24* | 405 |
|---|---|---|---|---|---|
| 2 | Schwartz, J. | *J. Am. Chem. Soc.* | **1974** | *96* | 8115 |
| 3 | Schwartz, J. | *J. Am. Chem. Soc.* | **1980** | *102* | 1333 |
| 4 | Schwartz, J. | *Angew. Chem. Int. Ed.* | **1976** | *15* | 333 |
| 5 | Srebnik, M. | *J. Org. Chem.* | **1995** | *60* | 1912 |
| 6 | Negishi, Ei-ichi | *Aldrichimica Acta* | **1985** | *18* | 31 |
| 7 | Schwartz, J. | *Chimica Scripta* | **1989** | *29* | 411 |

**3-(1-Octen-1-yl)cyclopentanone 4.**[3]  Chlorobis($\eta^5$-cyclopentadienyl)hydrozirconium **1** (38.68 g, 0.15 mol) in THF (50 mL) under Ar was treated with 1-octene **2** (23.6 mL, 0.16 mol) at 15-25°C. After 18 h stirring at 20°C, 2-cyclopentenone **3** (10.9 mL, 0.13 mol) was added and the mixture kept for 10 min in an ice bath. Nickel acetylacetonide (3.34 g, 13 mmol) was added in three portions at 10 min interval below 40°C. After 2 h stirring at 5°C and 2 h at 20°C the mixture was quenched with HCl-ice water. Extraction with hexane followed by chromatography (silica gel, 2% EtOAc in hexane) gave 15.43 g of **4** (61.2%).

**Octylamine 6.**[5]  A suspension of **1** (258 mg, 1 mmol) in THF (1 mL) was stirred at 20°C under Ar. 1-Octene **5** (134 mg, 1.2 mmol) was added, the mixture was cooled in an ice bath and MSH (O-mesitylsulfonyl hydroxylamine) 220 mg, 1.2 mmol) in Et$_2$O (1 mL) was added. After 10 min stirring, 1 M HCl (10 mL) was added. Usual work up and distillation of the solvent gave 99 mg of **6** (77%).

## S C H W E I Z E R Allylamine Synthesis

Synthesis of E-allylamines from vinylphosphonium salts, phthalimide aldehydes (via a Wittig reaction) (see 1st edition).

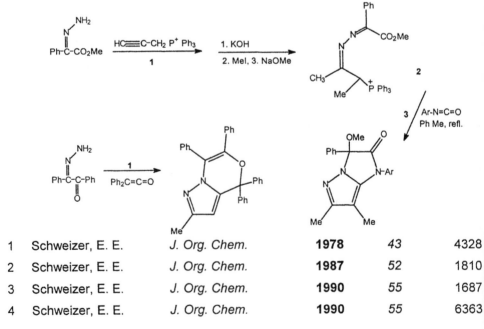

| 1 | Schweizer, E. E. | J. Org. Chem. | 1966 | 31 | 467 |
|---|---|---|---|---|---|
| 2 | Rouhut, M. M. | J. Org. Chem. | 1963 | 28 | 2565 |
| 3 | Evans, D. A. | J. Org. Chem. Soc. | 1978 | 100 | 1548 |
| 4 | Meyers, A. I. | J. Org. Chem. | 1981 | 46 | 3119 |

**(E)-3-Phenylpropenylamine (4).**[4] NaH (1.5 mmol) was washed (pentane), treated in THF with PhCHO (0.3 g, 1 mmol), **2** (0.4 g, 1.3 mmol) and phthalimide **1** (0.19 g, 1.3 mmol) and heated at 60 °C (TLC CHCl$_3$:Et$_2$O:hexane 5:1:4). Treatment with 5% citric acid in water and extraction with Et$_2$O gave 281 mg of **3** (80%), mp 150-151 °C. To **3** (174 mg, 0.6 mol) in anh. EtOH (19 mL) was added 95% hydrazine ( 60 µL, 1.8 mmol). 4.5 h reflux, acidification to pH=2, heating for 1 h, filtration, dilution of the filtrate, extraction with Et$_2$O and basification gave 93 mg of **4** (83%), mp 101-102 °C, 100% E.

## S C H W E I Z E R Rearrangement

Thermal reaction of "allenyl azines", derived from propargylphosphonium salts with ketenes, isocyanates, CS$_2$ or phthalic anhydride to form bi- and tricyclic fused pyrazolo heterocycles (see 1st edition).

| 1 | Schweizer, E. E. | J. Org. Chem. | 1978 | 43 | 4328 |
|---|---|---|---|---|---|
| 2 | Schweizer, E. E. | J. Org. Chem. | 1987 | 52 | 1810 |
| 3 | Schweizer, E. E. | J. Org. Chem. | 1990 | 55 | 1687 |
| 4 | Schweizer, E. E. | J. Org. Chem. | 1990 | 55 | 6363 |

## SCHWESINGER Bases

Very strong uncharged polyaminophosphazene bases with good chemical and thermal stability, their pKa ranging from 24 to 47 in the absolute MeCN scale and relatively non-nucleophilic. Useful in alkylation of enolates, in enantioselective α-alkylation of amino acids, in Ullmann synthesis.

BEMP

$P_2$Et

$P_4$ t-Bu (1)

| 1 | Schwesinger, R. | *Chimia* | **1985** | *39* | 269 |
|---|---|---|---|---|---|
| 2 | Schwesinger, R. | *Angew.Chem.Int.Ed.* | **1987** | *26* | 1165 |
| 3 | Schwesinger, R. | *Liebigs Ann.* | **1996** | | 1055 |
| 4 | Palomo, C. | *J.Chem.Soc.Chem.Commun.* | **1998** | | 2091 |
| 5 | Schwesinger, R. | *Chem.Ber.* | **1994** | *127* | 2435 |
| 6 | Solladie-Cavallo, A. | *J.Org.Chem.* | **1996** | *61* | 2690 |
| 7 | O' Donnell, M.J. | *Tetrahedron Lett.* | **1998** | *39* | 8775 |
| 8 | Prinzbach, H. | *J.Org.Chem.* | **2001** | *66* | 5744 |

**(-)8-Phenylmenthyl 2-phenylbutyrate 4**. To a stirred solution of ester **2** (0.5 mmol) in THF (2.5 mL) was added an excess of iodide **3** (234 mg, 1.5 mmol) and then, after cooling to –100°C, a solution of **1** (1M in hexane, 0.55 mmol, 0.55 mL) in dry THF (1.55 mL) so that the temperature of the mixture did not rise above –95°C. After being stirred for 1 h at 95°C, the mixture was warmed to 20°C. The solvent was removed in vacuum, and to the residual oil was added $Et_2O$. A precipitate formed which was filtered. Concentration of the filtrate afforded crude **4**. Flash chromatography (hexane:$Et_2O$) gave **4** in 95% yield, $[\alpha]_D$=-18° (c=0.5).

## S E Y F E R T H   Acyllithium Reagent

Direct nucleophilic acylation of electrophiles (ketones, esters) by acyllithium reagents.

$$n\text{--BuLi} \;+\; CO \xrightarrow[-110^\circ C]{THF/Et_2O} n\text{--Bu--}\overset{O}{\overset{\|}{C}}\text{Li} \xrightarrow[-110^\circ]{Me\text{--}\overset{O}{\overset{\|}{C}}\text{--}CH_2\text{--}CH_2\text{--}CH_2\text{--}Cl}$$

$$n\text{--Bu--}\underset{\underset{Me}{|}}{\overset{O}{\overset{\|}{C}}}\text{--}\overset{OLi}{\overset{|}{C}}\text{--}CH_2\text{--}CH_2\text{--}CH_2\text{--}Cl \xrightarrow{Me_3SiCl} n\text{--Bu--}\underset{\underset{Me}{|}}{\overset{O}{\overset{\|}{C}}}\text{--}\overset{OSiMe_3}{\overset{|}{C}}\text{--}CH_2\text{--}CH_2\text{--}CH_2\text{--}C$$

**2 (67%)**

$$n\text{--BuLi} \;+\; Ph\text{--COOMe} \xrightarrow[-110^\circ C]{CO} n\text{--Bu--}\overset{O}{\overset{\|}{C}}\text{--}\overset{O}{\overset{\|}{C}}\text{--}Ph$$

| 1 | Seyferth, D. | *J. Am. Chem. Soc.* | **1982** | *104* | 5534 |
|---|---|---|---|---|---|
| 2 | Seyferth, D. | *J. Am. Chem.* | **1983** | *48* | 114, 3367 |
| 3 | Seyferth, D. | *Tetrahedron Lett.* | **1984** | *25* | 1651,5251 |
| 4 | Seyferth, D. | *J. Org. Chem.* | **1992** | *57* | 5620 |
| 5 | Seyferth, D. | *Isr. J. Chem.* | **1984** | *24* | 167 |
| 6 | Seyferth, D. | *Org. Synth.* | **1990** | *69* | 114 |
| 7 | Seyferth, D. | *J. Org. Chem.* | **1991** | *56* | 5768 |

**4-(Trimethylsiloxy)-4-methyl-1-chloro-5-nonanone (2).**[4] To a mixture of THF (130 mL), Et$_2$O (130 mL) and pentane (40 mL) was added 5-chloro-2-pentanone **1** (2.1 mL, 18 mmol). This solution was cooled to –110 °C and CO was bubbled in for 30 min. BuLi in hexane (2.1 M, 4.0 mL, 8.2 mmol) was added (at a controlled rate of 0.5 mmol/min) under vigorous stirring. The mixture was stirred for 2 h at –110 °C, under a CO stream. Me$_3$SiCl (4.0 mL, 32 mmol) was added at the same temperature and finally the reaction mixture was graduallly warmed to 20 °C under N$_2$. Usual work up afforded 1.52 g of **2** (67%). GLC (100-240 °C, 6 °C/min, IS=C$_{12}$ ) showed the presence of one product.

## S E Y F E R T H Dihalocarbene Reagent

Phenyl (trihalomethyl) mercury compounds as versatile dihalocarbene precursors, useful in synthesis of halocyclopropanes.

**3 (93%)**

**6 (75%)**

(61%)

| 1 | Seyferth. D. | *J. Am. Chem. Soc.* | **1965** | 87 | 681, 4259 |
|---|---|---|---|---|---|
| 2 | Seyferth. D. | *J. Am. Chem. Soc.* | **1967** | 89 | 959 |
| 3 | Seyferth. D. | *J. Org. Chem.* | **1967** | 32 | 2980 |
| 4 | Seyferth. D. | *J. Am. Chem. Soc.* | **1969** | 91 | 6536 |
| 5 | Seyferth. D. | *J. Org. Chem.* | **1970** | 35 | 1297 |
| 6 | Seyferth. D. | *J. Org. Chem.* | **1972** | 37 | 4070 |
| 7 | Seyferth. D. | *Acc. Chem. Res.* | **1972** | 5 | 65 |
| 8 | Kang, Jahyo | *Synlett.* | **1990** | | 611 |

**1,1-Difluoro-trans-2,3-diethylcyclopropane (3).[6]** Trans-3-hexene **1** (5.04 g, 60 mmol) (98 % isomerically pure), PhHgCF$_3$ (6.54 g, 20 mmol), NaI (7.5 g, 50 mmol) in PhH (50 mL) were heated at 60-70 °C for 48 h. The cooled mixture was filtered from insoluble salts and distilled to afford **3** in 94% yield.

**3-Oxa-6-fluoro-6-chlorobicyclo [3.1.0] hexane (6).[5]** A mixture of PhHgCCl$_2$F-Ph$_2$Hg (11.85 g) containing PhHgCCl$_2$F (25 mmol) and 2,5-dihydrofuran **4** (5.18 g, 70 mmol) in PhH (50 mL) was refluxed for 48 h. Filtration, evaporation of the solvent and distillation afforded **6** in 75% yield as a mixture of syn and anti isomers.

## SEYFERTH-GILBERT  Diazoalkane Reagent

Dimethyl (diazomethyl)phosphonate **5** in reaction with olefins to form cyclopropanes or 1,3-dipolar addition products; also in synthesis of alkynes.

$$NaN_3 \ + \ p\text{-}AcNH\text{-}C_6H_4\text{-}SO_2Cl \longrightarrow p\text{-}AcNH\text{-}C_6H_4\text{-}SO_2N_3$$

$$\mathbf{1} \qquad\qquad\qquad\qquad \mathbf{2}$$

**3**             **4**             **5**

**6**          **7 (39%)**

(100%)

$$Ph\text{-}CO\text{-}Ph \ + \ (MeO)_2P(O)CHN_2 \xrightarrow{\ KOtBu\ } Ph\text{-}C\equiv C\text{-}Ph$$

$$\mathbf{8} \qquad\qquad \mathbf{5} \qquad\qquad\qquad \mathbf{9\ (92\%)}$$

| | | | | | |
|---|---|---|---|---|---|
| 1 | Seyferth, D. | *J. Am. Chem. Soc.* | **1967** | *89* | 4811 |
| 2 | Seyferth, D. | *Tetrahedron Lett.* | **1970** | | 2493 |
| 3 | Seyferth, D. | *J. Org. Chem.* | **1971** | *36* | 128; 1379 |
| 4 | Gilbert, J.C. | *J. Org. Chem.* | **1979** | *44* | 4994 |
| 5 | Gilbert, J.C. | *J. Org. Chem.* | **1982** | *47* | 1837 |
| 6 | Brisbois, R.G. | *J. Org. Chem.* | **1996** | *61* | 2540 |

**7-(Dimethyloxyphosphono)norcarane (7).**[3] To a stirred mixture of cyclohexene **6** (100 mL), $CH_2Cl_2$ (30 mL) and Cu powder (3.8 g) in an ice bath under $N_2$ was added **5** (10 mmol). The reaction mixture was stirred for 8 h at 0°C and 16 h at 20°C. After filtration through Celite and evaporation of the solvent, distillation afforded **7** (39%), bp 61-63°C/0.02 mm.

**1,2-Diphenylethyne (9).**[5] To a slurry of KOtBu (0.8 mmol) in THF (1.5 mL) under $N_2$ at -78°C was added a solution of **5** (0.8 mmol) in THF (2 mL) during 1 min. After another 5 min stirring, benzophenone **8** (0.7 mmol) in THF (2 mL) was added and the mixture was stirred for 16 h in a mixture THF:$Et_2NH$ 2:1. Usual work up afforded **9** in 92% yield.

# SHARPLESS Asymmetric Epoxidation

Enantioselective epoxidation of allyl alcohols by means of titanium alkoxide, (+) or (-) diethyl tartarate (DET) and t-butyl hydroperoxide (TBHP). In the presence of molecular sieves, a catalytic amount of Ti alkoxide suffices [7] (see 1st edition).

| 1 | Sharpless, K. B. | *J. Am. Chem. Soc.* | **1980** | *102* | 5974 |
|---|---|---|---|---|---|
| 2 | Sharpless, K. B. | *J. Am. Chem. Soc.* | **1981** | *103* | 464; 6237 |
| 3 | Sharpless, K. B. | *J. Org. Chem.* | **1985** | *50* | 1557 |
| 4 | Canali, L. | *J. Chem. Soc. Chem.* *Commun.* | **1997** | | 123 |
| 5 | Sharpless, K. B. | *Aldrichimica. Acta.* | **1983** | *16* | 67 |
| 6 | Katsuki,T.,Sharpless,K.B | *Org. React.* | **1996** | *48* | 1 |
| 7 | Sharpless, K. B. | *J. Am. Chem. Soc.* | **1987** | *109* | 5765 |

**2(S), 3(S)-Epoxygeraniol (2).**[1] To CH$_2$Cl$_2$ (200 ml) at −23 °C was added sequentially under stirring titanium tetraisopropoxide (5.68 g, 5.94 ml, 20 mmol). L (+) DET (4.12 g, 3.43 ml, 20 mmol) and after 5 min geraniol **1** (3.08 g, 3.47 ml, 20 mmol) and 3.67 M of (TBHP) 40 mmol in CH$_2$Cl$_2$. After 18 h at −20 °C, 10% aqueous tartaric acid (50 ml) was added under stirring and after 30 min the mixture was heated to 20 °C and stirred for 1 h. The organic layer was washed, dried and evaporated. The oily residue was diluted with Et$_2$O (150 ml) washed with 1 N NaOH (60 ml), brine, dried and the solvent evaporated. Chromatography on silica gel afforded 2.6 g of **2** (77%), 95 % ee, [α]$^{24}$$_D$ = − 6.36 °C(c 1.5, CHCl$_3$).

## SHARPLESS Asymmetric Dihydroxylation

Enantioselective syn dihydroxylation (also aminohydroxylation)[8] of olefins using AD-mix-α and AD-mix-β from phthalazine-dihydroquinidine or phthalazine- dihydroquinine and $OsO_4$ or by a new ligand $(DHQ)_2$ PYR or (DHQD)PYR respectively (see 1st edition).

(DHQ)₂PYR

**1**

(DHQD)₂PYR

**2**

**3**

2, $K_3Fe(CN)_6$

$OsO_4$

**4 (80%, 90% ee)**

| | | | | | |
|---|---|---|---|---|---|
| 1 | Sharpless, K. B. | *J. Am.Chem. Soc.* | **1989** | *110* | 1968· |
| 2 | Sharpless, K. B. | *Tetrahedron. Lett.* | **1990** | *31* | 2999; 3817 |
| 3 | Sharpless, K. B. | *J. Org. Chem.* | **1991** | *56* | 4585 |
| 4 | Soderquist, J. A. | *J. Org. Chem.* | **1992** | *57* | 5844 |
| 5 | Sharpless, K. B. | *J. Org. Chem.* | **1993** | *58* | 3785 |
| 6 | Soo Y. Ko. | *J. Org. Chem.* | **1994** | *59* | 2570 |
| 7 | Sharpless, K. B. | *Org. Synth.* | **1991** | *70* | 47 |
| 8 | Sharpless, K. B. | *Angew. Chem. Int.* | **1999** | *38* | 1080 |
| 9 | Philips, G.M.G. | *Synthesis* | **2000** | | 127 |

**(R)-3,3-Dimethy1-1,2-butandiol 4.** [5] To a well stirred solution of **2** (8.8 mg,1.0 mol%), $K_3Fe(CN)_6$ (990 mg, 3 mmol), $K_2CO_3$ (420 mg, 3 mmol) and $OsO_4$ (42 mL of a 0.25 M solution in PhMe 1.0 mol%) in 1:1 t-BuOH:$H_2O$ (5 mL of each) at 0 °C was added 3,3-dimethy1-1-butene **3** (84 mg, 1.0 mmol), After 3 h stirring, $Na_2S_2O_5$ was added (1.5 g) and the mixture was warmed to 20 °C. Extraction with $CH_2Cl_2$ was followed by drying ($MgSO_4$) and evaporation of the solvent. The crude product was flash chromatographed (silica gel, 7:3 EtOAc; hexane) to afford 94 mg of **4** (80%) as a clear oil.

## SHERADSKY–COATES–ENDO Rearrangement

Thermal hetero Cope [3,3] - rearrangement of O-arylated oximes (Sheradsky) or acid catalyzed anionic hetero [3,3] and [3,5] - rearrangement of hydroxylamines with N-O bond cleavage.(see 1st edition).

| 1 | Sheradsky, T. | *Tetrahedron Lett.* | **1966** | | 5225 |
|---|---|---|---|---|---|
| 2 | Sheradsky, T. | *Israel. J. Chem.* | **1968** | 6 | 859 |
| 3 | Sheradsky, T. | *J. Org. Chem.* | **1971** | 36 | 1061 |
| 4 | Laronze, J. Y. | *Tetrahedron Lett.* | **1989** | 30 | 2229 |
| 5 | Coates, R. M. | *J. Am. Chem. Soc.* | **1977** | 99 | 2355 |
| 7 | Endo, Y. | *Synthesis* | **1983** | | 471 |
| 8 | Endo, Y. | *Synthesis* | **1984** | | 1096 |

**3-(1-Oxo-2-tetralyl)-2-pyridone (2).**[3] A solution of **1** (0.1 g, 4.1 mmol) in ethylene glycol (20 mL) was refluxed for 20 h under $N_2$, and poured into water (100 mL). The precipitate was crystallized from EtOH to yield 0.73 g of **2** (73%), mp 206-207 °C.

**N-Methyl-N-(2-hydroxyphenyl) urea (4).**[6] To a solution of **3** (166 mg, 1 mmol) in $CH_2Cl_2$ (5 mL), TFA (3.8 mL, 50 mmol) was added under stirring at 0 °C. After 4 h stirring at 20 °C, the solvent and TFA were evaporated in vacuum and water (5 mL) was added to the residue. Extraction with EtOAc, drying ($MgSO_4$), evaporation and chromatography of the residue (sillica gel, EtOAc) afforded 141 mg of **4** (85%), mp 134-135 °C (PhH).

## SHEVERDINA–KOCHESHKOV Amination

Electrophilic amination of organolithium compounds with methyllithium-methoxamine or amination of higher order cuprates by N,O-bis(trimethylsilyl)hydroxylamine.    Also amination of aryllithium by vinyl azides.[3]

Ph$_2$CuCNLi  +  **2**  $\longrightarrow$  Ph-NH$_2$ (90%)

| | | | | | |
|---|---|---|---|---|---|
| 1 | Sheverdina, N.I.; Kocheshkov, Z. | *J. Gen. Chem. USSR* | **1938** | 8 | 1825 |
| 2 | Beak, P. | *J. Org. Chem.* | **1982** | 47 | 2822 |
| 3 | Hassner, A. | *Tetrahedron Lett.* | **1982** | 23 | 699 |
| 4 | Beak, P. | *J. Am. Chem. Soc* | **1986** | 108 | 6061 |
| 5 | Ricci, A.; Seconi, G. | *Synthesis* | **1991** | | 1201 |
| 6 | Ricci, A. | *Synlett* | **1992** | | 329 |
| 7 | Ricci, A.; Seconi, G. | *Synlett* | **1992** | | 981 |
| 8 | Ricci, A.; Seconi, G. | *J. Org. Chem.* | **1993** | 58 | 5620 |
| 9 | Erdik, E. | *Chem. Rev.* | **1989** | 89 | 1947 |

**2-Aminopyridine 3.**[8]  To a solution of n-BuLi (2 mmol-2.5 M) in hexane was added THF (10 mL) cooled to below 0°C.    The solution was cooled to -100°C and 2-bromopyridine **1** (0.388 L, 4 mmol) was added dropwise over 15 min with magnetic stirring under N$_2$.  The temperature was then allowed to rise to -80°C and the reaction mixture was kept at this temperature for 2 h.  To the deep orange solution CuCN (178 mg, 2 mmol) was added and after 30 min stirring at -80°C the temperature was allowed to rise to -60°C and N,O-bis(trimethylsilyl)hydroxylamine **2** (0.426 mL) was added.  The reaction solution was filtered through a pad of Celite, the solvent evaporated in vacuum and the residue chromatographed (silica gel 0-100, hexane-EtOAc gradient elution) to afford 110 mg of **3** (60%), mp 58-60°C.

## SHIBASAKI Cyclization

Introduction of nitrogen into organic molecules (primary enamine formation from ketones) in the presence of a titanium complex and Pd.

$$TiCl_4 \xrightarrow[\text{THF}]{\text{Mg / N}_2} (THF.Mg_2Cl_2TiN) \dashrightarrow (3THF,Mg_2Cl_2OTiNCO)$$

**1**

3 (48%)

(80%)

85%

| | | | | | |
|---|---|---|---|---|---|
| 1 | Mori, M.; Shibasaki, M. | *Tetrahedron Lett.* | **1987** | *28* | 6187 |
| 2 | Mori, M.; Shibasaki, M. | *J.Am.Chem.Soc.* | **1989** | *111* | 3725 |
| 3 | Mori, M.; Shibasaki, M. | *J.Chem.Soc.Chem.Comm.* | **1991** | | 81 |
| 4 | Mori, M.; Shibasaki, M. | *J.Synth.Org.Chem.Jpn.* | **1991** | *49* | 937 |

**Titanium complex (1).**[2] To Mg (7.0 g, 0.29 at.g) in THF (50 mL) was added TiCl$_4$ (1.9 g, 10 mmol) at -78°C under Ar. After degassing, the mixture was stirred at 20°C under N$_2$ for 16 h with a change of color and exothermicity. The unreacted Mg was removed by filtration under N$_2$ and the filtrate was stirred for 1 h at 20°C under CO$_2$. The reaction mixture under ice cooling was treated with hexane (1 mL) and the precipitate **1** was filtered, washed with Et$_2$O and dried in vacuum.

**3-Methyleneisoindoline (3).** A mixture of o-bromoacetophenone **2** (40 mg, 0.2 mmol), K$_2$CO$_3$ (55 mg, 0.4 mmol), Pd(Ph$_3$P)$_4$ (11.5 mg, 0.01 mmol) and **1** (265 mg, 0.6 mmol) in N-methylpyrrolidone (2 mL) was degassed and heated to 100°C for 16 h under a CO (1 atm) (TLC monitoring). The cooled mixture was diluted with EtOAc, stirred with water a few hours, filtered through cellite, the organic phase washed with water and the solvent evaporated in vacuum. Chromatography afforded 13 mg of **3** (48%).

## S I E G R I S T  Stilbene Synthesis

Synthesis of stilbenes by base catalyzed condensation of reactive toluenes with benzalanilines (see 1st edition).

| 1 | Siegrist, A.E. | *Helv. Chim. Acta* | **1967** | *50* | 906 |
| 2 | Siegrist, A.E. | *Helv. Chim. Acta* | **1969** | *52* | 1282; 2521 |
| 3 | Martin, R.H. | *Helv. Chim. Acta* | **1971** | *54* | 358 |
| 4 | Newman, M.S. | *J. Org. Chem.* | **1978** | *54* | 524 |

## S H E S T A K O V  Hydrazino Acid Synthesis

Synthesis of $\alpha$-hydrazino acids from $\alpha$-amino acids via ureas (see 1st edition).

| 1 | Shestakov, P. | *Z. Angew. Chem.* | **1903** | *16* | 1061 |
| 2 | Karady, S. | *J. Org. Chem.* | **1971** | *36* | 1949 |
| 3 | Viret, J. | *Tetrahedron* | **1987** | *43* | 891 |
| 4 | Kost, A.N. | *Russ. Chem. Rev.* | **1964** | *33* | 159 |

## SIMCHEN Azaheterocycle Synthesis

Cyclization of 2-cyano substituted benzoic acid chlorides to five, six and seven membered aza, diaza, and thiazabenzoheterocycles in aprotic solvents.

| 1 | Simchen, G. | *Angew. Chem. Int. Ed.* | **1966** | *5* | 663 |
| 2 | Simchen, G. | *Chem. Ber.* | **1969** | *102* | 3666 |
| 3 | Simchen, G. | *Chem. Ber.* | **1970** | *103* | 413 |
| 4 | Simchen, G. | *Angew. Chem. Int. Ed.* | **1973** | *12* | 119 |

**1,3-Dichloroisoquinoline 2.²**    2-Oximino-1-indanone **1** (1.61 g, 10 mmol) in POCl₃ (30 mL) was treated with PCl₅ (2.28 g, 11 mol) under stirring at 0°C. The clear solution was saturated with HCl and heated to 60-70°C for 2 h. A second portion of PCl₅ (2.28 g, 11 mmol) was added and heating was continued for 3-6 h at 80-100°C. After distillation of POCl₃ in vacuum, the residue was sublimed in vacuum (10⁻³ Torr), the crude product (1.98 g) was washed with NaHCO₃ solution and recrystallized to give 1.22 g of **2** (61%).

**1,3-Dichloroisoquinoline 2.²**    2-Cyanomethylbenzoic acid **3** (4.0 g, 27 mmol) and PCl₅ (10.4 g, 50 mmol) in POCl₃ (20 mL) were stirred at 20°C for 4 h and at 90°C for 5 h. Vacuum distillation of the POCl₃ gave a residue which after washing with NaHCO₃ solution and recrystallization from EtOH afforded 3 g of **2** (61%), mp. 120°C.

## SIMMONS-SMITH Cyclopropanation

Cyclopropanation from alkenes and carbenes with alkyl gem dihalides and Zn-Cu couple (Simmons-Smith) or Et$_2$Zn (Furukawa); Et$_3$Al (Yamamoto) or Sm (Molander) with high diastereoselectivity (see 1st edition).

1 → (Zn-Cu, CH$_2$I$_2$) → 2 (68%)

3 → (Sm(Hg), ICH$_2$Cl, THF, -78° to 20°C) → 4 (96%)

46 : 1

yield = 81%

| 1 | Simmons, H.E.; Smith, R.D. | *J.Am.Chem.Soc.* | **1958** | 80 | 5323 |
|---|---|---|---|---|---|
| 2 | Furukawa, J. | *Tetrahedron* | **1968** | 24 | 53 |
| 3 | Yamamoto, N. | *J.Org.Chem.* | **1985** | 50 | 4412 |
| 4 | Taguchi, T. | *Chem.Pharm.Chem.* | **1992** | 40 | 3189 |
| 5 | Molander, G.A. | *J.Org.Chem.* | **1987** | 52 | 3942 |
| 6 | Molander, G.A. | *J.Org.Chem.* | **1989** | 54 | 3525 |
| 7 | Lautens, M. | *J.Org.Chem.* | **1992** | 57 | 798 |
| 8 | Simmons, H.E. | *Org.React.* | **1973** | 20 | 1 |

**1-Fluoro-1-hydroxymethyl-2-(2-phenyllethyl)cyclopropane (2).**[4] A mixture of Zn-Cu couple (361 mg, 5.6 at g), CH$_2$I$_2$ (446.7 mg, 1.67 mmol) and olefin 1 (100 mg, 0.56 mmol) in Et$_2$O was stirred and refluxed for 10 h. The cooled mixture was diluted with Et$_2$O, quenched (aq. NH$_4$Cl), the organic phase concentrated and the residue chromatographed to give 73.4 mg of 2 (68%).

**cis-Bicyclo[4.1.0]heptan-2-ol (4).**[6]   To Sm metal (316 mg, 2.1 mmol) in THF (5 mL) was added, under Ar, a solution of HgCl$_2$ (54 mg, 0.2 mmol) in THF. After 10 min stirring, cyclohexenol 3 (49 mg, 0.5 mmol) was added followed by dropwise addition of ClCH$_2$I (483 mg, 2 mmol) maintaining a temperature of -78°C. The mixture was allowed to warm to 20°C, stirred for an additional 2 h, quenched with aq. sat. K$_2$CO$_3$ sol. Extraction with Et$_2$O, evaporation of the solvent and chromatography (silica gel, hexane:EtOAc 2:1) afforded 107 mg of 4 (96%), purity 99% (GC).

## S K A T T E B Ø L   Dihalocyclopropane Rearrangement

Rearrangement of gem-dihalocyclopropanes to allenes or of vinyl dihalocyclopropanes to cyclopentadienes and fulvenes by MeLi (see 1st edition).

| 1 | Skattebol, L. | *J. Org. Chem.* | **1964** | *29* | 2951 |
|---|---|---|---|---|---|
| 2 | Skattebol, L. | *Tetrahedron* | **1967** | *23* | 1107 |
| 3 | Skattebol, L. | *Tetrahedron Lett.* | **1977** | | 2347 |
| 4 | Skattebol, L. | *Acta Chem. Scand. B* | **1984** | *39* | 549 |
| 5 | Paquette, L.A. | *J. Am. Chem. Soc.* | **1984** | *106* | 8225 |
| 6 | Paquette, L.A. | *J. Org. Chem.* | **1987** | *52* | 2951 |

## S I M O N I S   Benzopyrone Synthesis

Benzopyrone synthesis from phenols and β-ketoesters.

| 1 | Simonis, H. | *Chem. Ber.* | **1913** | *46* | 2014 |
|---|---|---|---|---|---|
| 2 | Simonis, H. | *Chem. Ber.* | **1914** | *47* | 2229 |
| 3 | Lacey, R.N. | *J. Chem. Soc.* | **1954** | | 854 |
| 4 | Sethna, S.M. | *Chem. Rev.* | **1945** | *36* | 14 |
| 5 | Sethna, S.M. | *Org. React.* | **1953** | *7* | 15 |

**2,3,5-Trimethyl-1,4-benzopyrone 3.**[3]    A mixture of m-cresol **1** (15.0 g, 138 mmol), ethyl 2-methylacetoacetate **2** (10.0 g, 69 mmol) and $P_2O_5$ (20.0 g) was heated on a water bath for 2 h. After 45 min **1** (12.0 g, 114 mol) and $P_2O_5$ (20.0 g) was added. The cooled mixture was basified with NaOH. Work up gave 2.0 g of **3** (10%), mp. 96°C.

## S K R A U P  Quinoline Synthesis

Quinoline synthesis from anilines and acrolein or glycerol (see 1st edition).

| 1 | Skraup, Z.H. | *Chem. Ber.* | **1880** | *13* | 2086 |
|---|---|---|---|---|---|
| 2 | Yale, H.L. | *J. Am. Chem. Soc.* | **1948** | *70* | 254 |
| 3 | Wahren, M. | *Tetrahedron* | **1964** | *20* | 2773 |
| 4 | Bergstrom | *Chem. Rev.* | **1944** | *35* | 152 |
| 5 | Manske, R.H.F. | *Org. React.* | **1953** | *7* | 59 |

## S M I L E S  Aromatic Rearrangement

Rearrangement by nucleophilic aromatic substitution and aryl migration from one hetero atom to another (O to N or S to O) (see 1st edition).

| 1 | Smiles, S. | *J. Chem Soc.* | **1931** | | 2364 |
|---|---|---|---|---|---|
| 2 | Hauser, Ch.R. | *J. Org. Chem.* | **1968** | *33* | 2228 |
| 3 | Bayles, R. | *Synthesis* | **1977** | | 77 |
| 4 | Hyrota, T. | *Heterocycles* | **1995** | *41* | 1307 |
| 5 | Peet, N.P. | *J. Heterocyclic Chem.* | **1997** | *34* | 1857 |
| 6 | Bunnet, J. | *Chem. Rev.* | **1951** | *49* | 362 |
| 7 | Huisgen, R. | *Angew. Chem.* | **1960** | *72* | 314 |
| 8 | Truce, W.E. | *Org. React.* | **1970** | *18* | 100 |

**N-(p-Nitrophenyl)-2-hydroxyacetamide 2.**[3]   A solution of p-nitrophenoxyacetamide **1** (2.7 g, 13 mmol) in DMF (20 mL) was treated with a 50% suspension of NaH (330 mg). The mixture was stirred for 1 h at 50°C, water was added and the product recrystallized from EtOAc to give 1.2 g of **2** (45%), mp. 194°C.

**2-Benzylbenzenesulfinic acid 4.**[2]   To a stirred suspension of NaH (858 mg, 20 mmol) in liq. NH₃ (400 mL) was added phenyl o-tolyl sulfone **3** (4.64 g, 20 mmol) and the ammonia was replaced by THF.  After 7 h reflux the cooled solution was filtered to afford 2.73 g of **4** (54%); more **4** was recovered from the mother liquor to give a total yield of 70%.

## SMITH–MIDDLETON–ROZEN Fluorination

Conversion of carbonyls to $CF_2$ compounds by $SF_4$ (Smith) or diethylaminosulfur trifluoride (DAST) (Middleton) or by IF on hydrazones (Rozen) (see 1st edition).

(+) Ph—CH—CO$_2$H + SF$_4$ $\xrightarrow[\text{9 d}]{40°}$ (+) Ph—CH—CF$_3$
    |                                      |
   Et                                Et  (54%)

| 1 | Berg, M.A. | *Bull. Soc. Chim. Fr.* | **1925** | *37* | 637 |
| 2 | Smith, W.C. | *J. Am. Chem. Soc.* | **1959** | *81* | 3165 |
| 3 | Rozen, S. | *J. Am. Chem. Soc.* | **1987** | *109* | 896 |
| 4 | Middleton, W.J. | *J. Org. Chem.* | **1975** | *40* | 574 |
| 5 | Boswell, G.A. | *Org. React.* | **1974** | *21* | 1 |

**Iodine fluoride (IF).** A suspension of well-ground iodine (25 g), in CFCl$_3$ (500 mL) was sonicated for 30 min, cooled to –78°C and agitated with a vibromixer. Nitrogen-diluted F$_2$ (10% v/v) was bubbled through (1.1 equiv.) to give a light brown suspension of IF.

**4-tert-butyl-1,1-difluorocyclohexane (4).**[3] Ketone 1 (5 g, 33 mmol) in EtOH (15 mL) was added to hydrazine hydrate 2 (10 g) in EtOH (40 mL) and heated to reflux, then diluted with water, extracted with CHCl$_3$, dried (MgSO$_4$) and the solvent evaporated to give 5.5 g of 3 (100%). A solution of 3 (2 g, 11 mmol) in CHCl$_3$ (20 mL) at –78°C was treated with IF (6.42 g, 44 mmol) and the reaction was monitored by GC (5% SE-30 column). There was obtained 1.23 g of 4 (65%), and 10-15% of 2-iodo derivative 5.

## S N I E C K U S  Carbamate Rearrangement

Direct ortho lithiation of O-aryl carbamates and O to C carbamoyl migration to give salicylamides.

| 1 | Snieckus, V. | *Heterocycles* | **1980** | *14* | 1649 |
|---|---|---|---|---|---|
| 2 | Snieckus, V. | *J. Org. Chem.* | **1983** | *48* | 1935 |
| 3 | Snieckus, V. | *J. Am. Chem. Soc.* | **1985** | *107* | 6312 |
| 4 | Snieckus, V. | *J. Org. Chem.* | **1991** | *56* | 3763 |
| 5 | Snieckus, V. | *Acc. Chem. Res.* | **1982** | *15* | 306 |
| 6 | Snieckus, V. | *Chem. Rev.* | **1990** | *90* | 879 |

**N,N-Diethyl-2-methoxy-3-carboxybenzamide**    **2.**[3]    A    solution    of 0-(2-carboxyphenyl)-N,N-diethylcarbamate **1** (2.06 g, 8.7 mmol) in THF (10 mL) was added to sec-BuLi (13.8 mL, 19.14 mmol) (1.39 M sol) and TMEDA (2.9 mL, 19.14 mmol) in THF (170 mL) under $N_2$ at –78°C under stirring. After slow heating for 12 h to 20°C, a 25% $NH_4Cl$ solution was added, the solvent was removed in vacuum and the residue extracted with $Et_2O$. The aqueous layer was acidified, extracted with $Et_2O/CH_2Cl_2$, the residue (1.7 g) was heated with MeI (10 mL) and $K_2CO_3$ (3 g) in $Me_2CO$ (30 mL) for 20 h. Chromatography (silica gel EtOAc : hexane 1:1) afforded 976 mg of ester which after hydrolysis (NaOH 3 g, MeOH 60 mL and water 10 mL) (24 h) gave after recrystallization from $CH_2Cl_2$; hexane 84 mg of **2** (89%) mp. 123-125°C.

## S O M M E L E T  Aldehyde Synthesis

Aldehyde synthesis from primary alkyl halides with hexamethylene tetramine (see 1st edition).

| 1 | Sommelet, M. | *C.R.* | **1913** | *157* | 852 |
| 2 | Sommelet, M. | *Bull. Soc. Chim. Fr.* | **1913** | *13* | 1085(4) |
| 3 | Zaluski, M.C. | *Bull. Soc. Chim. Fr.* | **1970** | | 1445 |
| 4 | Angyal, S.J. | *Org. React.* | **1954** | *8* | 198 |

## S O M M E L E T – H A U S E R  Ammonium Ylid Rearrangement

Rearrangement of quaternary ammonium ylids to amines by aryl transfer (see 1st edition).

| 1 | Sommelet, M. | *C.R.* | **1937** | *205* | 56 |
| 2 | Hauser, C.R. | *J. Am. Chem. Soc.* | **1951** | *73* | 4122 |
| 3 | Sato, Y. | *J. Org. Chem.* | **1987** | *52* | 1844 |
| 4 | Pine, S.H. | *Org. React.* | **1970** | *18* | 404 |
| 5 | Jonczyk, A. | *Tetrahedron Lett.* | **1995** | *36* | 1355 |

## S O N N – M Ü L L E R  Aldehyde Synthesis

Aldehyde synthesis from amides or ketoximes, by reduction of imino chlorides.

$$C_6H_5-\underset{\underset{\textbf{1}}{NOH}}{\overset{\|}{C}}-C_6H_5 \xrightarrow{PCl_5} \left( C_6H_5-\underset{\underset{OPCl_4}{N}}{\overset{\|}{C}}-C_6H_5 \right) \xrightarrow[-(POCl_3)]{rearr.} C_6H_5-\underset{\underset{\textbf{2}\; C_6H_5}{N}}{\overset{\|}{C}}-Cl$$

$$\xrightarrow[HCl]{SnCl_2} C_6H_5-\underset{\underset{H}{|}}{C}=N-C_6H_5 \xrightarrow{4N\;HCl} \underset{\textbf{3}\;(85\%)}{C_6H_5-CHO}$$

| | | | | | |
|---|---|---|---|---|---|
| 1 | Sonn, A.; Müller, E. | *Chem. Ber.* | **1919** | 52 | 1929 |
| 2 | Coleman, C.R. | *J. Am. Chem. Soc.* | **1946** | 68 | 2007 |
| 3 | Ferguson, L.N. | *Chem. Rev.* | **1946** | 38 | 244 |
| 4 | Mossetig, E. | *Org. React.* | **1954** | 8 | 240 |

## S P E N G L E R – P F A N N E N S T I E L  Sugar Oxidation

Oxidation of reductive sugars in alkaline solution with molecular $O_2$ (see 1st edition).

maltose 1          2 (100%)

| | | | | |
|---|---|---|---|---|
| 1 | Spengler, O.; Pfannenstiel, A. | *DR Pat.* | 618164 | |
| 2 | Hardegger, E. | *Hel. Chim. Acta.* | **1952** 35 | 618 |
| 3 | Hardegger, E. | *Hel. Chim. Acta.* | **1951** 34 | 2343 |

**3-(α-D-Glucosido)-D-arabonic acid (2).**[2]  A solution of maltose 1 (18.0 g, 53 mmol) in water (200 mL) was added dropwise to a very well stirred solution of Ba(OH)$_2$cryst. (20 g) in water (150 mL), under a flow of $O_2$.  In 22 h there were absorbed 1250 mL of $O_2$ (calculated 1250 mL).  The mixture was saturated with $CO_2$ and filtered through Celite and 120 mL of Wofatit KS.  Concentration under vacuum afforded 17 g of crude 2 (100%).  Separation of 2 was carried out as the brucinate, mp = 152-154°C, ([α]$_D$ = 50° (c = 0.5 water).

## S O U L A  Phase Transfer Catalyst

Solid-liquid phase transfer catalyst **2** for aliphatic and aromatic nucleophilic substitution; synergistic effect with Cu in Ullmann synthesis; as ligand in homogeneous hydrogenation catalysis (see 1st edition).

| 1 | Pederson, C.J. | *J. Am. Chem. Soc.* | **1967** | 89 | 7017 |
|---|---|---|---|---|---|
| 2 | Lehn, J.M. | *Tetrahedron Lett.* | **1969** | | 2885 |
| 3 | Vogtle, F. | *Angew. Chem. Int. Ed.* | **1974** | 13 | 814 |
| 4 | Soula, G. | *Eur. Pat.* | **1978** | | 5094 |
| 5 | Soula, G. | *French Pat.* | **1979** | | 16673 |
| 6. | Soula, G. | *J. Org. Chem.* | **1985** | 50 | 3717; 3721 |
| 7 | Petrignani, J.F. | *Tetrahedron Lett.* | **1986** | 27 | 5979 |
| 8 | McKillop, L.S. | *Synth. Commun.* | **1987** | 17 | 647 |
| 9 | Bose, A.K. | *Tetrahedron Lett.* | **1987** | 28 | 2503 |
| 10 | Loupy, A. | *Synth. Commun.* | **1990** | 20 | 2833 |

**Anhydride 9.**[10] Anhydride **8** (348 mg, 2 mmol) in $CH_2Cl_2$ (3 mL), followed by $PdCl_2$ (17.7 mg, 0.1 mmol) and tris(3,6-dioxaheptyl)amine (0.5 mL) was stirred at 25°C under $H_2$ (1 atm) until absorption ceased. $Et_2O$ (100 mL) was added, the organic phase was filtered on Celite, washed with AcOH and then with water to neutrality. Drying ($MgSO_4$) and evaporation gave 334.6 mg of **9** (94%), mp 128°C.

# S P E C K A M P  Ring Closure

N-Acyliminium ions in ring closure with $\pi$-nucleophiles.

| 1 | Speckamp, W.N. | *Tetrahedron* | **1975** | *31* | 1437 |
| 2 | Speckamp, W.N. | *Tetrahedron* | **1978** | *34* | 163 |
| 3 | Speckamp, W.N. | *Tetrahedron* | **1980** | *36* | 143 |
| 4 | Speckamp, W.N. | *Rec.Trav.Chim.Pay Bas* | **1981** | *100* | 345 |
| 5 | Speckamp, W.N.; Hiemstra, H. | *Tetrahedron* | **1985** | *41* | 4367 |
| 6 | Hiemstra, H. | *J.Org.Chem.* | **1997** | *62* | 8862 |

**(1S,6R,9S)-7-Benzyl-9-hydroxy-7-azabicyclo[4.2.1]nonane-4,8-dione 2.**[6] A solution of **1** (3.19 g, 11.1 mmol) in HCOOH (55 mL) was stirred at 85°C for 2.5 days. The residue obtained after evaporation in vacuum was stirred for 1 h at 20°C in 50% methanolic NH₃ (50 mL). After evaporation of the MeOH the residue was chromatographed (CH₂Cl₂:Me₂CO 1:1) to give **2**. Recrystallization from EtOAc afforded 2.26 g of **2** (79%), mp 208-210°C, $[\alpha]_D^{20}$ = +16.9 (c 0.99, CHCl₃).

## S T A A B   Reagent

1,1'-Carbonyldiimidazole **2** an activating reagent for carboxylic acids in formation of esters, amides, peptides, aldehydes and ketones via acylimidaroles **3** (see 1st edition).

| 1 | Staab, H.A. | *Chem. Ber.* | **1956** | *89* | 1927 |
|---|---|---|---|---|---|
| 2 | Staab, H.A. | *Liebigs Ann.* | **1957** | *609* | 75, 83 |
| 3 | Staab, H.A. | *Liebigs Ann.* | **1962** | *654* | 119 |
| 4 | Komives, T. | *Org. Prep. Proc. Int.* | **1989** | *21* | 251 |
| 5 | Ley, S.V. | *Synlett* | **1990** | | 255 |
| 6 | Staab, H.A. | *Angew. Chem. Int. Ed.* | **1962** | *7* | 351 |

## S T A U D I N G E R   Azide Reduction

Conversion of organic azides with phosphines or phosphites to iminophosphoranes (phosphazo compounds) and their conversion to amines (see 1st edition).

| 1 | Staudinger, H. | *Helv. Chim. Acta.* | **1919** | *2* | 635 |
|---|---|---|---|---|---|
| 2 | Marsh, F.D. | *J. Org. Chem.* | **1972** | *37* | 2966 |
| 3 | Cooper, R.D.G. | *Pure and Appl. Chem.* | **1987** | *59* | 485 |
| 4 | Gololobov, Yu G. | *Tetrahedron* | **1981** | *37* | 437 |
| 5 | Carrie, R. | *Bull. Chem. Soc. Fr.* | **1985** | | 815 |
| 6 | Wipf, P. | *Synlett* | **1997** | | 1 |

## STAUDINGER  Ketene Cycloaddition

Cycloaddition of ketenes to olefins.

| 1 | Staudinger, H. | *Chem. Ber.* | **1908** | *41* | 594, 1516 |
| 2 | Chick, F.; Wilsmore, N.T. | *J. Chem. Soc.* | **1908** | | 946 |
| 3 | Barton, D.H.R. | *J. Chem. Soc.* | **1962** | | 2708 |
| 4 | Corey, E.J. | *J. Am. Chem. Soc.* | **1973** | *95* | 6832 |
| 5 | Hassner, A. | *J. Org. Chem.* | **1978** | *43* | 3173 |
| 6 | Hassner, A. | *J. Org. Chem.* | **1983** | *48* | 3382 |
| 7 | Hyatt, J.A.; Raynolds, P.W. | *Org. React.* | **1994** | *45* | 159 |

**2,2-Dichloro-3-trimethylsiloxy-4,4-dimethylcyclobutanone 2.**[5] To a mixture of silyl enol ether **1** (2.0 g, 13.8 mmol) and activated Zn (13.45 g, 20.7 mmol) in Et$_2$O (100 mL) under N$_2$ and stirring was added Cl$_3$CCOCl (32.6 g, 18 mmol) in Et$_2$O (45 mL) dropwise over 45 min.  Stirring under N$_2$ was continued until (NMR or GC) all **1** was consumed. The unreacted Zn was removed by filtration, the solutions concentrated in vacuo and the Zn salts precipitated with hexane.  After washing and evaporation of the solvent there were obtained 3.2 g of **2** (92%).

**2,2-Dichloro-3-hydroxy-4,4-dimethylcyclobutanone 3.**  **2** (1.0 g, 3.9 mmol) in THF (40 mL) and a few drops of 10% HCl was stirred for 1 h at 20°C followed by usual work up and distillation to give 590 mg of **3** (83%).

## S T E G L I C H - H A S S N E R  Direct Esterification

Direct room temperature esterification of carboxylic acids with alcohols, including tert. alcohols with the help of dicyclohexylcarbodiimide (DCC) and 4-diakylaminopyridine catalysts 3.

3, R : Me or $(CH_2)_4$

| 1 | Steglich, W. | *Angew. Chem. Int. Ed.* | **1969** | 9 | 981 |
| 2 | Steglich, W. | *Angew. Chem. Int. Ed.* | **1978** | 17 | 522 |
| 3 | Hassner, A. | *Tetrahedron* | **1978** | 34 | 2069 |
| 4 | Hassner, A. | *Tetrahedron Lett.* | **1978** | | 4475 |
| 5 | Steglich, W. | *Angew. Chem. Int. Ed.* | **1978** | 17 | 569 |

## S T A U D I N G E R – P F E N N I N G E R  Thiirane Dioxide Synthesis

Thiirane dioxide (episulfone) synthesis by reaction of diazomethane with sulfenes or $SO_2$ (see 1st edition).

| 1 | Staudinger, H.; Pfenninger, F. | *Chem. Ber.* | **1916** | 42 | 1941 |
| 2 | Hesse, G. | *Chem. Ber.* | **1957** | 90 | 1166 |
| 3 | Opitz, G. | *Z. Naturforschung.* | **1963** | b18 | 775 |
| 4 | Opitz, G. | *Angew. Chem.* | **1961** | 77 | 41 |
| 5 | Fischer, N.H. | *Synthesis* | **1970** | | 396 |

## S T E P H E N  Aldehyde Synthesis

Synthesis of aldehydes from nitriles and $SnCl_2 \cdot HCl$ (see 1st edition).

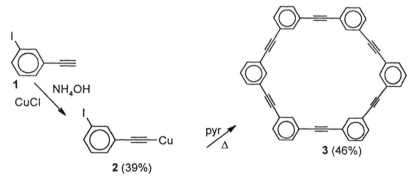

| 1 | Stephen, J. | *J. Chem. Soc.* | **1925** | *127* | 1874 |
|---|---|---|---|---|---|
| 2 | Stephen, T.W. | *J. Chem. Soc.* | **1956** | | 4695 |
| 3 | Tolbert, J. | *J. Org. Chem.* | **1963** | *28* | 696 |
| 4 | Ferguson, L.W. | *Chem. Rev.* | **1946** | *38* | 243 |
| 5 | Mosettig, E. | *Org. React.* | **1954** | *8* | 246 |

## S T E P H E N S – C A S T R O  Acetylene Cyclophane Synthesis

Polyacetylene cyclophane synthesis from an iodophenyl copper acetylide (see 1st edition).

| 1 | Stephens, R.D.; Castro, C.E. | *J. Org. Chem.* | **1963** | *28* | 3313 |
|---|---|---|---|---|---|
| 2 | Campbell, I.D. | *Chem. Commun.* | **1966** | | 87 |
| 3 | Staab, H.E. | *Chem. Ber.* | **1970** | *103* | 1157 |
| 4 | Staab, H.E. | *Synthesis* | **1974** | | 424 |

**STETTER**  1,4-Dicarbonyl Synthesis

Michael addition of aromatic or heterocyclic aldehydes (via cyanohydrins) to α,β-unsaturated systems.    Also addition of an aliphatic aldehydes catalyzed by thiazolium ylids (see 1st edition).

| 1 | Stetter, H. | *Angew. Chem.* | **1973** | *85* | 89 |
|---|---|---|---|---|---|
| 2 | Stetter, H. | *Chem. Ber.* | **1974** | *107* | 210 |
| 3 | Stetter, H. | *Synthesis* | **1975** | | 379 |
| 4 | Stetter, H. | *Angew. Chem. Int. Ed.* | **1976** | *15* | 639 |
| 5 | Stetter, H. | *Org. Synth.* | **1985** | *65* | 26 |
| 6 | Enders, D. | *Helv. Chim. Acta.* | **1996** | *79* | 1899 |

**4-oxo-4-phenylbutanenitrile 3.[4]**  A solution of Ph-CHO **1** (10.6 g, 0.1 mol) in DMF (50 mL) was added over 10 min to a stirred NaCN (2.45 g, 0.05 mol) in DMF (50 mL) at 35°C.  After 5 min acrylonitrile **2** (4 g, 0.075 mol) in DMF (100 mL) is added over 20 min at 35°C.   After 3 h stirring and work up, one obtained 9.5 g of **3** (80%), bp 114°C/0.3 torr, mp 70°C.

**Chroman-4-one 6.[6]**  To a stirred solution of 4-(2-formylphenoxy)-but-2-enoate **4** (275 mg, 1.25 mmol) and chiral catalyst **5** (118 mg. 0.25 mmol) in THF (40 mL) were added $K_2CO_3$ (17.5 mg) at 20°C.  After 24 h the mixture was diluted with water, extracted with $CH_2Cl_2$ and the solvent evaporated.   The residue after chromatography (silica gel $Et_2O$:pentane 1:1) gave 200 mg, of **6** (73%) yield, 60% ee, config R, $\alpha_D^{20} = -4.6°$.

# STEVENS Rearrangement

Base catalyzed migration of one alkyl group from a quaternary nitrogen atom to the $\alpha$-carbon atom of a second alkyl group or to an ortho aromatic position (see also Sommelet-Hauser) (via ammonium ylids, or oxonium ylids)[7].

| 1 | Stevens, T.S. | *J.Chem.Soc.* | 1928 | | 3193 |
|---|---|---|---|---|---|
| 2 | Stevens, T.S. | *J.Chem.Soc.* | 1932 | | 1932 |
| 3 | Hauser, C.R. | *J.Org.Chem.* | 1958 | 23 | 354 |
| 4 | Stevens, T.S. | *Prog.Org.Chem.* | 1960 | 7 | 48 |
| 5 | Sato, Y. | *J.Org.Chem.* | 1988 | 53 | 194 |
| 6 | Joncze, A. | *J.Org.Chem.* | 1991 | 56 | 6933 |
| 7 | West, F.G. | *J.Org.Chem.* | 1992 | 57 | 3479 |
| 8 | Sato, Y. | *J.Org.Chem.* | 1992 | 57 | 5034 |
| 9 | Coldham, J. | *Synlett* | 2000 | | 236 |
| 10 | Pedrosa, R. | *Synlett* | 2000 | | 893 |
| 11 | Pine, S.M. | *Org.React.* | 1970 | 57 | 5034 |

**2-Methyl-1-dimethylaminomethylnaphthalene 2.**[3] To a rapidly stirred suspension of sodamide (31.2 g, 0.8 mol) in liquid ammonia (1200 mL) were added in 30 min 2-naphthymethyltrimethylammonium chloride **1** (94.3 g, 0.4 mol). After 2 h stirring, the mixture was quenched with NH$_4$Cl. The ether solution was concentrated to afford 66.5 g of **2** (84%), bp 152-153°C/10 mm.

**Tetrahydro-2-benzyl-4,4-dimethylfuran-3-one 4.**[7] A 0.04M solution of **3** (92 mg, 0.4 mmol) in CH$_2$Cl$_2$ (10 mL) was added dropwise (0.5 mmol/h) to a $4 \times 10^{-4}$m solution of Rh(OAc)$_2$ (5.3 mg, 0.012 mmol) in CH$_2$Cl$_2$ (30 mL) under N$_2$. The mixture was stirred for an additional 30 min then washed (0.5M K$_2$CO$_3$, brine) and dried (MgSO$_4$). Evaporation and flash chromatography (silica gel EtOAc:hexane 15:85) afforded 49 mg of **4** (65%).

# STILES–SISTI Formylation

Synthesis of aldehydes by formylation of Grignard reagents with p-dimethylaminobenzaldehyde and a diazonium salt (see 1st edition).

| 1 | Stiles, M.; Sisti, A. | *J. Org. Chem.* | **1960** | *25* | 1691 |
| 2 | Sisti, A. | *J. Org. Chem.* | **1962** | *27* | 279 |
| 3 | Sisti, A. | *J. Chem. Eng. Data* | **1964** | *9* | 108 |

**Cyclohexanecarboxaldehyde (4).** [2] A solution of sulfanilic acid (100 g, 0.31 mol) in water (200 mL) and $Na_2CO_3$ (18.4 g, 0.18 mmol) was diazotized with HCl (64 mL) and $NaNO_2$ (24.4 g, 0.35 mol) in water (75 mL) at 0-5 °C. NaOAc (70 g) in water (200 mL) was added to pH=6. A solution of **3** (45.8 g, 0.2 mol) in acetone, obtained from **1** and **2**, was added. The red solution was stirred for 30 min at 0-5 °C, for 30 min at 20 °C diluted with water, extracted with $Et_2O$ and distilled gave 15.45 g of **4** (69%), bp 50-53 °C/20 mm.

# SZARVASY–SCHÖPF Carbomethoxylation

Carboxylation of activated CH groups with MMMC (methoxy magnesium methyl carbonate) **1** (Szarvasy) and addition of the resulting activated groups to C=N bonds (Schöpf) (see 1st edition).

| 1 | Szarvasy, S. | *Chem. Ber.* | **1887** | *30* | 1836 |
| 2 | Finkbeiner, H. L. | *J. Org. Chem.* | **1963** | *28* | 215 |
| 3 | Schöpf, C. | *Liebigs Annl.* | **1959** | *626* | 123 |
| 4 | Schöpf, C. | *Angew. Chem.* | **1949** | *61* | 31 |
| 5 | Grisar, J. M. | *Synthesis* | **1974** | | 284 |

**(2).** [2] A 2 N solution of MMMC **1** (100 mL) was heated to 60 °C under a flow of $CO_2$, 1-nitropropane (9.80 g, 0.2 mol) was added and $CO_2$ was replaced by $N_2$. After 6 h at 60 °C, 32% HCl (60 mL) and ice (75 g) were added, the acid was extracted with $Et_2O$, the solvent evaporated and the residue esterified by MeOH-HCl to afford 6.5 g of **2** (44%), bp 77 °C (2.5 mm).

## STILL – GENNARI  Z-Olefin Synthesis

A modified Horner-Wadsworth-Emmons reagent with high Z stereoselectivity using trifluoroethyl phosphonates in reaction with saturated, unsaturated or aromatic aldehydes.

| | | | | |
|---|---|---|---|---|
| 1 | Still, W.C., Gennari, C. | *Tetrahedron Lett.* | **1983** | 24 | 4405 |
| 2 | Bodnarchuk, N.D. | *Zh. Obshch. Khim.* | **1970** | 40 | 1210 |

**Z- Methyl cinnamate 3.** A solution of **2** (318 mg, 1 mmol), 18-crown-6 (5 mmol) $CH_3CN$ complex in THF (20 mL) was cooled under $N_2$ at -78°C and treated with K-N(TMS)$_2$ in PhMe (1 mmol, 0.6M). 4-Methoxy-benzaldehyde **1** (136 mg, 1 mmol) was then added and the mixture was stirred for 30 min at -78°C. Quenching with saturated $NH_4Cl$, extraction ($Et_2O$) and chromatography gave 182.4 mg of **3** (95%), Z:E = 50:1.

## S T I L L E  Carbonyl Synthesis

Synthesis of aryl ketones or aldehydes from aryl triflates or iodides and organo stannanes in the presence of CO and a palladium catalyst (see 1st edition).

$$p\text{-MeO}-C_6H_4-I \quad + \quad CO \quad + \quad Bu_3SnH \quad \xrightarrow[\text{CO, 50°}]{(Ph_3P)_4Pd} \quad p\text{-MeO}-C_6H_4-CHO$$

**1**                                                                    **2** (77%)

| 1 | Stille, J.K. | *J. Am. Chem. Soc.* | **1983** | *105* | 7175 |
| 2 | Stille, J.K.. | *J. Am. Chem. Soc.* | **1987** | *109* | 5478 |
| 3 | Stille, J.K. | *J. Am. Chem. Soc.* | **1988** | *110* | 1557 |

**p-Methoxybenzaldehyde (2).**[1]  p-Methoxyiodobenzene **1** (234 mg, 1 mmol) in PhH (4.0 mL) and tetrakis(triphenylphosphine)palladium (0) (35.6 mg) were maintained under 1 atm. of CO at 50°C. A solution of tributyltin hydride (350 mg, 1.1 mmol) was added via syringe pump over 2.5 h. Tributyltin halide was removed and purification by chromatography afforded 104 mg of **2** (77%) (CG yield 100%).

**(E)-1-(p-Methoxyphenyl)-3-phenyl-2-propen-1-one (5).**[3]  To 4-methoxyphenyl triflate **3** (390 mg, 1.52 mmol) in DMF (7 mL) were added (E)-phenyl-tri-n-butylstyrylstannane **4** (645 mg, 1.64 mmol) LiCl (200 mg, 4.72 mmol), dichloro-1,1'-bis(diphenylphosphino) ferrocene palladium (II)/PdCl$_2$(dppf)/(45 mg, 0.06 mmol), a few crystals of 2,6-di-tert-butyl-4-methylphenol and 4 Å molecular sieves (100 mg). The mixture was heated at 70°C under CO (1 atm). Work up after 23 h and chromatography (hexane: EtOAc 20:1) afforded 246 mg of **5** (58%), mp. 105-106°C.

# STILLE Cross Coupling

Coupling of organotin reagents (and Pd catalyst) with aryl or vinyl halides or triflates, acyl chlorides or allyl acetates. (see 1st edition).

| | Stille, J.W. | *J. Am. Chem. Soc.* | **1984** | *106* | 4630 |
|---|---|---|---|---|---|
| 1 | Stille, J.W. | *J. Am. Chem. Soc.* | **1984** | *106* | 4630 |
| 2 | Stille, J.W. | *Angew. Chem. Int. Ed.* | **1985** | *25* | 508 |
| 3 | Stille, J.W. | *J. Org. Chem.* | **1990** | *55* | 3019 |
| 4 | Sucholeiki, I. | *J. Org. Chem.* | **1995** | *60* | 523 |
| 5 | Farina, V. | *Org. React.* | **1997** | *50* | 1 |

**1-Vinyl-4-tert-butylcyclohexene (3).**[1]    To LiCl (0.56 g, 13 mmol) and tetrakis (triphenylphosphine) palladium (0) (0.032 g, 0.028 mol, 1.6 mol%) under Ar was added THF (10 mL) followed by a solution of vinyl triflate **1** (0.51 g, 1.8 mmol) and tributylvinyltin **2** (0.56 g, 1.8 mmol) in THF (10 mL). The slurry was heated to reflux for 17 h, cooled to 20°C and diluted with pentane (60 mL). The mixture was washed with 10% NH₄OH solution, dried (MgSO₄), filtered through a short pad of silica gel and the solvent evaporated in vacuum to afford 0.26 g of **3** (91%).

**(2E,6E)-1-(4-Methyl-2,6-dimethoxyphenyl)-3,7,11-trimethyl-2,6,10-dodecatriene (6).**[2]   From **5** (1.57 g, 5 mmol), trans, trans-farnesyl acetate **4** (1.32 g, 5 mmol), LiCl (0.632 g, 15 mmol) and (bis(dibenzylideneacetone)palladium (0.144 g, 0.25 mol, 5 mol%). Column chromatography (silica gel, 5% EtOAc/hexane) afforded 1.26 g of **6** (71%), R$_f$ = 0.53.

## S T O R K  Enamine Alkylation

α-Alkylation and acylation of ketones via enamines or imines.  Also Michael addition via enamines (see 1st edition).

| 1 | Stork, G. | *J. Am. Chem. Soc.* | **1954** | *76* | 2029 |
|---|---|---|---|---|---|
| 2 | Hunig, S. | *Chem. Ber.* | **1962** | *95* | 2493 |
| 3 | Hickmott, P.W. | *Tetrahedron* | **1982** | *38* | 1975 |
| 4 | Knebine, M.F. | *Synthesis* | **1970** | | 510 |
| 5 | Stork, G. | *J. Am. Chem. Soc.* | **1971** | *93* | 5939 |
| 6 | Willemin, D. | *Synth. Commun.* | **1996** | *26* | 2901 |
| 7 | Nasi, R. | *Tetrahedron* | **1998** | *54* | 10851 |

**N-(1-Cyclohexen-1-yl)morpholine (3).**      Cyclohexanone **1** (19.6 g, 0.2 mol), morpholine **2** (19.08 g, 0.22 mol) and TsOH (catalyst) in PhH or PhMe was refluxed with a Dean-Stark unit.  After water was removed azeotropically, distillation afforded 28 g of **3** (85%), bp. 117-118°C.

**2-(Δ[10]-Undecenoyl)cyclohexanone (5).**[2]   To **3** (18.4 g, 0.11 mol) and TEA (15.3 mL, 0.11 mol) in CHCl$_3$ (130 mL) was added **4** (20.2 g, 0.1 mol) in CHCl$_3$ (90 mL) at 35°C during 2.5 h.  After 12 h the red solution was refluxed with 32% HCl (50 mL) for 5 h. Separation of water, washing and distillation afforded 18.4 g of **5** (70%), bp. 132-136°C (0.003 mm).

## S T O R K  Reductive Cyclization

Cyclization of acetylenic ketones to allyl alcohols by one electron reduction with Li/NH₃; also electrochemically (Shono) or by $SmI_2$ (Kagan-Molander) (see 1st edition).

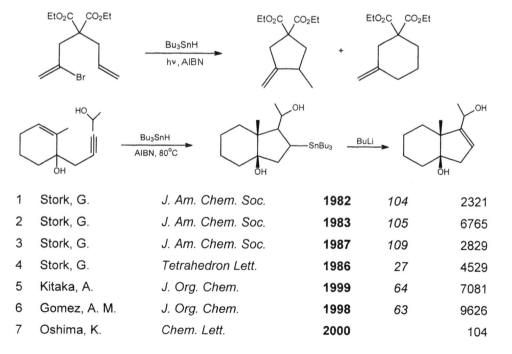

60%

| | | | | | |
|---|---|---|---|---|---|
| 1 | Stork, G. | *J. Am. Chem. Soc.* | **1965** | *87* | 1148 |
| 2 | Pradhan, S. K. | *J. Org. Chem.* | **1976** | *41* | 1943 |
| 3 | Stork, G. | *J. Am. Chem. Soc.* | **1979** | *101* | 7107 |
| 4 | Shono, T. | *Chem. Lett.* | **1976** | | 1233 |
| 5 | Molander, G. A. | *J. Am. Chem. Soc.* | **1989** | *111* | 8236 |
| 6 | Cossy, J. | *Pure Appl. Chem.* | **1992** | *64* | 1883 |
| 7 | Cossy, J. | *J. Org. Chem.* | **1998** | *63* | 3141 |

## S T O R K  Radical Cyclization

Free radical cyclization with preferential formation of cyclopentanes (see 1st edition).

| | | | | | |
|---|---|---|---|---|---|
| 1 | Stork, G. | *J. Am. Chem. Soc.* | **1982** | *104* | 2321 |
| 2 | Stork, G. | *J. Am. Chem. Soc.* | **1983** | *105* | 6765 |
| 3 | Stork, G. | *J. Am. Chem. Soc.* | **1987** | *109* | 2829 |
| 4 | Stork, G. | *Tetrahedron Lett.* | **1986** | *27* | 4529 |
| 5 | Kitaka, A. | *J. Org. Chem.* | **1999** | *64* | 7081 |
| 6 | Gomez, A. M. | *J. Org. Chem.* | **1998** | *63* | 9626 |
| 7 | Oshima, K. | *Chem. Lett.* | **2000** | | 104 |

## STORK–HÜNIG  Cyanohydrin Alkylation

Conversion of aldehydes to ketones via cyanohydrin, derivatives (ethers) by alkylation or Michael addition; also via cyanohydrin silyl ethers, or via α-dialkylaminonitriles (see also Stetter reaction), (see 1st edition).

| 1 | Stork, G. | *J. Am. Chem. Soc.* | **1971** | 93 | 5286 |
| 2 | Stork, G. | *J. Am. Chem. Soc.* | **1974** | 96 | 5272 |
| 3 | Hünig, S. | *Synthesis* | **1973** | | 777 |
| 4 | Hünig, S. | *Chem. Ber.* | **1979** | 112 | 2062 |
| 5 | Hünig, S. | *Chem. Ber.* | **1981** | 114 | 959 |
| 6 | Albright, J.D. | *Tetrahedron* | **1983** | 39 | 3207 |
| 7 | Hünig, S. | *Synthesis* | **1975** | | 180 |
| 8 | Watt, D.S. | *Org. React.* | **1984** | 31 | 47 |

**O-(Trimethylsilyl)-benzaldehyde cyanohydrin 9.**  To a mixture of **8** (14.85 g, 0.15 mol) and AlCl₃ (0.3 g), was added PhCHO **7** (10.6 g, 0.1 mol). After 1 h at 40-50°C and distillation, one obtains 14.5 g of **9** (72%), bp 64°CC/0.5 Torr.

**1,2,2-Triphenyl-2-(trimethylsiloxy)-1-ethanone 10.**[4]  A solution of **9** (10 mmol) and LDA in DME was treated with benzophenone (1.7 g, 10 mmol) at -78°C. After warming to 20°C, 3.67 g of **10** (98%) were isolated. Hydrolysis of **10** gave 3.17 g of **11** (89%), mp 84°C.

## STRECKER Aminoacid Synthesis

Synthesis of α–amino acids from aldehydes or ketones via cyanohydrins (see 1st edition).

| | | | | | |
|---|---|---|---|---|---|
| 1 | Strecker, A. | *Liebigs Ann.* | **1850** | 75 | 27 |
| 2 | Pollard, C. B. | *J. Am. Chem. Soc.* | **1955** | 77 | 40 |
| 3 | Weinges, K. | *Chem. Ber.* | **1971** | 104 | 3594 |
| 4 | Stoul, D. | *J. Org. Chem.* | **1983** | 48 | 5369 |
| 5 | Georgiades, M. P. | *Synthesis* | **1989** | | 616 |
| 6 | Mowry, D. T. | *Chem. Rev.* | **1948** | 42 | 236 |

**(S)-α-Methyl-3,4-dimethoxyphenylalanine (5).**[3] (4S,5S) **1** (20.7 g, 0.1 mol), ketone **2**(19.4 g, 0.1 mol) and NaCN (5.4 g, 0.11 mol) in MeOH (70 mL) was heated to 60 °C and HOAc (9 mL) was added dropwise. The mixture was cooled, filtered, stirred with water (100 mL) for 1 h and filtered. Crystallization from MeOH afforded 33.6 g (82%) of **3**, mp 127-128 °C, $(\alpha)_D$=+85.7°. **3** (14 g, 40 mmol) was added to cooled conc HCl (100 mL). After stirring for 1 h at -5 °C, 1 h at 20 °C and 4 h at 50 °C, the mixtrue was cooled for 2 h, filtered and recrystallized from MeOH to give 11.6 g (83%) of **4**, mp 208 °C, $(\alpha)_D$=-8.4 °. Heating of **4** with Raney-nickel and 2 N NaOH at 120 °C for 29 h gave **5** as **5.HCl** in 98% yield, mp 174-175 °C, $(\alpha)_D$=-4.3 °.

## STRYKER Regioselective Reduction

Regioselective conjugate reduction and reductive silylation of α,β-unsaturated ketones, esters, and aldehydes, also of acetylenes using a stable copper (I) hydride cluster $[(Ph_3P)CuH]_6$ (see 1st edition).

| | | | | | |
|---|---|---|---|---|---|
| 1 | Stryker, J. M. | *Tetrahedron Lett.* | **1988** | 29 | 3749 |
| 2 | Stryker, J. M. | *J. Am. Chem. Soc.* | **1988** | 110 | 291 |
| 3 | Stryker, J. M. | *Tetrahedron Lett.* | **1989** | 30 | 5677 |
| 4 | Stryker, J. M. | *Tetrahedron Lett.* | **1990** | 31 | 3237 |
| 5 | Stryker, J. M. | *J. Am. Chem. Soc.* | **1989** | 111 | 8818 |

## **SUAREZ** Photochemical Iodo Functionalization

Photochemical radical reaction of alcohols, amines in the presence of iodine and hypervalent iodine reagents: $PhI(OAc)_2$ (diacetoxyiodo)benzene (DIB), PhIO iodoxylbenzene leading to decarboxylation, transannular functionalization, radical amidation, fragmentation.

| | | | | | | |
|---|---|---|---|---|---|---|
| 1 | Suarez, E. | *Tetrahedron Lett.* | **1984** | *25* | 1953 |
| 2 | Suarez, E. | *J.Org.Chem.* | **1986** | *51* | 402 |
| 3 | Stork, G. | *Tetrahedron Lett.* | **1989** | *30* | 3609 |
| 4 | Millan, J. | *Tetrahedron Lett.* | **1991** | *32* | 7493 |
| 5 | Suarez, E. | *Angew.Chem.Int.Ed.* | **1992** | *31* | 772 |
| 6 | Lusztik, J. | *Tetrahedron Lett.* | **1994** | *35* | 1003 |
| 7 | Danishefsky, S.J. | *Tetrahedron Lett.* | **1993** | *34* | 3989 |
| 8 | Yokoyama, M. | *J.Chem.Soc.Perkin 1* | **1997** | | 787 |
| 9 | Yokoyama, M. | *J.Org.Chem.* | **1998** | *63* | 5193 |
| 10 | Suarez, E. | *J.Am.Chem.Soc.* | **1993** | *115* | 8865 |
| 11 | Suarez, E. | *J.Org.Chem.* | **1998** | *63* | 2099, 4697 |
| 12 | Suarez, E. | *Tetrahedron Lett.* | **2000** | *41* | 2495, 7869 |

**N-(Trifluoromethanesulfonyl)-1,2,3,4-tetrahydroquinoline (2).[9]** To sulfonamide 1 (135.5 mg, 0.5 mmol) in 1,2 dichloroethane were added DIB (also called iodobenzenediacetate (257.6 mg, 0.8 mmol) and iodine (127 mg, 0.5 mmol). The mixture was irradiated with a tungsten lamp (500W) at 60-70°C for 2 h under Ar. The mixture was poured into saturated aq. $Na_2SO_3$ and extracted with EtOAc. Evaporation of the solvent and preparative TLC (silica gel, hexane:EtOAc 2:1 -> 8:1) afforded 94 mg of 2 (71%).

## SUZUKI (KYODAI*)   Nitration

Nitration of aromatic compounds by oxides of nitrogen and ozone.
*Symbolic abbreviation of Kyoto University.

$$C_6H_6 + NO_2 + O_3 \xrightarrow{-10°, 1h} C_6H_5\text{-}NO_2$$
$$\mathbf{1} \qquad\qquad\qquad\qquad \mathbf{2}\ (100\%)$$

$$C_6H_5\text{-CHO} + NO_2 + O_3 \xrightarrow[0°]{MeSO_3H} O_2N\text{-}C_6H_4\text{-CHO} \qquad o\ :\ m:p$$
$$\mathbf{3} \qquad\qquad\qquad\qquad \mathbf{4}\ (100\%) \qquad 32:64:4$$

| | | | | |
|---|---|---|---|---|
| 1 | Suzuki, H. | *Chem. Let.* | **1991** | 817 |
| 2 | Suzuki, H. | *J. Chem. Soc. Perkin 1* | **1993** | 1591 |
| 3 | Suzuki, H. | *J. Chem. Soc. Perkin 1* | **1994** | 903 |
| 4 | Suzuki, H. | *Synthesis* | **1994** | 841 |
| 5 | Suzuki, H. | *Synlett* | **1995** | 383 |

**Nitrobenzene 2.**[2]   Benzene **1** (780 mg, 10 mmol) in $CH_2Cl_2$ (50 mL) was cooled to $-10°C$. A flow of ozonized oxygen and nitrogen dioxide was slowly introduced for 1 h. Quenching with aq. $NaHCO_3$ and work up afforded **2** in nearly quantitative yield.

**Nitrobenzaldehydes 4.**[3]   Benzaldehyde **3** (1.06 g, 10 mmol) and methane-sulfonic acid (530 mg, 5 mmol) in $CH_2Cl_2$ (50 mL) were cooled below 0°C and treated in the same manner as shown for **2**. Progress of the reaction was monitored by TLC. After 3 h the reaction mixture was worked up to give a mixture of **4** (o:m:p 32:64:4) in quantitative yield.

## S U Z U K I  Coupling

Pd catalyzed cross-coupling reactions of aryl, alkynyl or vinyl halides with aryl or vinyl boronic acids (see 1st edition).

| 1 | Suzuki, A. | Tetrahedron Lett. | **1979** | *20* | 3437 |
| 2 | Suzuki, A. | J. Am. Chem. Soc. | **1985** | *107* | 972 |
| 3 | Kishi, Y. | J. Am. Chem. Soc. | **1987** | *109* | 4756 |
| 4 | Suzuki, A. | Pure. Appl. Chem. | **1991** | *63* | 419 |
| 5 | Novak, B.M. | J. Org. chem. | **1994** | *59* | 5034 |
| 6 | Beller, M. | Angew. Chem. Int. Ed. | **1995** | *34* | 1848 |

**Arylmethyl ketone 4.**[6]   A mixture of p-bromoacetophenone **1** (1.99 g, 10 mmol), phenylboronic acid **2** (1.83 g, 15 mmol), $K_2CO_3$ (2.76 g, 20 mmol) and palladacycle catalyst **3** (4.24 mg, 0.5 mmol), was heated in o-xylene (100 mL) to reflux. Washing with water, evaporation of the solvent and chromatography, afforded 1.8 g of **4** (92%).

## S U Z U K I  Selective Reduction

Selective reduction of nitriles or amides with $NaBH_4$ in the presence of transition metal salts.

$$C_6H_5-CN \quad \xrightarrow[CoCl_2]{NaBH_4} \quad C_6H_5-CH_2-NH_2$$
$$\textbf{1} \qquad\qquad\qquad \textbf{2} \ (72\%)$$

$$p-O_2N-C_6H_4-CH_2-CN \quad \xrightarrow[-10°C]{NaBH_4/Co^{++}} \quad p-O_2N-C_6H_4-CH_2-CH_2-NH_2$$

$$CH_2{=}CH-CN \quad \xrightarrow[MeOH/20°C]{NaBH_4/Co^{++}} \quad CH_2{=}CH-CH_2-NH_2$$

$$C_6H_5-CONH_2 \quad \xrightarrow[MeOH/30°C]{NaBH_4/Co^{++}} \quad C_6H_5-CH_2-NH_2$$

| | | | | | |
|---|---|---|---|---|---|
| 1 | Suzuki, S.; Suzuky, Y. | *Tetrahedron Lett.* | **1969** | | 4555 |
| 2 | Suzuki, S. | *Chem. and Ind.* | **1970** | | 1626 |
| 3 | Atta-ur Rahman | *Tetrahedron Lett.* | **1980** | | 1773 |
| 4 | Sung-eun Yoo | *Synlett* | **1990** | | 419 |
| 5 | Paquette, L.A. | *J. Am. Chem. Soc.* | **1994** | *116* | 4689 |

**Benzyl amine 2.**[1]  To a solution of benzonitrile 1 (5.0 g, 50 mmol) and $CoCl_2{\cdot}6H_2O$ (23.8 g, 100 mmol) in MeOH (300 mL), were added $NaBH_4$ (19.0 g, 500 mmol) in portions under stirring at 20°C. After addition was complete, the stirring was continued for 1 h at the same temperature. The reaction mixture was acidified with 3N HCl (100 mL) under stirring until the black precipitate was dissolved. The solvent was removed by distillation and the unreacted 1 by extraction with $Et_2O$. The aqueous layer was basified with conc. $NH_4OH$ and extracted with $Et_2O$. After washing (brine) and drying ($MgSO_4$) of the ether extract, the solvent was removed and the residue distilled to afford 3.6 g of 2 (72%), bp. 90°C/11 mm Hg. 2 hydrochloride, mp. 254°C.

## SUZUKI–MIYAURA Coupling

Palladium or nickel-catalyzed coupling of organoboron compounds with unsaturated halides or triflates.

| 1 | Suzuki, A. | *Tetrahedron Lett.* | **1979** | *20* | 3437 |
|---|---|---|---|---|---|
| 2 | Miyaura, N.; Suzuki, A. | *J. Am. Chem. Soc.* | **1989** | *111* | 314 |
| 3 | Miyaura, N.; Suzuki, A. | *J. Org. Chem.* | **1993** | *58* | 2201 |
| 4 | Johnson, C. R. | *J. Am. Chem. Soc.* | **1993** | *115* | 11014 |
| 5 | Miyaura, N. | *J. Org. Chem.* | **1997** | *62* | 8024 |
| 6 | Beller, M. | *Angew. Chem. Int. Ed.* | **1995** | *34* | 1848 |
| 7 | Miyaura, N.; Suzuki, A. | *Chem. Rev.* | **1995** | *95* | 2457 |

**2-[(*E*)-3-(*t*-Butyldimethylsiloxy)-1-octenyl]-2-carboethoxycyclopentene (3).**[2] A mixture of triflate **1** (288 mg, 1 mmol), boronate (396 mg, 1.1 mmol), Pd(PPh₃)₄(29 mg, 0.025 mmol), and K₃PO₄·H₂O (fine power, 318 mg) in dioxane (5 mL) was stirred at 85 °C for 10 h under N₂. The mixture was diluted with toluene (10 mL), washed with brine, and dried over MgSO₄. Chromatography over silica gel gave **3** (96% based on **1**).(TBS=SiMe₂ *t*-Bu).

## S W A R T S  Fluoroalkane Synthesis

Substitution of chlorine atoms with fluorine atoms by means of SbF$_5$ (see 1st edition).

$$Cl_2FC-CCIF-CCl_3 + SbF_5 \xrightarrow{\Delta} Cl_2FC-CCIF-CCIF_2$$

**1**                                              **2** (70%)

$$p-Cl-C_6H_4-CCl_3 + SbF_5 \xrightarrow{\Delta} p-Cl-C_6H_4-CF_3$$

**3**                                         **4** (95%)

| | | | | | |
|---|---|---|---|---|---|
| 1 | Swarts, F. | *Bull. Acad. Royal Belge* | **1892** | *24* | 309 |
| 2 | Swarts, F. | *Rec. Trav. Chim.* | **1915** | *35* | 131 |
| 3 | Henne, A.I. | *J. Am. Chem. Soc.* | **1941** | *63* | 3478 |
| 4 | Finger, G.C. | *J. Am. Chem. Soc.* | **1956** | *78* | 6034 |
| 5 | Finger, G.C. | *Org. React.* | **1994** | *2* | 49 |

**1,1,2,3-Tetrachloro-1,2,3,3-tetrafluoropropane**                                        **(2)**.[3]
1,1,1,2,3-Pentachloro-2,3,3-trifluoropropane **1** (213 g, 1 mol) was heated in a steel vessel with SbF$_5$ (10.8 g, 0.05 mol). From the reactor **2** (bp. 112°C) was distilled and **1** was refluxed back (bp. 152°C) by raising the temperature slowly and progressively from 125°C to 170°C. Finally the temperature was raised to force out the organic material with a small amount of SbF$_5$. The distillate was steam distilled from a 10% NaOH solution to give 117.8 g of **2** (70%) and 15% recovery of **1**.

**p-Chloro-α,α,α-trifluorotoluene (4).**[4] A mixture of p-chloro-α,α,α-trichlorotoluene **3** (23.0 g, 0.1 mol) and SbF$_5$ (29.58 g, 0.11 mol) was heated until the reaction started. After completion of the reaction, the mixture was washed with 6 N HCl and dried on BaO. Distillation afforded 17.1 g of **4** (95%), bp. 136-138 °C, mp. -36°C, $n_D^{20}$ = 1.4463, D = 1.353.

## S W E R N - P F I T Z N E R – M O F F A T Oxidation

Oxidation of alcohols to aldehydes or ketone by DMSO activated with DCC (Pfitzner-Moffat), $Ac_2O$, $(COCl)_2$, TFA, (Swern), $P_2O_5$, or pyridine-$SO_3$ (see 1st ed).

$$p\text{-}O_2N\text{-}C_6H_4\text{-}CH_2\text{-}OH \xrightarrow{\text{DMSO-DCC}} p\text{-}O_2N\text{-}C_6H_4\text{-}CHO$$

**1**                                              **2** (92%)

| 1  | Pfitzner, K.E.; Moffat, J.G. | *J.Am.Chem.Soc.*     | **1963** | *85*  | 3027       |
|----|------------------------------|----------------------|----------|-------|------------|
| 2  | Moffat, J.G.                 | *J.Org.Chem.*        | **1971** | *36*  | 1909       |
| 3  | Albright, J.D.; Goldman, L.  | *J.Am.Chem.Soc.*     | **1965** | *87*  | 4214       |
| 4  | Albright, J.D.               | *J.Org.Chem.*        | **1974** | *39*  | 1977       |
| 5  | Onodera, K.                  | *J.Am.Chem.Soc.*     | **1965** | *87*  | 4651       |
| 6  | Taber, D.F.                  | *J.Org.Chem.*        | **1988** | *53*  | 2984       |
| 7  | Doering, v. W.               | *J.Am.Chem.Soc.*     | **1967** | *89*  | 5505       |
| 8  | Nicolaou, K.C.               | *J.Am.Chem.Soc.*     | **1989** | *111* | 6676       |
| 9  | Swern, D.                    | *J.Org.Chem.*        | **1976** | *41*  | 957, 3329  |
| 10 | Swern, D.                    | *Synthesis*          | **1981** |       | 165        |
| 11 | Maycock, C.D.                | *J.Chem.Soc.Perk. 1* | **1987** |       | 1221       |
| 12 | Liu, H.J.                    | *Tetrahedron Lett.*  | **1988** | *29*  | 3167, 5467 |
| 13 | Tidwell, T.T.                | *Synthesis*          | **1990** |       | 857        |
| 14 | Tidwell, T.T.                | *Org.React.*         | **1990** | *39*  | 297        |

**P-Nitrobenzaldehyde 2.**[1] To a solution of p-nitrobenzyl alcohol **1** (135 mg, 1 mmol) in DMSO was added dicyclohexylcarbodiimide (DCC) (618 mg, 3 mmol). The reaction is quantitative (TLC) and **2** was isolated as the DNPH derivative in 92% yield, mp 316-317°C.

**Methyl 12-Hydroxy-3-oxodeoxycholanate 4.**[11] To oxalyl chloride (240 mg, 1.93 mmol) and DMSO (0.28 mL, 3.94 mmol) in $CH_2Cl_2$ at –60°C was added rapidly **3** (1.07 g, 1.93 mmol) in $CH_2Cl_2$, and the temperature was allowed to rise to –40°C during 15 min and maintained for 30 min at –40°C. $Et_3N$ (0.89 mL, 6.38 mmol) was added and after 5 min the temperature was allowed to rise to 20°C. The TMS group was removed with 5% HCl in MeOH (TLC). Usual work up and chromatography gave 580 mg of **4** (74%), mp 137-140°C.

## T A M A O – F L E M I N G Stereoselective Hydroxylation

Stereoselective conversion of alkyl silanes to alcohols by means of peracids.

| 1 | Kumada, M.; Tamao, K. | *Tetrahedron* | **1983** | *39* | 983 |
|---|---|---|---|---|---|
| 2 | Kumada, M.; Tamao, K. | *J. Org. Chem.* | **1983** | *48* | 2120 |
| 3 | Fleming, I. | *J. Chem. Soc. Chem. Commun.* | **1984** | | 29 |
| 4 | Tamao, K. | *J. Organometallic Chem.* | **1984** | *269* | C37-C39 |
| 5 | Fleming, I. | *Tetrahedron Lett.* | **1987** | *28* | 4229 |
| 6 | Fuchs, P. L. | *Tetrahedron Lett.* | **1991** | *32* | 7513 |

**(SR,RS)-4-Hydroxy-3-methyl-4-phenylbutan-2-one (2).[5]** To a stirred solution of β–silylketone **1** (79 mg, 0.27 mmol) in $MeCO_3H$ (3 mL, of 15% solution in $MeCO_2H$, containing 1% $H_2SO_4$, 7.2 mmol) was added $Hg(OAc)_2$ (130 mg, 0.41 mmol) and the mixture was maintained for 3 h at 20 °C. The mixture was diluted with $Et_2O$ (60 m L) and washed with $Na_2S_2O_3$, $NaHCO_3$ solution, brine and dried ($MgSO_4$). Evaporation and preparative TLC (hexane: EtOAc 1:1) gave 43 mg of **2** (88%).

## T E B B E – G R U B B S – P E T A S I S  Olefination

Methylenation of carbonyl groups from aldehydes, ketones, esters, lactones, amides by Ti reagents (see 1 st edition).

Tebbe **1**          Grubs **2**          Petasis **3**          Cp$_2$TiMe$_2$

**5** (96%)

**7** (70%)

**9** (80%)

| 1 | Tebbe, F.N. | *J.Am.Chem.Soc.* | **1978** | *100* | 3611 |
|---|---|---|---|---|---|
| 2 | Pine, S.K. | *Synthesis* | **1991** | | 165 |
| 3 | Grubbs, R.H. | *Tetrahedron Lett.* | **1984** | *25* | 5733 |
| 4 | Petasis, N.A. | *J.Am.Chem.Soc.* | **1990** | *112* | 6392 |
| 5 | Pine, S. K. | *Org. React.* | **1993** | *43* | 1 |

**2-t-Butyl-1-methylenecyclohexane 5.[2]** To a solution of 2-t-butylcyclo- hexanone **4** (154 mg, 1mmol) in THF (3 mL) at 0°C was added a toluene solution of **1** (2 mL, of 0.5 M sol 1 mmol). The reaction mixture was allowed to warm to 20°C, Et$_2$O (20 mL) was added followed by 0.1 N NaOH (5-10 drops). Evaporation of the organic layer and chromatography (alumina 2% Et$_2$O in pentane) afforded 146 mg of **5** (96%).

**Olefin 7.[3]** To a solution of **2** (304 mg, 1.1mmol) in Et$_2$O (4 mL) under Ar at 0°C was added ketone **6** (152 mg, 1 mmol). The mixture was allowed to warm to 20°C over 20 min. Dilution with pentane (50 mL), filtration (Celite) and chromatography afforded 106 mg of **7** (70%).

**Enol ether 9.[4]** A solution of **3** (427 mg, 3 mmol) in THF was stirred with **8** (148 mg, 1 mmol) under Ar at 60-65°C for 12-26 h. Dilution with petroleum ether, filtration of insoluble matter and chromatography on silica gel gave **9** (80%).

## T E U B E R   Quinone Synthesis

Oxidation of phenols or anilines to quinones by means of potassium nitrosodisulfonate (Fremy's salt) (see 1st edition).

| 1 | Teuber, H.I. | Chem. Ber. | 1952 | 85 | 95 |
| 2 | Teuber, H.I. | Chem. Ber. | 1953 | 86 | 1036 |
| 3 | Teuber, H.I. | Chem. Ber. | 1955 | 88 | 802 |
| 4 | Teuber, H.I. | Angew. Chem. Int. Ed. | 1969 | 8 | 218 |
| 5 | Roth, R.A. | J. Org. Chem. | 1966 | 31 | 1014 |
| 6 | Zimmer, H. | Chem. Rev. | 1971 | 71 | 229 |
| 7 | Kozikowski, A. | J. Org. Chem. | 1981 | 46 | 2426 |

## T H I L E - W I N T E R   Quinone Acetoxylation

Synthesis of triacetoxyaryl derivatives from quinones (see 1st edition).

| 1 | Thile, J. | Chem. Ber. | 1898 | 31 | 1247 |
| 2 | Thile, J.; Winter, E. | Liebigs Ann. | 1900 | 311 | 341 |
| 3 | Fieser, L.F. | J. Am. Chem. Soc. | 1948 | 70 | 3165 |
| 4 | Blatchly, J.M. | J. Chem. Soc. | 1963 | | 5311 |
| 5 | McOmie, J.F.W. | Org. React. | 1972 | 19 | 200 |

## TIETZE Domino or Cascade Reactions

One pot domino (cascade) reactions like tandem Knoevenagel-hetero Diels-Alder, Knoevenagel-ene, Pictet-Spengler-ene, Sakurai carbonyl-ene reactions.

| 1 | Tietze, L.F. | *Chem.Ber.* | **1989** | *122* | 997, 1955 |
|---|---|---|---|---|---|
| 2 | Tietze, L.F. | *Synthesis* | **1989** | | 439 |
| 3 | Tietze, L.F. | *Angew.Chem.Int.Ed.* | **1992** | *331* | 1079 |
| 4 | Tietze, L.F. | *Angew.Chem.Int.Ed.* | **1993** | *32* | 131 |
| 5 | Tietze, L.F. | *J.Org.Chem.* | **1994** | *59* | 192 |
| 6 | Tietze, L.F. | *Synthesis* | **1994** | | 1185 |
| 7 | Tietze, L.F. | *Chem.Rev.* | **1996** | *96* | 115 |
| 8 | Tietze, L.F. | *Synlett* | **1997** | | 35 |
| 9 | Tietze, L.F. | *Curr.Opin.Chem.Biol.* | **1998** | *2* | 363 |

**Octahydroindolo[2,3]quinolizine**   **4**.[6]    (1RS)-Benzyloxycarbonyl-1-(2-oxoethyl)-1,2,3,4-tetrahydro-β-carboline **1** (49.2 mg, 0.14 mmol), N,N-dimethylbarbituric acid **2** (26.5 mg, 0.17 mmol) and the enol ether **3** (360 mg, 2.22 mmol) in the presence of a few crystals of ethylenediammonium diacetate (EDDA) in an ultra sound bath ($H_2O$, 50-60°C)gave after 4 h a clear red solution. Flash chromatography (hexane) afforded 89 mg of a cycloaduct. Pd/C 10% (90 mg) in anh. EtOH was stirred under $H_2$ for 30 min. The cycloaduct was added and stirring was continued for 24 h at 20°C. Chromatography ($CHCl_3$:MeOH 5:1) gave 51 mg of **4** (98%) as an α:β mixture (2.7:1).

## T I F F E N E A U   Aminoalcohol Rearrangement

Cationic rearrangement (ring enlargement) of 1,2-aminoalcohols by diazotization (see 1st edition).

| | | | | | |
|---|---|---|---|---|---|
| 1 | Tiffeneau, M. | *C. R.* | **1937** | *205* | 54 |
| 2 | Tiffeneau, M. | *C. R.* | **1941** | *212* | 195 |
| 3 | Cox, R.H. | *J. Am. Chem. Soc.* | **1952** | *74* | 2924 |
| 4 | Parham, W.E. | *J. Org. Chem.* | **1972** | *37* | 1975 |
| 5 | Smith, P.A.S. | *Org. React.* | **1960** | *11* | 157 |

**Suberone (2).**[3] Aminomethylcyclohexane **1** (129 g; 1 mol) at pH=4 and maintained at 0-5°C was treated with NaNO$_2$ (83.0 g; 1.20 mol) in water on 2 h. The mixture was stirred 2 h at 20°C at pH=5-6 and finally refluxed on a water bath for 1 h. Usual work up gave 56-64 g of **2** (50-57%), bp 66-70°C/16 mm.

## T I M M I S   Pteridine Synthesis

Base catalyzed condensation of 4-amino-5-nitrosopyrimidines with cyano acetic derivatives to afford pteridines.

| | | | | | |
|---|---|---|---|---|---|
| 1 | Timmis, G.M. | *Nature* | **1949** | *163* | 2032 |
| 2 | Timmis, G.M. | *J. Chem. Soc.* | **1954** | | 2881; 2995 |
| 3 | Timmis, G.M. | *J. Chem. Soc.* | **1955** | | 2032 |
| 4 | Timmis, G.M. | *Nature* | **1964** | *178* | 139 |

**4,7-Diaminopteridine-6-carboxylic acid (3).**[3] To a solution of Na (1.1 g; 48 mat) in 2-ethoxyethanol (200 mL) was added 4,6-diamino-5-nitroso pyrimidine **1** (3.2 g; 23 mmol) and cyanoacetic acid **2** (2.0 g; 23 mmol). The mixture was refluxed for 15 min and the brown precipitate filtered and acidified with AcOH. Recrystallization from water (charcoal) gave 2.1 g of **3** (43%), mp 292°C.

## TISCHENKO–CLAISEN Dismutation

Conversion of aldehydes to esters in the presence of metal alcoholates, involving oxidation-reduction (see 1st edition)

| 1 | Claisen, L. | *Chem.Ber.* | **1887** | *20* | 648 |
| 2 | Tischenko, W. | *J.Russ.Phys.Chem.Soc.* | **1906** | *39* | 335, 542 |
| 3 | Lin, I. | *J.Am.Chem.Soc.* | **1957** | *74* | 5133 |
| 4 | Stapp, P.R. | *J.Org.Chem.* | **1973** | *38* | 1433 |
| 5 | Mahrwald, R. | *Synthesis* | **1996** | | 1087 |
| 6 | Iashutaka, I. | *J.Org.Chem.* | **1997** | *62* | 3409 |

**(1RS, 2RS, 3RS)-3-Hydroxy-2-methyl-1-phenylpentyl benzoate 3[5] and (1SR, 2RS, 3RS)-1-Ethyl-3-hydroxy-2-methyl-3-phenylpropyl benzoate 3a.** To a solution of Ti(OiPr)$_4$ (0.32 mL, 1 mmol) in 1-tert-butoxy-2-methoxy ethane (1.5 mL) was carefully added BuLi (0.64 mL, 1 mmol) in hexane under Ar. After stirring for 30 min at 20°C, ketoalcohol **1** (0.5 mL, 5 mmol) and then Ph-CHO **2** (1 mL, 10 mmol) were added. The solution was stirred for 24 h at 20°C and after usual work up the product was chromatographed (silica gel, hexane:iPrOH 95:5) to afford **3** and **3a** in ratio 5:95.

**3-Hydroxybutyl acetate 6.[6]** To a solution of Cp$_2$ZrH$_2$ (50 mmol) in THF (0.25 mL) were added β-hydroxy ketone **4** (76 mg, 1 mmol) followed by Me-CHO **5** (176 mg, 4 mmol) under Ar at 20°C. After 5 h stirring the reaction mixture was quenched with wet Et$_2$O. Purification by chromatography (silica gel, EtOAc:hexane 1:10) gave **6** (92%).

## T O D A  Solid State Reactions

Organic reactions in the solid state, e. g. Baeyer-Villiger, Reformatsky, Luche, Glaser, Eglington, Wittig, Brown. Michael, Robinson often more efficient than in solution.

$$R_1R_2N-(CH_2)n-O-C_6H_4-C-C_6H_5 \xrightarrow[\text{stirr, 5 d}]{NaBH_4} R_1R_2N-(CH_2)n-O-C_6H_4-CH-C_6H_5$$

**1**  2 (70%)[3]

$$Ph-CHO + Br-CH_2-CH=CH_2 \xrightarrow{Zn/NH_4Cl} Ph-CH-CH2-CH=CH_2$$

(90%)[6]

mCPBA

FeCl₃

(95%)[4]

| 1 | Toda, F. | *J. Chem. Soc. Chem.Commun.* | **1988** | | 958 |
|---|----------|------------------------------|----------|----|------|
| 2 | Toda, F. | *J. Org. Chem.* | **1988** | *54* | 3007 |
| 3 | Stummer, C. | *Unpublished results* | | | |
| 4 | Toda, F. | *J. Org. Chem.* | **1989** | *54* | 3007 |
| 5 | Seebach, D. | *Angew. Chem. Int. Ed.* | **1990** | *29* | 1320 |
| 6 | Toda, F. | *Chem. Lett.* | **1990** | | 373 |
| 7 | Toda, F. | *J. Org. Chem.* | **1993** | *58* | 6208 |
| 8 | Toda, F. | *J. Chem. Soc. Perkin 1* | **1999** | | 3069 |
| 9 | Toda, F. | *Chem. Lett.* | **2000** | | 888 |
| 10 | Toda, F. | *Acc. Chem. Res.* | **1995** | *28* | 480 |
| 11 | Toda, F. | *Chem. Rev.* | **2000** | *100* | 1025 |

**4-(Dialkylaminoalkoxy)benzhydrol (2).**[3] A mixture of 4-(dialkylaminoalkoxy) benzophenone **1** (10 mmol) and NaBH₄ (3.783 g, 100 mmol) were mixed in a mortar and pestle in a glove dry-box under N₂ at 20 °C. The operation was repeated once a day for five days. Extraction with Et₂O, drying, filtration through a pad of Celite and evaporation afforded **2** in 60-72% yield.

# T O R G O V   Vinyl Coupling

$SN_2$ type condensation of vinyl carbinols with $\square$-diketones (without additional acid).

**1**     **2**     xylene, 90 min   tBuOH; refl.     **3 (70%)**

| 1 | Torgov, I.V. | *Dokl. Akad. Nauk SSSR* | **1959** | 127 | 553 |
|---|---|---|---|---|---|
| 2 | Torgov, I.V. | *Isv. Akad. Nauk SSSR Otd. Khim.* | **1962** | | 298 |
| 3 | Weyl Reiner, J. | *Bull. Soc. Chim. Fr.* | **1969** | | 4561 |
| 4 | Kuo, C.H. | *J. Org. Chem.* | **1968** | 33 | 3126 |
| 5 | Blazejewsky, J.C. | *Tetrahedron Lett.* | **1994** | 35 | 2021 |

**Dione 3.**[4] A mixture of 1-vinyl-6-methoxytetralol **1** (700 mg; 3.7 mmol) and 2-methyl-cyclopentane-1,3-dione **2** (420 mg; 3.7 mmol) in xylene (4 mL) and t-butyl alcohol (2 mL) was refluxed with stirring and under $N_2$ for 90 min. $Et_2O$ was added and **2** was removed by filtration (115 mg). The filtrate after washing (water, 5% $NaHCO_3$, brine) and drying ($MgSO_4$) was concentrated. The residue after recrystallization from MeOH gave 575 mg of **3** (70%) from two crops, mp 76-78°C.

## T R A H A N O V S K Y  Ether Oxidation

Oxidation of aromatic ethers to carbonyl compounds or of dimethoxy aromatics to quinones with cerium ammonium nitrate (see 1st edition).

| 1 | Trahanovsky, W.S. | *J. Chem. Soc.* | **1965** | | 5777 |
|---|---|---|---|---|---|
| 2 | Jacobs, P. | *J. Org. Chem.* | **1976** | *41* | 3627 |
| 3 | Lepage, L. & Y. | *Can. J. Chem.* | **1980** | *58* | 1161 |
| 4 | Lepage, L. & Y. | *Synthesis* | **1983** | | 1018 |

## T R A U B E  Purine Synthesis

Pyrimidine synthesis from guanidine and cyanoacetic ester and purine synthesis from aminopyrimidines (see 1st edition).

| 1 | Traube, W. | *Chem. Ber.* | **1900** | *33* | 1371; 3035 |
|---|---|---|---|---|---|
| 2 | Traube, W. | *Liebigs Ann.* | **1904** | *331* | 641 |
| 3 | Katritzky, A. | *Quart. Rev. (London)* | **1956** | *10* | 397 |

**Guanine (5).**[1] A suspension of guanidine·HCl **1** (40.0 g; 0.4 mol) in EtOH was treated with NaOEt (from Na 9.2 g; 0.4 at g). **2** (48.0 g; 0.4 mol) was added and the mixture was refluxed for 6 h. The salts were filtered and the filtrate was concentrated to dryness. Pyrimidine **3** after nitrosation and reduction with $(NH_4)_2S$ gave 2,4,5-triamino-6-oxypyrimidine **4**. By refluxing **4** (10.0 g; 74 mmol) with HCOOH (190 mL) for 4-5 h there are obtained 7-8 g of **5** (60-67%).

## **T R A U B E**   Reducing Agent

$CrCl_2$ reduction of alkyl halides to alkanes, of acetylenes to trans olefins, of epoxides to olefins, or of nitro compounds to oximes.

$$\xrightarrow[\text{Bu-SH; DMSO}]{\text{Cr(OAc)}_2;\ 28°}$$

**1**            **2**

$$\xrightarrow{\text{CrCl}_2}$$

$$\xrightarrow{\text{CrCl}_2}$$

| 1 | Traube, W. | *Chem. Ber.* | **1916** | *49* | 1692 |
|---|---|---|---|---|---|
| 2 | Traube, W. | *Chem. Ber.* | **1925** | *58* | 2466 |
| 3 | Barton, D.H.R. | *J. Am. Chem. Soc.* | **1965** | *87* | 4601 |
| 4 | Hanson, J.R. | *J. Chem. Soc. (C)* | **1969** | | 1201 |
| 5 | Hanson, J.R. | *Synthesis* | **1974** | | 1 |

**3β-Acetoxy-6β-hydroxyandrostan-17-one (2).**[3] To a solution of Cr$^{II}$ acetate (5.3 g; 5 equiv.) in DMSO (75 mL) under $N_2$ are added n-butyl mercaptan (1.6 mL; 8 equiv), followed by 3β-acetoxy-5α-bromo-6β-hydroxyandrostan-17-one **1** (4.07 g; 9.55 mmol). After 2 h stirring at 28°C, the mixture was poured into water (200 mL) and the solution extracted with $CH_2Cl_2$. Chromatography on alumina afforded **2**, mp 183-184°C, $[\alpha]_D^{24} = +42°$.

## T R E I B S  Allylic Oxidation

Allylic oxidation of alkenes using mercuric trifluoroacetate with possible allylic rearrangement (see 1st edition).

| 1 | Treibs, W. | *Naturwissenschaften* | **1948** | *35* | 125 |
|---|---|---|---|---|---|
| 2 | Treibs, W. | *Liebigs Ann.* | **1953** | *581* | 59 |
| 3 | Treibs, W. | *Chem. Ber.* | **1960** | *93* | 1234 |
| 4 | Wiberg, K.B. | *J. Org. Chem.* | **1964** | *29* | 3353 |
| 5 | Arzoumanian, N. | *Synthesis* | **1971** | | 527 |
| 6 | Halpern, J. | *J. Am. Chem. Soc.* | **1972** | *94* | 1985 |
| 7 | Bloosey, E.C. | *J. Chem. Soc. Chem. Commun.* | **1973** | | 56 |
| 8 | Husson, H.P. | *Synthesis* | **1974** | | 722 |

**17-Oxo-$\Delta^4$–androsten-3β,6β–diol-3-acetate 2.**[8] A solution of 17-oxo-$\Delta^5$–androsten-3β-ol acetate **1** (1,03 g, 31 mmol) and mercury(II) trifluoro acetate (3,1 g, 72 mmol) in dichloromethane (100 mL) were stirred for 24 h at 20°C. Part of the solvent (66mL) was evaporated in vacuum and the reaction mixture was filtered over glass fiber filter paper. The filtrate was washed with 5% $Na_2CO_3$ aqueous solution, water and again filtered. After evaporation of the solvent the residue (720 mg) was dissolved in MeOH. After crystalization there was obtained a first crop of 411 mg of **2** (40%), mp 148-150°C $[\alpha]_D^{25} = +25°$ (CHCl$_3$).

## T R O S T  Cyclopentanation

Methylenecyclopentane    formation    from    siloxychloromethylallylsilane    or
acetoxymethyl-allylsilane **2** with Michael acceptor olefins and Pd catalysts (via a
trimethylene methane equivalent) (see 1st edition).

| 1 | Trost, B.M. | *J. Am. Chem. Soc.* | **1979** | *101* | 6429 |
|---|---|---|---|---|---|
| 2 | Trost, B.M. | *J. Am. Chem. Soc.* | **1983** | *105* | 2315 |
| 3 | Trost, B.M. | *Angew. Chem. Int. Ed.* | **1986** | *25* | 1 |
| 4 | Trost, B.M. | *J. Org. Chem.* | **1988** | *53* | 4887 |
| 5 | Trost, B.M. | *J. Am. Chem. Soc.* | **1989** | *111* | 7487 |

**2-Methylene-4-(methoxymethoxy)-8aβ-(phenylsulfonyl)-3aβ-decahydroazulene
(3).**[5] To Pd(OAc)$_2$ (15 mg; 0.06 mmol) and P(OiPr)$_3$ (101 mg; 0.487 mmol) in PhMe
(2 mL) was added **1** (1.05 g; 3.54 mmol) in PhMe (2 mL) followed at 60°C by **2** (0.95 g;
5.3 mmol). After 40 h at 80°C, chromatography (3:1 hexane:EtOAc, R$_f$ = 0.33) gave
1.15 g of **3** (93%).

## T R O S T – C H E N  Decarboxylation

Ni complex catalyzed decarboxylation of dicarboxylic acid anhydrides to form alkenes
(see 1st edition).

| 1 | Trost, B.M.; Chen, F. | *Tetrahedron Lett.* | **1971** | | 2603 |
|---|---|---|---|---|---|
| 2 | Cramer, R. | *J. Org. Chem.* | **1975** | *40* | 2267 |
| 3 | Jennings, P.W. | *J. Org. Chem.* | **1975** | *40* | 260 |
| 4 | Flood, T.C. | *Tetrahedron Lett.* | **1977** | | 3861 |
| 5 | Rose, J.D. | *J. Chem. Soc.* | **1950** | | 69 |
| 6 | Grunewald, G.L. | *J. Org. Chem.* | **1978** | *43* | 3074 |

## TSUJI–TROST Allylation

Direct C-allylation of enolizable ketones or of tin enol ethers with allyl esters using Pd(O) catalysts (see 1st edition).

| | | | | | |
|---|---|---|---|---|---|
| 1 | Tsuji, J. | *Tetrahedron Lett.* | **1965** | | 4387 |
| 2 | Tsuji, J. | *J. Org. Chem.* | **1985** | *50* | 1523 |
| 3 | Tsuji, J. | *Acc. Chem. Res.* | **1969** | *2* | 144 |
| 4 | Trost, B. | *J. Am. Chem. Soc.* | **1973** | *95* | 292 |
| 5 | Trost, B. | *J. Am. Chem. Soc.* | **1980** | *102* | 5699 |
| 6 | Trost, B. | *Acc. Chem. Res.* | **1980** | *13* | 385 |
| 7 | Ukai, T. | *J. Organomet. Chem.* | **1974** | *65* | 235 |

## TSCHUGAEF Olefin Synthesis

Olefin formation (preferentially less substituted) from alcohols via xanthate pyrolysis (see 1st edition).

| | | | | | |
|---|---|---|---|---|---|
| 1 | Tschugaef, J. | *Chem. Ber.* | **1898** | *31* | 1775 |
| 2 | de Groote, A. | *J. Org. Chem.* | **1968** | *33* | 2214 |
| 3 | De Puy, C.H. | *Chem. Rev.* | **1960** | *60* | 444 |
| 4 | Nace, H.R | *Org. React.* | **1962** | *12* | 58 |

### U G I   Multicomponent Condensation

Peptide synthesis via three or four component condensation (amino acid, imine and isocyanide) (see 1st edition).

| 1 | Ugi, I. | *Angew.Chem.* | **1977** | *89* | 267 |
|---|---|---|---|---|---|
| 2 | Yamada, T. | *J.Chem.Soc.Chem.Commun* | **1984** | | 1500 |
| 3 | Yamada, T. | *Chem.Lett.* | **1987** | | 723 |
| 4 | Yamada, T. | *J.Chem.Soc.Chem.Commun* | **1990** | | 1640 |
| 5 | Marcaccini, S. | *Synthesis* | **1994** | | 765 |
| 6 | Wipf, P.; Curran, D.P. | *J.Org.Chem.* | **1997** | *62* | 2917 |
| 7 | Bossio, R. | *Heterocycles* | **1999** | *50* | 463 |
| 8 | Byk, G. | *J.Comb.Chem.* | **2000** | *2* | 732 |

**Peptide (4).[4]** A mixture of N-carbobenzoxy-L-valine **1** (1.104 g, 4.4 mmol), Schiff base **2** (1.139 g, 4.4 mmol), methyl isocyanidoacetate **3** (433 mg, 4.4 mmol) in $CH_2Cl_2$ (4 mL) was compressed for 14 days at 9 kbar. Evaporation of the solvent and chromatography afforded 1.675 g of **4** (63%), mp 126-127°, $\alpha_D$ = -16.0° (c 1.0 CHCL$_3$)

**N-Benzoyl-N-benzylphenylglycine tert-butylamide (8).[6]** 4-Tris(2-perfluorodecyl) silylbenzoic acid **5** (26.2 mg, 0.015 mmol), benzylidenebenzylamine **6** (51 mg, 0.25 mmol), and t-butyl isocyanide **7** (30 L, 0.25 mmol) were heated in a sealed tube with $CF_3CH_2OH$ (0.3 mL) under Ar to 90°C for 48 h. After evaporation of the solvent, the residue in THF (2 mL) was stirred with TBAF in THF (22 µL) for 30 min at 25°C. Evaporation of the solvent, extraction with PhH, washing and evaporation of the solvent gave 5 mg of **8** (83% yield and 85% purity).

## U L L M A N N – F E D V A D J A N  Acridine Synthesis

Synthesis of polynuclear pyridines from anilines, phenols and paraformaldehyde (see 1st edition).

| 1 | Ullmann, F.; Fedvadjan, A. | *Chem.Ber.* | **1903** | *36* | 1027 |
| 2 | Buu Hoi, N.P. | *Bull.Soc.Chim.Fr.Mem.* | **1944** | *11* | 406 |
| 3 | Buu Hoi, N.P. | *J.Chem.Soc.(C)* | **1967** | | 213 |

## U L L M A N N – L A  T O R R E  Acridine Synthesis

Cyclization of o-methyldiarylamines with anilines by heating with PbO to provide acridines.

| 1 | Ullmann, F.; Torre, A.L. | *Chem.Ber.* | **1904** | *37* | 2922 |
| 2 | Buu Hoi, N.P. | *J.Chem.Soc.* | **1949** | | 670 |
| 3 | Motohashi, N. | *Org.Prep.Proc.Int.* | **1993** | *25* | 259 |

**8-Methyl-1,2-benzacridine 4.**[2] A mixture of **3** (10 g, 40 mmol) and lead oxide (100 g) was heated slowly to boiling. The distillate was dissolved in hot EtOH and treated with picric acid. The crude picrate (2 g) after recrystallization from PhCl melted at 239-240°C (decomp), free base **4**, mp 148°C.

## ULLMANN–GOLDBERG Aromatic Substitution

Substitution of aromatic halides or recrystes in the synthesis of diaryls, diaryl ethers, diaryl amines, phenols etc catalyzed by Cu and other catalysts

| 1 | Ullmann, F. | *Chem.Ber.* | **1903** | *36* | 2389 |
|---|---|---|---|---|---|
| 2 | Goldberg, I. | *Chem.Ber.* | **1906** | *39* | 1691 |
| 3 | Yamamoto, T. | *Can.J.Chem.* | **1983** | *61* | 86 |
| 4 | Renger, B. | *Synthesis* | **1985** | | 856 |
| 5 | Percec, V. | *J.Org.Chem.* | **1995** | *60* | 176,1066 |
| 6 | Liebeskind, L.S. | *J.Org.Chem.* | **1997** | *62* | 2312 |
| 7 | Bunnell, J.F. | *Chem.Rev.* | **1951** | *49* | 392 |
| 8 | Schulenburg, J.W. | *Org.React.* | **1965** | *14* | 19 |

**2,2'-Dibenzoyl-4,4'-dimethoxybiphenyl 5.**[5] A Schlenk tube was charged with $NiCl_2$ $(PPh_3)_2$ (65.3 mg, 0.1 mmol), triphenylphosphine (104.8 mg, o.4 mmol), Zn powder (110.5 mg, 1.7 at g.) and tetraethylammonium iodide (385.5 mg 1.5 mmol). Under nitrogen, was added dry THF (0.5mL) and after stirring for 5 min at 20°C was added 2-benzoyl-4-methoxyphenyl mesylate 4 (322 mg, 1 mmol) in THF (0.5 mL). After 24 h reflux, the cooled mixture was filtered, diluted with water and extracted with $CHCl_3$. The organic phase after washed, dried ($MgSO_4$) and evaporated in vacuum and the residue chromatographed (silica gel, Hexane-ethyl acetate). Recrystallization from hexane-chloroform afforded 68.5 mg of 5 (65%), mp 138-140°C.

## ULLMANN–HORNER Phenazine Synthesis

Synthesis of dibenzo(a, h)phenazine from 1-phenylazo-2-naphthylamine and 2-naphthol (Ullmann) or by autooxidation of 1-aminonaphthalene (Horner).

| 1 | Ullmann, F. | *Chem. Ber.* | **1905** | 38 | 1811 |
|---|---|---|---|---|---|
| 2 | Horner, L. | *Chem. Ber.* | **1963** | 96 | 786 |
| 3 | Itoho, K. | *Daiichi Yakka Daigaku Kenkyu Nepo* | **1993** | 24 | 19 |
| | | *C. A.* | **1994** | 121 | 83285 n |

**Dibenzo-(a,h)-phenazine (3).**[3] To melted **1** (4.0 g, 27 mmol) was added **2** (2.0 g, 8.1 mmol). When generation of steam and aniline subsided, the mixture was cooled to 20 °C and the product was recrystallized from PhH and chromatographed on silica gel to give 1.25 g of **3** (87%), mp 291 °C.

**Synthesis of 3 from (4).** A mixture of KOtBu (23 g) and **4** (14.3 g, 0.1 mol) in PhMe (500 mL) after auto-oxidation with oxygen, was neutralized with 2 N $H_2SO_4$. Work up and chromatography afforded 9.7 g of **3** (72%), mp 291 °C.

## VAN BOOM Phosphorylating Reagent

Phosphorylation of sugars or nucleosides by means of salicylchlorophospite **2** (see 1st edition).

R = 4,4'-Dimethoxytrityl

**1**

dioxane, 20°, 5 min

3 (88%)[4]

| | | | | | | |
|---|---|---|---|---|---|---|
| 1 | Anshutz, R. | *Liebigs Ann.* | **1887** | 239 | | 301 |
| 2 | Young, R. W. | *J. Am. Chem. Soc.* | **1952** | 74 | | 1672 |
| 3 | Van Boom, J. H. | *Rec. Trav. Chim.* | **1986** | 105 | | 510 |
| 4 | Van Boom, J. H. | *Tetrahedron Lett.* | **1986** | 27 | | 2661,6271 |

## VAN LEUSEN Reagent

A one-step synthesis of nitriles from carbonyls by a reductive cyanation with tosylmethyl isocyanide **2** (TosMIC); also synthesis of 1,3-azole or of ketones (see 1st edition).

$R_1$-CO-R ← $F_1$-$\overset{R}{\underset{R_1}{C}}$-N≡C ← Tos-$\overset{R}{CH}$-N≡C

**4**

| | | | | | |
|---|---|---|---|---|---|
| 1 | van Leusen, A. M. | *Tetrahedron Lett.* | **1973** | | 1357 |
| 2 | van Leusen, A. M. | *J. Org. Chem.* | **1977** | 42 | 3114 |
| 3 | van Leusen, A. M. | *Synth. Commun.* | **1980** | 10 | 399 |
| 4 | van Leusen, A. M. | *Lect. Heteroc. Chem.* | **1980** | V | S-111 |
| 5 | van Leusen, A. M. | *Org. Synth.* | **1977** | 57 | 102 |

## VARVOGLIS - MORIARTY Hypervalent Iodine Reagents

Iodobenzene diacetate PhI(OAc)$_2$ **1**; bis(trifluoroacetoxy)iodobenzene PhI(OCOCF$_3$)$_1$ **2**; hydroxy(tosyloxy)iodobenzene PhI$_4$(OH)OTs **3** in oxidation, dehydrogenation, Hofmann rearrangement, dethioacetalization, -carbonyl functionalization (Moriarty-Prakash).

| 1 | Varvoglis, A. | *Synthesis* | **1975** | | 445 |
|----|----|----|----|----|----|
| 2 | Varvoglis, A. | *J.Chem.Res.* | **1982** | | 150 |
| 3 | Stork, G. | *Tetrahedron Lett.* | **1989** | 30 | 287 |
| 4 | Loudon, G.M. | *J.Org.Chem.* | **1984** | 49 | 4273 |
| 5 | Varvoglis, A. | *Synthesis* | **1984** | | 709 |
| 6 | Moriarty, R.M.; Prakash, C. | *Acc.Chem.Res.* | **1986** | 19 | 244 |
| 7 | Varvoglis, A. | *Tetrahedron Lett.* | **1997** | 53 | 1179 |
| 8 | Moriarty, R.M.; Prakash, C. | *Synlett* | **1997** | | 1255 |
| 9 | Varvoglis, A. | *Synlett* | **1998** | | 221 |
| 10 | McKillop, | *J.Chem.Soc.Perkin 1* | **1994** | | 2047 |
| 11 | Moriarty, R.M.; Prakash, C. | *Org.React.* | **1999** | 54 | 273 |
| 12 | Xian Huang | *Synth.Commun.* | **2000** | 30 | 9 |

**Aldehyde (5).**[3] To a stirred solution of thioacetal **4** (10 mmol) in MeOH/H$_2$O (10 mL) was added **2** (15 mmol) at 20°C. After the reaction was completed (10 min), the mixture was poured into saturated aqueous NaHCO$_3$ (20 mL). Extraction (Et$_2$O), evaporation of the solvent and chromatography (silica gel, petroleum ether:EtOAc) gave **5** in 91% yield.

**Methyl 2-phenyl, 2-thiocyanatoethanoate (7).**[8] (Dichloroiodo)benzene (660 mg, 2.4 mmol) was added to a suspension of Pb(SCN)$_2$ (970 mg, 3mmol) in CH$_2$Cl$_2$ (20 mL) at 0°C under Ar. After 15 min, silyl keten acetal **6** (436 mg, 2 mmol) in CH$_2$Cl$_2$ (10 mL) was added. The mixture was stirred for 2 h at 0-5°C. Work up and chromatography of the residue (EtOAc:hexane afforded 289 mg of **7** (70%).

# VASELLA – BERNET

## Chiral Cyclopentane Synthesis From Sugars

Transformation of monosaccharides into enantiomerically pure penta- substituted cyclopentanes via fragmentation and nitrone-olefin dipolar cycloaddition.

| 1 | Bernet, B.; Vasella, A. | *Helv.Chim.Acta* | **1979** | 62 | 1900 |
|---|---|---|---|---|---|
| | | | | | 2400, 2411 |
| 2 | Bernet, B.; Vasella, A. | *Helv.Chim.Acta* | **1984** | 67 | 1328 |

**1D-(1,2,5/3,4)-1¹,2¹,-Anhydro-3,4,5-tri-O-benzyl-1-(hydroxymethyl)-2-(hydroxymeth ylamino)-3,4,5-cyclopentantriol (3).**[1] A solution of methyl 2,3,4-tri- O-benzyl-6-bromo -6-desoxy- α-D-mannopyranoside **1** (788 mg, 1.49 mmol) in PrOH (13 mL) and water (1 mL) was refluxed with active Zn (968 mg, 14.9 mmol) for 30 min. After filtration through Celite, the solution was stirred for 30 min with Amberlite IR-45 (OH⁻) and charcoal. The solution was filtered through Celite, the filtrate evaporated, and the residue dried in vacuum. The residue of **2** in MeOH was refluxed for 30 min with N-methylhydroxylamine (1.13 g, 13.6 mmol), NaOMe (784 mg, 14.52 mmol) and NaHCO₃ (120 mg, 1.42 mmol). After usual work up and chromatography there was obtained 428 mg of **3** (64.2 g), $\alpha_D$ = -53.9 (c=0.7).

## V E D E J S Hydroxylation

Oxidation of ketones to α-hydroxyketones by means of oxodiperoxymolybdenum (pyridine) (hexamethylphosphoric triamide) (MoOPH) prepared and MoO₃, 30% H₂O₂, HMPA and pyridine (see 1st edition).

2 (53%)[1]

| 1 | Vedejs, E. | *J. Org. Chem.* | **1978** | *43* | 194 |
| 2 | Krohn, K. | *Chem. Ber.* | **1989** | *122* | 2323 |

## V I L S M E I E R – H A A C K – V I E H E Reagent

Formylation of aromatics, alkenes, activated H compounds by $Me_2N^+=CHClCl^-$ (Vilsmeier-Haack) or $Me_2N^+=CCl_2Cl^-$(Viehe) reagent. (see 1st edition).

2 (81%)[2]

| 1 | Vilsmeier, A.; Haack, A. | *Chem. Ber.* | **1927** | *60* | 119 |
| 2 | Krishna-Rao, G. S. | *J. Org. Chem.* | **1981** | *46* | 5371 |
| 3 | Konvar, D. | *Tetrahedron Lett.* | **1987** | *28* | 955 |
| 4 | Ferguson, L. N. | *Chem. Rev.* | **1946** | *38* | 230 |
| 5 | Grundmann, A. | *Angew. Chem.* | **1966** | *78* | 747 |
| 6 | Viehe, H. G. | *Angew. Chem. Int. Ed.* | **1971** | *10* | 575 |
| 7 | Bergmann, J. | *Tetrahedron Lett.* | **1986** | *27* | 1939 |

**Aldehyde (2).**[2] To **1** (1 g, 3.7 mmol) in DMF (4 mL) at 0 °C was added dropwise POCl₃ (0.5 mL) . After 10 h at 95 °C, POCl₃ (0.5 mL) was again added at 25 °C and heating was continued for 5 h. Quenching with aq.NaOAc and preparative TLC (PhH) gave 0.85 g of **2** (81%), mp 116-117 °C.

## V O I G H T α-Aminoketone Synthesis

Synthesis of α-aminoketones from α-hydroxyketones (see 1st edition).

| 1 | Voight, K. | *J. Prakt.Chem.* | **1886** | *34* | 1(2) |
| 2 | Lutz, R. E. | *J. Am. Chem. Soc.* | **1948** | *70* | 2015 |
| 3 | Kay, J. A. | *J. Am. Chem. Soc.* | **1953** | *75* | 746 |
| 4 | Lutz, R. E. | *J. Org. Chem.* | **1956** | *21* | 49 |

## V O L H A R D T – E R D M A N N Thiophene Synthesis

Thiophene synthesis from succinic acids (see 1st edition).

| 1 | Volhardt, J. | *Chem. Ber.* | **1885** | *18* | 454 |
| 2 | Lindstead, R. | *J. Chem. Soc.* | **1937** | | 915 |
| 3 | Wolff, E. W. | *Org. React.* | **1951** | *6* | 412 |

**3,4-Dimethylthiophene (2).**[2]    Disodium salt **1** (195 g, 1 mol) and phosphorus pentasulfide ~(245 g) was distilled dry under a stream of $CO_2$ to give 83 g of crude **2**, which after 15 h contact with NaOH and 6 h reflux over Na was distilled to afford 50 g of **2** (44.6 %), bp 145-148 °C.

## **V O R B R U G G E N** Nucleoside Synthesis

Synthesis of nucleosides by condensation of sugars with silyl heterocycles and Lewis acids such as AlCl$_3$, SnCl$_4$, Me$_3$SiSO$_3$CF$_3$ (see 1$^{st}$ edition)

**1**         **2**         **3 (95%)**

| 1 Vorbrüggen, H. | *Angew. Chem. Int. Ed.* | **1969** | *8* | 976 |
|---|---|---|---|---|
| 2 Vorbrüggen, H. | *Angew. Chem. Int. Ed.* | **1970** | *9* | 461 |
| 3 Vorbrüggen, H. | *Chem. Ber.* | **1973** | *106* | 3039 |
| 4 Vorbrüggen, H. | *J. Org. Chem.* | **1974** | *39* | 3654; 3660 |
| | | | | 3664; 3668 |
| 5 Vorbrüggen, H. | *Chem. Ber.* | **1981** | *114* | 1234 |
| 6 Shreiber, S. L. | *J. Am. Chem. Soc.* | **1990** | *112* | 9657 |
| 7 Danishefsky, S. | *J. Org. Chem.* | **1990** | *55* | 4211 |
| 8 Vorbrüggen, H. | *Org. React.* | **2000** | *55* | 1 |

**1-(2,3,5-Tri-O-benzoyl-β-D-ribofuranosyl)-5-ethyl-1,2,3,4-tetrahydropyrimidine-2, 4-dione (3).**[4] To 1-O-acetyl-2,3,5-tri-O-benzoyl-β-D-ribofuranose **2** (4.27 g, 8.47 mmol) in 1,2-dichloroethane (150 mL) was added 5-ethyl-2,4-bis (trimethylsilyloxy) pyrimidine **1** (3.0 g, 10.5 mmol) and SnCl$_4$ (0.71 mL, 6 mmol) in 1,2-dichloroethane (10 mL). After 20 h at 22$^0$C (TLC PhMe:AcOH:H$_2$O 5:5:1), the reaction mixture was stirred with NaHCO$_3$ solution, filtered (Celite), the organic layer separated, dried, and the solvent evaporated. Crystallization afforded 4.7 g of **3** (95%), mp 159-160$^0$C, [α]$^{23}$$_D$ –96.7$^0$ (c= 0.6 CHCl$_3$)

## W A C K E R – T S U J I Olefin Oxidation

Oxidation of olefins to ketones by a Pd (II) catalyst (see 1st edition).

| 1 | Phillips, F. C. | *Amer. Chem. J.* | **1894** | *16* | 255 |
|---|----------------|------------------|----------|------|-----|
| 2 | Tsuji, J. | *Tetrahedron Lett.* | **1982** | *23* | 2679 |
| 3 | Smidt, J. (Wacker) | *Angew. Chem.* | **1959** | *71* | 176 |
| 4 | Tsuji, J. | *Synthesis* | **1984** | | 369 |
| 5 | Wayner, D. D. M. | *J. Org. Chem.* | **1990** | *55* | 2924 |
| 6 | Tamasu, Y. | *J. Org. Chem.* | **1997** | *62* | 2113 |

**Cyclohexanone (2).**[3] A mixture of Pd(OAc)$_2$ (44.8 mg, 0.2 mmol), benzoquinone (1.06 g, 9 mmol) and an inorganic acid such as (HCl, HClO$_4$, HBF$_4$, H$_2$SO$_4$ or HNO$_3$ 0.1 M) were dissolved in MeCN (43 ml) and water (7 ml). The solution was deoxygenated with Ar (minimum 30 min) and stirred until Pd(OAc)$_2$ had dissolved. Cyclohexene **1** (0.82 g, 10 mmol) was added and the mixture was stirred for 10 min. Extraction of **2** with hexane or Et$_2$O followed by washing with 30% NaOH and evaporation of the solvent afforded 0.98 g of **2** (100% by capillary GC).

**Oxazolidinone 4.**[6] Into a flask containing PdCl$_2$ (17.7 mg, 0.1 mmol) and CuCl$_2$ (308 mg, 2.3 mmol) purged with CO, was added **3** (356 mg, 1 mmol) in trimethylortho acetate 5 ml). The mixture was stirred for 7 h at 35 °C (monitored by TLC PhH:EtOAc 1:1 ,R$_{f1}$=0.55, R$_{f2}$=0.65). The reaction mixture was diluted with EtOAc, and washed with NH$_4$Cl and 5% NH$_3$. Evaporation of the solvent and chromatography (SiO$_2$ hexane: EtOAc gradient), gave a mixture of **4** trans:cis 1:1, 272 mg (68%). Recrystallized from CH$_2$Cl$_2$/hexane, mp 108.5°-109°.

# WAGNER–MEERWEIN–NAMETKIN Rearrangement

Skeletal rearrangement via carbocations (see 1st edition).

| 1 | Wagner, G. | *J. Rus. Phys. Chem. Soc.* | **1899** | *31* | 680 |
|---|---|---|---|---|---|
| 2 | Meerwein, H. | *Liebigs Ann.* | **1914** | *405* | 129 |
| 3 | Nametkin,S. S. | *J. Russ. Phys. Chem. Soc.* | **1925** | *57* | 80 |
| 4 | Coates, R. M. | *J. Org. Chem.* | **1971** | *36* | 3277 |
| 5 | Zefirov, N. S. | *Tetrahedron* | **1975** | *31* | 2948 |
| 6 | Cristol, S. I. | *J. Org. Chem.* | **1986** | *51* | 4326 |
| 7 | Pliemingen, H. | *Angew. Chem. Int. Ed.* | **1976** | *15* | 293 |

# WALLACH Azoxybenzene Rearrangement

Acid catalyzed rearrangement of azoxybenzenes to p-hydroxyazobenzenes.

| 1 | Wallach, O. | *Chem. Ber.* | **1880** | *13* | 525 |
|---|---|---|---|---|---|
| 2 | Hahr, C. S. | *J. Am. Chem. Soc.* | **1962** | *84* | 946 |
| 3 | Brigelow, H. E. | *Chem. Rev.* | **1931** | *9* | 139 |
| 4 | Oae, S. | *Bull. Chem. Soc. Jpn.* | **1963** | *36* | 601 |

**P- Hydroxyazobenzene 2.**[2] Azoxybenzene 1 (198 mg, 1 mmol) was heated in $H_2SO_4$ (5 ml). The cooled mixture was diluted with water, purified via the sodium salt and recrystallized from MeOH, mp 152-153 °C.

## WAKAMATSU (SYNGAS) Amino Acid Synthesis

Synthesis of amidoalkylcarboxylic acids from olefins or aldehydes catalyzed by cobalt carbonyl. Amidocarbonylation as an alternative to the Strecker synthesis.

$$H_2C{=}CH{-}CO_2Et \quad + \quad H_3C{-}CONH_2 \quad \xrightarrow[CO/H_2]{Co_2(CO)_8} \quad \begin{array}{c} HO_2C \\ \diagdown \\ CH{-}CH_2{-}CH_2{-}CO_2Et \\ \diagup \\ Ac{-}NH \end{array}$$

**1**                **2**                                                    **3** (77%)

$$Ph{-}CH_2{-}CHO \quad + \quad H_3C{-}CONH_2 \quad \xrightarrow[CO/H_2]{Co_2(CO)_8} \quad Ph{-}CH_2{-}\underset{\underset{NHAc}{|}}{CH}{-}CO_2H$$

| | | | | |
|---|---|---|---|---|
| 1 | Wakamatsu, H. | *J. Chem. Soc. Chem. Comm.* | **1971** | | 1540 |
| 2 | Parnaud, J.J. | *J. Mol. Catal.* | **1979** | *6* | 341 |
| 3 | Ojima, I. | *J. Organomet. Chem.* | **1985** | *279* | 203 |
| 4 | Lin, J.J.; Knifton, J.F. | *J. Organomet. Chem.* | **1991** | *147* | 99 |
| 5 | Ichikawa, M. | *J. Chem. Soc. Chem. Comm.* | **1985** | | 321 |

**N-Acetylglutamic acid 3.[4]** A mixture of dicobalt octacarbonyl (5.1 g, 15 mmol), acetamide **2** (53 g, 0.898 mol), ethyl acrylate **1** (75 g, 0.75 mol) and p-dioxane (150 g) was pressurized in an autoclave with a mixture of CO:H$_2$ to 500 psi. The system was heated to 130-153°C, repressurized to 2000 psi and maintained in this condition for 2 h. The mixture was basified with K$_2$CO$_3$ to pH=10, the by-product extracted with EtOAc and the aqueous solution acidified with 85% H$_3$PO$_4$ and extracted with EtOAc. There are obtained 126 g of **3** (77.4%).

## W A L L A C H Imidazole Synthesis

Cyclization of N-alkylamides with PCl₅ to imidazoles.

| | 1 | | | | |
|---|---|---|---|---|---|
| 1 | Wallach, O. | *Liebigs Ann.* | **1887** | *184* | 51 |
| 2 | Wallach, O. | *Liebigs Ann.* | **1882** | *214* | 257 |
| 3 | Godefroi, E. F. | *J. Org. Chem.* | **1967** | *32* | 1259 |
| 4 | Sannicolo, F. | *J. Chem. Soc. Perkin I* | **1993** | | 675 |

**Imidazole (2).** [4] **1** (1.9 g, 6.3 mmol) was suspended in POCl₃ (15 mL) treated with PCl₅ (1.2 g, 5.7 mmol) and heated for 85 min at 100 °C. Work up (30% NH₄OH) and chromatography, yielded 1.1 g of **2** (70%), mp 96 °C (iPrOH).

## W A T A N A B E Heterocyclization

Synthesis of quinolines, indoles and carbazoles from aminoarenes with 1,2 or 1,3 diols in the presence of catalytic amounts of RuCl₂ or RuCl₃ and PR₃.

| | 1 | 2 | | 3 (46%)[7] | |
|---|---|---|---|---|---|
| 1 | Watanabe, Y. | *Bull. Chem. Soc. Jpn.* | **1983** | 56 | 2452 |
| 2 | Watanabe, Y. | *J. Organomet. Chem.* | **1984** | 286 | C44 |
| 3 | Watanabe, Y. | *J. Organomet. Chem.* | **1984** | 270 | 333 |
| 4 | Watanabe, Y. | *J. Org. Chem.* | **1984** | 49 | 3359 |
| 5 | Watanabe, Y. | *Bull. Chem. Soc. Jpn.* | **1984** | 57 | 435 |
| 6 | Watanabe, Y. | *J. Org. Chem.* | **1985** | 50 | 1365 |
| 7 | Watanabe, Y. | *J. Org. Chem.* | **1987** | 52 | 1673 |

## WASSERMAN – BORMANN Macrocyclic Lactam Synthesis

Ring expansion sequence of lactams by reaction with cyclic iminoethers followed by reductive ring opening to a macrocyclic lactam (see 1st edition).

| 1 | Bormann, D. | *Chem. Ber.* | **1970** | *103* | 1797 |
|---|---|---|---|---|---|
| 2 | Wasserman, H. H. | *J. Am. Chem. Soc.* | **1981** | *103* | 461 |
| 3 | Wasserman, H. H. | *Tetrahedron Lett.* | **1983** | *24* | 3669 |

## WEERMAN Degradation

Synthesis of lower homolog aldehydes from α,β–unsaturated carboxamides (via Hofmann degradation) (see 1st edition).

| 1 | Weerman, R .A. | *Rec. Trav.Chim.* | **1918** | *37* | 1 |
|---|---|---|---|---|---|
| 2 | Masson, C. D. | *J. Org. Chem.* | **1951** | *16* | 1869 |
| 3 | Lane, J. F. | *Org. React.* | **1946** | *3* | 276 |

**2-Thienylacetaldehyde (3).**[2] To a suspension of **1** (12.0 g, 80 mmol) in MeOH (100 mL) was added 0.8 N KOH (150 mL ) and 0.8 M KOCl. The temperature rose to 55-60 °C and after cooling crude **2** was filtered and recrystallized from EtOH to give 10 g of **2** (70%), mp 115-116 °C. **2** (22.0 g, 0.12 mol) in 50% EtOH (200 mL) was treated with oxalic acid dihydrate (20.0 g, 0.4 mol) and heated on a water bath for 15 min. The solvent was evaporated and the residue steam distilled. Extraction of the distillate with Et$_2$O and evaporation gave an oil, which after distillation gave 2.9 g of **4** (19%), bp 69-74 °C (8 mm); semicarbazone mp 131-132 °C.

## WEIDENHAGEN Imidazole Synthesis

Imidazole synthesis from α-ketols, formaldehyde and ammonia (see 1st edition).

$(10\%)^3$

| 1 | Weidenhagen, R. | *Chem. Ber.* | **1935** | *68* | 1953 |
| 2 | Weidenhagen, R. | *Chem. Ber.* | **1937** | *70* | 570 |
| 3 | Huebner, C. F. | *J. Am. Chem. Soc.* | **1951** | *73* | 4667 |
| 4 | Schobert, E. S. | *J. pr. Chem.* | **1962** | *18* | 192 |

## WEISS Annulation

Synthesis of fused cyclopentanones (bicyclo[3.3.0] octadiones) or of propellanes from α-dicarbonyl compounds via a double aldol condensation with β-ketoesters **2** (see 1st edition).

| 1 | Weiss, U. | *Tetrahedron Lett.* | **1968** | | 4885 |
| 2 | Cook, J. M. | *Can. J. Chem.* | **1978** | 56 | 189 |
| 3 | Cook, J. M. | *Tetrahedron Lett.* | **1991** | 47 | 3665 |
| 4 | Paquette, L. A. | *J. Org. Chem.* | **1995** | 60 | 353 |

## **W E I N R E B** Ketone Synthesis

**Synthesis of ketones and aldehydes from acid chlorides (or esters) via reaction of N-methoxy-N-methylamides with a Grignard or organolithium reagent (see 1$^{st}$ edition).**

$$C_6H_{11}\text{-}\overset{O}{\overset{\|}{C}}\text{-Cl} + \underset{NH\text{-}OMe}{\overset{Me}{|}} \xrightarrow[\text{EtOH}]{\text{pyr, }0^0} C_6H_{11}\text{-}\overset{O}{\overset{\|}{C}}\text{-}\underset{\overset{|}{OMe}}{N}\text{-OMe} \xrightarrow{\text{BuLi}} \underset{Bu}{\overset{C_6H_{11}}{\diagdown}} \underset{N\text{-}O}{\overset{O}{\diagup}}Li \longrightarrow C_6H_{11}\text{-}\overset{O}{\overset{\|}{C}}\text{-Bu}$$

90%

1

$$\xrightarrow{\text{ClMgN(OMe)Me}} \begin{bmatrix} \underset{R}{\overset{ClMgO}{\diagdown}}\underset{OMe}{\overset{N(OMe)Me}{\diagup}}C \end{bmatrix} \xrightarrow{\text{-MeOMgCl}} \underset{R}{\overset{O}{\overset{\|}{C}}}N(OMe)Me$$

$$\xrightarrow{\text{PhMgCl}} \underset{R}{\overset{ClMgO}{\diagdown}}\underset{Ph}{\overset{N(OMe)Me}{\diagup}}C \xrightarrow{H^+}$$

2

| | | | | | |
|---|---|---|---|---|---|
| 1 | Weinreb, S. M. | *Tetrahedron Lett.* | **1982** | 22 | 3818 |
| 2 | Fehrentz, J. A. | *Synthesis* | **1983** | | 676 |
| 3 | Goel, O. P. | *Org. Prep. Proced. Intn.* | **1987** | 19 | 75 |
| 4 | Williams, R. M. | *J. Org. Chem.* | **1987** | 52 | 2615 |
| 5 | Einhorn, J. | *Synth. Commun.* | **1990** | 20 | 1105 |
| 6 | Flippin, L. A. | *J. Org. Chem.* | **1993** | 58 | 2631 |
| 7 | Williams, J. M. | *Tetrahedron Lett.* | **1995** | 36 | 5461 |
| 8 | Sibi, M. P. | *Org. Prep. Proced. Intn.* | **1993** | 25 | 15 |
| 9 | Salvino, J. M. | *J. Org. Chem.* | **1999** | 64 | 1823 |
| 10 | Giacomelli, G. | *J. Org. Chem.* | **2001** | 66 | 2534 |

**17 β-Benzoyl-4-aza-5-ol-androst-1-ene-3-one (2).**[7] To a slurry of Me(MeO)NH.HCl (3.71 g, 37.9 mmol)and the methyl ester 1 (10.0 g, 30.2 mmol) in THF (400 mL) at −5°C under N$_2$ was added PhMgCl in THF (126 mL, 2.0 M) over 2 h maintaining the temperature between −2 and −5°C. After 1 h at −5°C the reaction mixture was warmed to 25°C over 1h, aged for 8 h then quenched into 1N HCl. The mixture was heated to 30-35°C, the organic layer was separated and the solvent evaporated in vacuum. The residue was crystallized by adding iPrOH and water and cooling to 0°C. Filtration, washing (iPrOH) and drying (80°C/vacuum) afforded 9.9 g of 2 (87%), [α]$_{408}$ +94.5° (25°C, c=1 glac. AcOH).

## **W E N C K E R** Aziridine Synthesis

Synthesis of aziridines from 1,2-aminoalcohols (see 1st edition).

$$H_2N-CH_2-CH_2-OH \xrightarrow{H_2SO_4} \overset{+}{H_3N}-CH_2-OSO_3^- \xrightarrow{NaOH} \text{(90%)[2]}$$

t-Bu-NH-CH₂-CH-CH₂Cl (OH) **1** $\xrightarrow[\text{MeSO}_2\text{Cl}]{\text{Pyr, MeSO}_3\text{H}}$ t-Bu-NH-CH₂-CH-CH₂Cl (O Ms) **2** $\xrightarrow{\text{Na}_2\text{CO}_3}$ t-Bu-N⟨⟩-CH₂Cl **3 (55%)[3]**

| 1 | Wencker, H. | *J. Am. Chem. Soc.* | **1935** | 57 | 2338 |
|---|---|---|---|---|---|
| 2 | Leighton, P. A. | *J. Am. Chem. Soc.* | **1947** | 69 | 1540 |
| 3 | Gaertner, V. R. | *J. Org. Chem.* | **1970** | 35 | 3952 |
| 4 | Nakagawa, Y. | *Bull. Chem. Soc. Jpn.* | **1972** | 45 | 1162 |

**1-tert-Butyl-2-chloromethylaziridine (3).**[3] 1-tert-Butylamino-3-chloro-2-propanol **1** (79.1 g, 0.47 mol) in CHCl₃ (250 mL) at 0-10 °C was treated with pyridine (71 g, 0.9 mol), MeSO₃H (36 g, 0.37 mol) and then dropwise with MeSO₂Cl (57.3 g, 0.46 mol). After 24 h stirring, the mixture was poured into ice-water and finally treated with Na₂CO₃ at pH=8. The CHCl₃ layer and the extract was dried (MgSO₄). Evaporation of the solvent (below 15 °C) gave **2**; cold **2** was added to a stirred solution of Na₂CO₃ (53 g), diethylenetriamine (10 g) (scavenger for epoxide impurities) and water (400 mL) under cooling in an ice bath. After 24 h of stirring the mixture was extracted (Et₂O) and the solvent distilled to afford 40.6 g of **3** (55%), bp 47-48 °C/10 mm.

## **W H A R T O N** Olefin Synthesis

Conversion of α-haloketones to olefins using hydrazine (via enediimides C=C-N=NH). Also reduction of α,β-epoxy ketones to allyl alcohos. (see 1st edition).

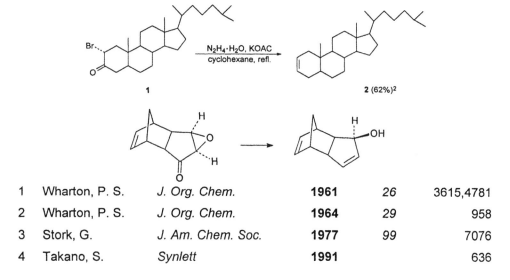

| 1 | Wharton, P. S. | *J. Org. Chem.* | **1961** | 26 | 3615,4781 |
|---|---|---|---|---|---|
| 2 | Wharton, P. S. | *J. Org. Chem.* | **1964** | 29 | 958 |
| 3 | Stork, G. | *J. Am. Chem. Soc.* | **1977** | 99 | 7076 |
| 4 | Takano, S. | *Synlett* | **1991** | | 636 |

## W E N D E R Homologous Diels-Alder Reaction

Synthesis of cycloheptenes by intermolecular 5+2 cycloaddition of singlcyclopropaves with alkyner catalyzed by $Rh(CO)_2Cl_2$.

| 1 | Sarel, S.; Breuer, E. | *J. Am. Chem. Soc.* | **1959** | *81* | 6522 |
|---|---|---|---|---|---|
| 2 | Wender, P. A. | *J. Am. Chem. Soc.* | **1995** | *117* | 4720 |
| 3 | Wender, P. A. | *J. Am. Chem. Soc.* | **1998** | *120* | 1940 |
| 4 | Wender, P. A. | *Tetrahedron,* | **1998** | *54* | 7203 |
| 5 | Gilbertson, S. R. | *Tetrahedron Lett.* | **1998** | *39* | 2075 |
| 6 | de Meijere, A. | *Eur. J. Org. Chem.* | **1998** | | 113 |
| 7 | Wender, P. A. | *J. Am. Chem. Soc.* | **1998** | *120* | 10976 |

**Dimethyl cyclohept-4-ene-1-one-4,5-dicarboxylate (5).**[7] To a solution of catalyst $Rh(CO)_2Cl_2$ **3** (115 mg, 0.05 mmol) in $CH_2Cl_2$ under Ar atmosphere were added (t-butyldimethylsilyloxy)-1-cyclopropylethene **1** (182 mg , 1 mmol) and the mixture was stirred under Ar for 1 min. Dimethyl acetylenedicarboxylate **2** (170 mg, 1.2 mmol) was added and the reaction mixture was heated to 40 °C for 2 h. Upon completion of the reaction (TLC), the reaction mixture was treated with 1% HCl in EtOH (0.2 mL). Filtration through a pad of silica gel evaporation of the solvent and chromatography afforded 205 mg of **5** (90.7%).

## WENZEL–IMAMOTO Reduction

Selective reduction of C=C or C=O with LaNi alloy prepared from La Ni$_5$ ingot and H$_2$ at 200 °, 30 atm[2]. (see 1st edition).

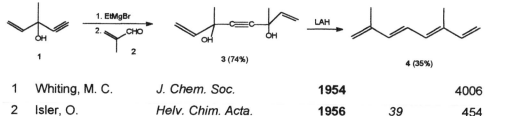

| 1 | Wenzel. H. | *Int. Met. Rev.* | **1982** | 27 | 140 |
| 2 | Imamoto, T. | *J. Org. Chem.* | **1987** | 52 | 5695 |

## WHITING Diene Synthesis

Diene synthesis from 2-alkyne-1,4,-diols with LAH (see 1st edition).

| 1 | Whiting, M. C. | *J. Chem. Soc.* | **1954** | | 4006 |
| 2 | Isler, O. | *Helv. Chim. Acta.* | **1956** | 39 | 454 |

**Cosmene (4).**[1] A solution of **3** (3 g, 20 mmol) in Et$_2$O (50 mL) was treated with LAH (2.5 g, 65 mmol) in Et$_2$O (200 mL) . After 4 h reflux, the cooled mixture was poured on ice and tartaric acid. The Et$_2$O layer was dried (MgSO$_4$), the solvent evaporated under N$_2$ at −4 °C and the residue extracted with petroleum ether and passed through alumina. After evaporation of the solvent at 0 °C, there was obtained 0.8-0.9 g of **4** (33-37%), bp 50 °C/0.2 mm. mp −7 °C, (stable for a few months under N$_2$ at −60 °C).

## WESSELY-MOSER Rearrangement

Acid catalyzed rearrangement of dihydroxyxanthones. (see 1st edition).

| 1 | Wessely, F.; Moser, G. H. | *Monatsh.* | **1930** | *56* | 97 |
|---|---|---|---|---|---|
| 2 | Wheeler, T. S. | *J. Chem. Soc.* | **1956** | | 4455 |
| 3 | Molho, D. | *Bull. Soc. Chim. Fr.* | **1963** | | 603 |
| 4 | Suschitzky, H. | *J. Chem. Soc. Chem. Commun.* | **1984** | | 2275 |

## WESTPHAL Heterocycle Condensation

Synthesis of azabicycles containing a quaternary aromatic bridgehead nitrogen, by condensation of α-methylcycloimmonium salts and 1,2-diketones.

| 1 | Westphal, O. | *Liebigs Ann.* | **1957** | *605* | 8 |
|---|---|---|---|---|---|
| 2 | Westphal, O. | *Arch. Pharm.* | **1961** | *294* | 37 |
| 3 | Alvarez-Builla, J. | *Liebigs Ann.* | **1992** | | 777 |
| 4 | Alvarez-Builla, J. | *J. org. Chem.* | **1994** | *59* | 8294 |
| 5 | Alvarez-Builla, J. | *Tetrahedron Lett.* | **1991** | *32* | 7575 |

**2,3-Dimethyldehydrochinolizinium bromide (5).** [2] 3 (2.5 g, 9.6 mmol), 4 (0.9 g, 10.46 mmol) and dibutylamine (1.29 g, 10 mmol) in EtOH (20 mL) was refluxed for 40 min. Work up and recrystallization from EtOH:Et$_2$O afforded 1.8 g of 5. H$_2$O (70%).

## WIDEQUIST Cyclopropane Synthesis

Tetracyanocyclopropane synthesis from bromomalonitrile and ketones (see 1st edition).

|   | Widequist, S. | *Arkiv. Kemi. Mineral. Geolog.* | **1937** | *12A* | (22) |
|---|---|---|---|---|---|
| 1 | Widequist, S. | *Arkiv. Kemi. Mineral. Geolog.* | **1937** | *12A* | (22) |
| 2 | Widequist, S. | *Arkiv. Kemi. Mineral. Geolog.* | **1945** | *20B* | 12(4) |
| 3 | Scribner, R. M. | *J. Org. Chem.* | **1960** | *25* | 1140 |
| 4 | Hart, H. | *J. Org. Chem.* | **1963** | *28* | 1220 |
| 5 | Hart, H. | *J. Am. Chem. Soc.* | **1963** | *85* | 1161 |
| 6 | Hart, H. | *J. Org. Chem.* | **1966** | *31* | 2784 |

## WILLGERODT-KINDLER Rearrangement

Rearrangement of ketones to amides by heating with sulfur and ammonia or amines (see 1st edition).

| 1 | Willgerodt, O. | *Chem. Ber.* | **1887** | *20* | 2467 |
|---|---|---|---|---|---|
| 2 | Kindler, K. | *Liebigs Ann.* | **1923** | *431* | 193 |
| 3 | De, Tar, D. F. | *J. Am. Chem. Soc.* | **1946** | *68* | 2028 |
| 4 | Cavallieri, L. | *J. Am. Chem. Soc.* | **1945** | *67* | 1755 |
| 5 | Brown, E. W. | *Synthesis* | **1975** | | 358 |
| 6 | Mayer, R.; Wehl, J. | *Angew. Chem.* | **1964** | *76* | 861 |
| 7 | Strauss. C.R. | *Org. Prep. Proced. Int.* | **1995** | *27* | 555 |
| 8 | Hundt, R. N. | *Chem. Rev.* | **1961** | *61* | 52 |
| 9 | Wolff, E. | *Org. React.* | **1951** | *6* | 439 |

**4-Methoxyphenylacetamide (2).**[7] A mixture of **5**. (13.66, 83.3 mmol), $S_8$ (15 g, 58.4 mmol), iPrOH (15 mL) and $NH_4OH$ (30 mL). was heated for 20 min at 210 °C. Trituration with $Et_2O$ and recrystallization from water gave 7.7 g of **6** (62%).

## **WILKINSON** Decarbonylation Rh Catalyst

Rh catalyst for carbonylation, decarbonylation, oxygenation, benzyl cleavage (see 1st edition).

| | | | | | |
|---|---|---|---|---|---|
| 1 | Wilkinson, G. | *J.Am.Chem.Soc.* | 1951 | *73* | 5501 |
| 2 | Wilkinson, G. | *J.Chem.Soc. (A)* | 1966 | | 1711 |
| 3 | Wilkinson, G. | *J.Chem.Soc.* | 1968 | | 1954 |
| 4 | Tsuji, J. | *J.Am.Chem.Soc.* | 1968 | *90* | 99 |
| 5 | Sakai, K. | *Tetrahedron Lett.* | 1972 | | 4375 |
| 6 | Trost, B.M. | *J.Am.Chem.Soc.* | 1973 | *95* | 7863 |
| 7 | Fraser-Reid, B. | *J.Am.Chem.Soc.* | 1975 | *97* | 2563 |
| 8 | Reuter, J.M. | *J.Org.Chem.* | 1978 | *43* | 2438 |
| 9 | Grigg, R. | *Synthesis* | 1983 | | 1009 |
| 10 | Kozikowski, A.P. | *Synthesis* | 1976 | | 562 |
| 11 | Hsing-Jang, Liu | *Synth.Commun.* | 1990 | *20* | 557 |

**Tris(triphenylphosphine)chlorrhodium 1.[2]** To PPh$_3$ (12 g, 19 mmol) in hot EtOH (350 mL) was added RhCl$_3$.3H$_2$O (2 g, 7.6 mmol) in EtOH (70 mL) and the solution was refluxed for 30 min. Hot filtration of the red crystals, washing (Et$_2$O) and vacuum drying afforded 6.25 g of 1 (88%), mp 157-158°C.

**3-(Trimethylsilyl)cyclopent-2-en-1-one 3.[8]** 2 (500 mg, 3.56 mmol) and 1 (33 mg, 0.035 mmol) were heated at 95°C for 1.5 h under a flow of O$_2$. Usual work up and distillation gave 455 mg of 3 (83%), bp 90°C/9.5 mm.

**1-Methylnaphthalene 7.[11]** To a solution of ester 6 (143 mg, 0.5 mmol) in PhH (5 mL), Et$_3$SiH (475 mg, 4.1 mmol) and 1 (100 mg, 0.1 mmol) were added. Under an Ar atmosphere, the mixture was refluxed for 30 min. After evaporation of the solvent the residue was chromatographed (silica gel, hexane) to afford 46 mg of 7 (65%).

## WILLIAMS – BEN ISHAI Amino Acid Synthesis

Asymmetric synthesis (Williams) of $\alpha$-amino acids through C-C bond construction on an electrophylic glycine template (see 1st edition).

| 1 | Ben-Ishai, D. | *J.Chem.Soc.Chem. Commun.* | **1975** | | 349, 905 |
| 2 | Williams, R.M. | *J.Am.Chem.Soc.* | **1986** | *108* | 1103 |
| 3 | Tishler, M. | *J.Am.Chem.Soc.* | **1951** | *73* | 1216 |
| 4 | Williams, R.M. | *J.Org.Chem.* | **1986** | *51* | 5021 |
| 5 | Williams, R.M. | *J.Am.Chem.Soc.* | **1988** | *110* | 1547 |
| 6 | Williams, R.M. | *J.Am.Chem.Soc.* | **1988** | *110* | 1553 |

**1,4-Oxazine-2-one (3).**[6] A suspension of **2** (51 g, 239 mmol) in THF (1200 mL) was treated with ethyl bromoacetate (60 g, 359 mmol) and TEA (49 g, 485 mmol). After 18 h stirring, TEA.HBr was filtered and the filtrate evaporated in vacuum. The residue was washed with water and recrystallized from EtOH (250 mL) to yield 60.3 g (84%) of products. To 239 g (80 mmol) of this product in 160 mL of $CHCl_3$ was added a mixture of di-tert-butyl dicarbonate (34.9 g, 160 mmol), NaCl (32.8 g, 560 mmol) and saturated solution of $NaHCO_3$ (160 mL) and the mixture was refluxed for 20 h. The aqueous phase was extracted with $CHCl_3$, and the combined organic layers washed (water), dried ($Na_2SO_4$), the solvent evaporated and the residue distilled to remove the di-tert-butyl carbonate. The crude product was used for the next step. Recrystallized (hexane:EtOAc 3:1), mp 60-62°C, $\alpha_D$ = -20.5° (c 5.5 $CH_2Cl_2$). To the crude product above (32 g), in PhH (750 mL) was added pTsOH (1.5 g, 8 mmol), the mixture was refluxed (Soxhlet extractor filled with $CaCl_2$ for 8 h, the solid after removal of solvent was dissolved in $CH_2Cl_2$, washed (water) and evaporated. Recrystallization from EtOH (750 mL) gave 20.7 g of **3** (73%).

**R-$\beta$-Ethyl aspartate (7).** A solution of **6** (86.5 mg, 0.18 mmol) in THF (2 mL) and anh. EtOH (2 mL) treated with $PdCl_2$ (19 mg) was hydrogenated 24 h at 20°C and 20 psi. After usual workup there are obtained 34.2 mg of **7** (85%), 96% ee.

## WILLIAMSON Ether Synthesis

Synthesis of ethers from alcoholates with alkyl halides, also with phase transfer catalysis (see 1st edition).

$$Me-(CH_2)_9-OH \ + \ n\text{-}BuBr \ + \ KOH \xrightarrow[\text{PEG 2000}]{20°, 5h} Me-(CH_2)_9-O-Bu$$

$$\underset{1}{} \quad\quad \underset{2}{} \quad\quad\quad\quad\quad\quad\quad\quad\quad \underset{3\ (84\%)}{}$$

| 1 | Williamson, A.W. | *J.Chem.Soc.* | **1852** | 4 | 229 |
|---|---|---|---|---|---|
| 2 | Dermer, O.C. | *Chem.Rev* | **1934** | 14 | 409 |
| 3 | Kalinowsky, K.O. | *Angew.Chem.Int.Ed.* | **1975** | 14 | 763 |
| 4 | Nakatsugi, T. | *Synthesis* | **1987** | | 280 |
| 5 | Soula, G. | *J.Org.Chem.* | **1985** | 50 | 3717 |
| 6 | Abribat, B. | *Syn.Commun.* | **1994** | 24 | 2091 |

**Decyl butyl ether (3).[6]** A mixture of decyl alcohol 1 (79g, 0.5mol), 5% PEG 2000 and powdered KOH (42g, 0.75mol) was treated with butyl bromide 2 (102.75g, 0.75mol) and the mixture was stirred for 5 h at 20°C. Removal of the solids by filtration and chromatography (hexane) gave 90 g of 3 (84%).

## WISSNER Hydroxyketone Synthesis

Conversion of acyl chlorides to functionalized (OH, $OCH_3$, $SCH_3$) ketones by means of tris(trimethylsilyloxy)ethylene 4 (see 1st edition).

| 1 | Wissner, A. | *Tetrahedron Lett.* | **1978** | | 2749 |
|---|---|---|---|---|---|
| 2 | Wissner, A. | *J.Org.Chem.* | **1978** | 44 | 4617 |

**1-Hydroxy-2-nonanone (6).[2]** Octanoyl chloride 5 (4.0 g, 24.6 mmol) and 4 (15.5 g, 53 mmol) were heated for 3 h at 95-100°C. Dioxane (25 mL) and 0.6N HCl (10 mL) was added (exothermic, gas evolution). The mixture was heated for 30 min at 95°C, extracted with $Et_2O$, and distilled to afford 3.28 g of 6 (84%), bp83-85°C/0.5 mm.

## WITTIG Olefin Synthesis

Olefin synthesis from phosphorane ylides (e.g. **2**) with aldehydes or ketones; cis olefins predominate in aliphatic systems, trans in conjugated olefins (see 1st edition).

| 1  | Wittig, G.       | *Liebigs Ann.*             | **1949** | 562 | 187  |
|----|------------------|----------------------------|----------|-----|------|
| 2  | Wittig, G.       | *Chem.Ber*                 | **1961** | 94  | 1373 |
| 3  | Ketcham, R.      | *J.Org.Chem.*              | **1962** | 27  | 4666 |
| 4  | Angeletti, E.    | *J.Chem.Soc.Perkin Tr.1*   | **1987** |     | 713  |
| 5  | Doudon, A.       | *Tetrahedron*              | **1988** | 44  | 2021 |
| 6  | Emmons, W.       | *Angew.Chem.Int.Ed.*       | **1966** | 5   | 126  |
| 7  | Murphy, P.B.     | *Chem.Soc.Rev.*            | **1988** | 17  | 1    |
| 8  | Maercker, A.     | *Org.React.*               | **1965** | 14  | 270  |
| 9  | Maryanoff, B.E.  | *Chem.Rev.*                | **1989** | 89  | 863  |
| 10 | Lorcharich, R.J. | *J.Org.Chem.*              | **1995** | 60  | 156  |
| 11 | Mikami, M.       | *Org. React.*              | **1994** | 46  | 2    |

**(E) and (Z) (6).**[10] To a mixture of ketone **4** (173 mg, 0.49 mmol) and (5-carboxypentyl)triphenylphosphonium bromide **5** (247.3 mg, 0.54 mmol) in THF (2.5 mL) was added dropwise 1M t-BuOK (1.1 mL, 1.1 mmol) in THF at 0°C. After stirring for 1 h the reaction was quenched with 10 drops of 1 N HCl at 0°C and the solvent was evaporated. Flash chromatography ($CH_2Cl_2$:EtOAc:AcOH 50:48:2) afforded 176.5 mg of **6** (80%), E:Z=26:74.

## WITTIG Rearrangement

A stereoselective base catalysed [2,3] sigmatropic rearrangement of allyl ethers to homoallylic alcohols (stereoselective) (see 1st edition).

| 1 | Wittig, G. | *Liebigs Ann.* | **1942** | *550* | 260 |
|---|---|---|---|---|---|
| 2 | Marshall, J.A. | *J.Org.Chem.* | **1988** | *53* | 4108 |
| 3 | Schöllkopf, U. | *Angew.Chem.Int.Ed.* | **1962** | *1* | 126 |
| 4 | Mikami, K. | *Chem.Lett.* | **1985** | | 1729 |
| 5 | Brückner, R. | *Contakte (Darmstadt)* | **1991** | *3* | 3 |
| 6 | Mikami, K. | *Org.React.* | **1994** | *46* | |
| 7 | Kurishima, M. | *Tetrahedron Lett.* | **2001** | *42* | 415 |

**Cyclohexenol (2).**[2] 2,2,6,6-Tetramethylpiperidine 3.2 mL and 2.5 M n-BuLi in hexane 7.9 mL was stirred in THF (20 mL) under $N_2$ at 0°C and for 30 min at 20°C. This lithium tetramethylpiperidide (LTMP) reagent was added to a solution of **1** (1.0 g, 658 mmol) in THF (20 mL) at 0°C. After 14 h at 20°C, work up and chromatography afforded 0.78 g of **2** (78%).

## WOHL - ZIEGLER Bromination

Allylic or benzylic bromination or chlorination with N-bromosuccinimide (NBS) or N-chlorosuccinimide (NCS), thermal or photochemical (see 1st edition).

| 1 | Wohl, A. | *Chem.Ber* | **1919** | *52* | 51 |
|---|---|---|---|---|---|
| 2 | Ziegler, K. | *Liebigs Ann.* | **1942** | *551* | 80 |
| 3 | Dauben, H. | *J.Am.Chem.Soc.* | **1959** | *81* | 4863 |
| 4 | Zalusky, M.C. | *Bull.Soc.Chim.Fr.* | **1970** | | 1447 |
| 5 | Djerassi, C. | *Chem.Rev.* | **1948** | *43* | 271 |
| 6 | Horner, L. | *Angew.Chem.* | **1959** | *71* | 349 |

**Furan (2).**[3] To **1** (96 g, 0.7 mol) in PhH (600 mL) under reflux and stirring, was added a mixture of N-bromosuccinimide (137 g, 0.76 mol) and benzoyl peroxide (3 g). The floating succinimide was filtered. Work up gave 151 g of **2** (98%), recrystallized from $Et_2O$-hexane or sublimed at 60°C (0.1 mm), mp 54°C.

## WOHL - WEYGAND Aldose Degradation

Degradation of sugar oximes via cyanohydrins by means of an acid chloride/pyridine (Wohl) or of fluorodinitrobenzene (Weygand).

| 1 | Wohl, A. | *Chem.Ber* | **1893** | 26 | 730 |
| 2 | Wohl, A. | *Chem.Ber* | **1899** | 32 | 3666 |
| 3 | Deferrari, J.O. | *J.Org.Chem.* | **1966** | 31 | 905 |
| 4 | Weygand, F. | *Chem.Ber* | **1950** | 83 | 559 |
| 5 | Weygand, F. | *Chem.Ber* | **1952** | 85 | 256 |

**1,1-Bis(benzamido)-1-deoxy-D-galactitol (3).**[3] **1** (3.92 g, 17 mmol) was added to 1:1 pyridine:PhCOCl (48 mL) and kept at 100°C by cooling. After 24 h at 20°C, all was poured into water (400 mL) and **2** was recrystallized from $Me_2CO$:EtOH (1:3), 9.45 g (65.6%), mp 190-191°C, ($\alpha$)$_D$=+19.5° (c 0.7 $CHCl_3$). **2** (4g, 4.8 mmol) was stirred with $NH_3$ in MeOH (100 mL) for 60 min. After 24 h evaporation afforded 770 mg of **3**, mp 194°C; from EtOH 700 mg (37%), mp 203-204°C, ($\alpha$)$_D$=-5.8° (c 0.85 pyr).

**Arabinose (6).**[4] $CO_2$ was bubbled through $NaHCO_3$ (1.5 g, 17 mmol) in water (70 mL) and glucose oxime **4** (1 g, 4.5 mmol). At 55-60°C **5** (1.8 g, 10 mmol) in iPrOH (30 mL) was added. After 1.5-2h, 420 mg of **6** (61%), mp 162°C, was isolated.

## **WOLFF** Rearrangement

Rearrangement (ring contraction) of $\alpha$-diazoketones to carboxylic acids or their derivatives (esters, amides) via ketenes (see also Arndt-Eistert) (see 1st edition).

| 1 | Wolff, L. | *Liebigs Ann.* | **1912** | *394* | 25 |
|---|---|---|---|---|---|
| 2 | Borch, R.F. | *J.Org.Chem.* | **1969** | *34* | 1481 |
| 3 | Wynberg, H. | *J.Org.Chem.* | **1968** | *33* | 4025 |
| 4 | Cissy, J. | *Synthesis* | **1988** | | 720 |
| 5 | Stoutland, P.O. | *J.Am.Chem.Soc.* | **1996** | *118* | 1551 |
| 6 | Podlech, J. | *J.Org.Chem.* | **1997** | *62* | 5873 |
| 7 | Meier, H. | *Angew.Chem.Int.Ed.* | **1975** | *14* | 32 |
| 8 | Bachmann, W.E. | *Org. React.* | **1942** | *1* | 39 |
| 9 | Smith, P.A.S. | *J.Org.Chem.* | **1961** | *26* | 27 |
| 10 | Maisden, S.P. | *Chem.Commun.* | **1999** | | 1199 |

**4-Carboxy-3,3,5,5-tetramethyl-1-thiacyclohexane (3).**[3] Hydrazoketone **1** (4.3 g, 20 mmol) in PhH (50 mL) was added to active $MnO_2$ (6 g) and $MgSO_4$ (10 g) in PhH (50 mL) over 30 min. After 24 h at 20°C, the mixture was filtered and the filtrate concentrated. 500 mg of the residue after reflux with dil. HCl gave 195 mg of **3** (48%), mp 149-152°C. The rest of the residue was chromatographed (silica gel, PhMe) to give 0.55 g (16.5%) of unreacted ketene **2**, bp 103-104°C/12 mm.

**Azetidinone 6a and 6b.**[6] A mixture of diazoketone **4** (579 mg, 2 mmol) and imine **5** (1.56 g, 8 mmol) in $Et_2O$ (300 mL) was irradiated for 90 min at $-15$°C. The mixture was stirred for 30 min at $-15$°C and warmed to 20°C. The solvent was evaporated and the nonpolar compounds were filtered. Chromatography gave 507 mg of **6a** (56%) and 59 mg of **6b** (6%).

## WOLFF – KISHNER – HUANG MINLON Reduction

Reduction of ketones to hydrocarbons by heating with $NH_2NH_2$ and aqueous KOH (Wolff-Kishner) or KOH in ethylene glycol (Huang Minlon) (see 1st edition).

| 1 | Kishner, J. | *J.Russ.Phys.Chem.Soc.* | **1911** | *43* | 582 |
|---|---|---|---|---|---|
| 2 | Wolff, C. | *Liebigs Ann.* | **1912** | *394* | 86 |
| 3 | Huang Minlon | *J.Am.Chem.Soc.* | **1946** | *68* | 2487 |
| 4 | Huang Minlon | *J.Am.Chem.Soc.* | **1949** | *71* | 3301 |
| 5 | Nickon, A. | *J.Org.Chem.* | **1981** | *46* | 4692 |
| 6 | Burgstahler, A.W. | *J.Org.Chem.* | **1969** | *34* | 1562 |
| 7 | Todd, D. | *Org.React.* | **1948** | *4* | 378 |

$\gamma$-(p-Phenoxyphenyl)butyric acid (2).[3] A mixture of $\gamma$-(p-phenoxybenzoyl)propionic acid 1 (500 g, 1.85 mol), KOH (350 g, 6.25 mol), 85% hydrazine hydrate (250 mL) and di (or tri) ethylene glycol (2500 mL) was refluxed for 2 h. The condensor was removed and heating continued until the temperature reached 195°C. Refluxing was continued for 4 h, the cooled mixture was diluted with water (2500 mL) and slowly poured into 6N HCl (1500 mL). Filtration and drying gave 451 g of 2 (95%), mp 64-66°C.

## WOLFRAM – SCHÖRNIG - HANSDORF Carboxymethylation

Carboxymethylation of aromatics in the presence of oxidants or photochemically (see 1st edition).

| 1 | Wolfram, A.; Schörnig, L. | *Germ. Pat.* | | 562.391 |
|---|---|---|---|---|
| 2 | Hansdorf, E. | *U.S. Pat.* | | 1.951.686 |
| 3 | Ogata, Y. | *J.Am.Chem.Soc.* | **1950** *72* | 4302 |
| 4 | Ogata, Y. | *J.Org.Chem.* | **1951** *16* | 239 |
| 5 | Southwick, P.L. | *Synthesis* | **1970** | 628 |

**Naphtylacetic acid (3).[4]** 1 (56.6 g, 0.44 mol), 2 (14.1 g, 0.149 mol), $Fe_3O_4$ (87.6 mg) and KBr (420 mg) were heated gently for 20 h so that 200°C was attained after 10 h and 218°C after 20 h. Unreacted 1 was distilled (43 g) and the residue extracted with hot NaOH solution, filtered and acidified with HCl to give 19.4 g of 3 (70%).

## WOODWARD Peptide Synthesis

Peptide synthesis mediated by N-ethyl-5-phenylisoxazolium-3'-sulfonate **3** (see 1st edition).

$$H_2N-CO-CH_2-\underset{\underset{NH-CO_2-CH_2-Ph}{|}}{CH}-COOH \quad + \quad H_2N-CH_2-CO_2Et \xrightarrow[Et_3N, 20°]{3, MeNO_2}$$

**1**                    **2**

$$H_2N-CO-CH_2-\underset{\underset{NH-CO_2-CH_2-Ph}{|}}{CH}-CO-NH-CH_2-CO_2Et$$

**4**

| | | | | |
|---|---|---|---|---|
| 1 | Woodward, R.B. | *Tetrahedron, Suppl. 7* | **1966** | 21 |
| 2 | Woodward, R.B. | *J.Am.Chem.Soc.* | **1961** | 83 | 1007 |
| 3 | Woodward, R.B. | *Tetrahedron, Suppl. 8* | **1966** | 22 | 321 |

## WÜRTZ Coupling

Coupling of alkyl halides with Na, supplanted by the coupling of alkyl halides or sulfonates with Grignard reagents or RLi in the presence of Cu(I) salts (see 1st edition).

$$\text{1} \xrightarrow[\text{dioxane}]{\text{Zn/Cu, }\triangle} \text{2(84\%)}^9 \quad n\text{-}C_8H_{17}\text{-OTS} \xrightarrow[\text{tBu-MgBr}]{\text{Li}_2\text{CuCl}_4,\ 25°} n\text{-}C_8H_{17}\text{-CMe}_3 \quad \text{4 (75\%)}$$

| | | | | | |
|---|---|---|---|---|---|
| 1 | Würtz, A | *Liebigs Ann.* | **1855** | 96 | 364 |
| 2 | Bailey, W.J. | *J.Org.Chem.* | **1962** | 27 | 3088 |
| 3 | Horner, L. | *Angew.Chem.* | **1962** | 74 | 586 |
| 4 | Buntrock, R.E. | *Chem.Rev.* | **1968** | 68 | 209 |
| 5 | Kosolapoff, G.M. | *Org.React.* | **1951** | 6 | 326 |
| 6 | Whitesides, G.M. | *J.Am.Chem.Soc.* | **1969** | 91 | 4871 |
| 7 | Erdile, E. | *Tetrahedron* | **1984** | 40 | 641 |
| 8 | Schlosser, M. | *Angew.Chem.Int.Ed.* | **1974** | 43 | 1863 |
| 9 | Furukawa, M. | *Heterocycles* | **1996** | 43 | 1863 |

**2,2-Dimethyldecane (4).**[8] To octyl tosylate **3** (40 mmol) in 50 mL of THF at –78°C was added tBuMgBr (56 mmol in 32 mL of Et$_2$O) and Li$_2$CuCl$_4$ (0.2 mmol in 2 mL of THF). The mixture was warmed to 25°C during 2 h and stirred for 12 h. Acidification with 2N H$_2$SO$_4$, washing (2x50 mL water), drying and distillation gave 5.1 g (85%) of pure **4**.

## Y A M A D A Peptide Coupling

Diethylphosphoryl cyanide 3 a reagent for amide bond formation and application to peptide synthesis free of racemization (see 1st edition).

| | | | | |
|---|---|---|---|---|
| 1 | Yamada, S. | *J. Am. Chem. Soc.* | **1971** | 3595 |
| 2 | Yamada, S. | *J. Am. Chem. Soc.* | **1972** | 94 6203 |
| 3 | Yamada, S. | *Tetrahedron Lett.* | **1973** | 1595 |
| 4 | Takamizawa, A. | *Yakugaku Zasshi* | **1965** | 851 298 |

**N-Benzoyl-L-leucylglycine ethyl ester (4).**[3] N-Benzoy-L-leucine 1 (0.235 g, 1 mmol) and 2 (0.1534 g, 1.1 mmol) in DMF (10 mL) under stirring was treated with 3 (0.179 g, 1.1 mmol) in DMF at 0 °C, followed by TEA (0.212 g, 2.1 mmol). The mixture was stirred for 30 min at 0 °C, and 4 h at 20 °C, diluted with PhH-EtOAc, washed with 5% HCl, water, 5% NaHCO$_3$ solution and brine. Evaporation and chromatography afforded 0.271 g of 4 (86%) (pure L), mp 158-160 °C.

## Y A M A G U C H I Lactonization Reagent

2,4,6-Trichlorobenzoyl chloride a reagent for esterification of acids via a mixed anhydride, also used for large ring lactonization with DMAP (see also Steglich-Hassner, (see 1st edition).

| | | | | |
|---|---|---|---|---|
| 1 | Yamaguchi, M. | *Bull Chem. Soc. Jpn.* | **1979** | 52 1989 |
| 2 | Yonemitsu, O. | *J. Org. Chem.* | **1990** | 55 7 |

**Lactone (3).**[1] Acid 1 (272 mg, 1 mmol) and TEA (0.153 mL, 1.1 mmol) in THF (10 mL) was stirred for 10 min at 20 °C, and then 2,4,6-trichlorobenzoyl chloride 2 (160 mL, 1 mmol) was added. After 2 h the precipitate was filtered and washed with THF. The filtrate was diluted with PhH (100 mL) and slowly added to a refluxing solution of 4-dimethylaminopyridine (DMAP) (732 mg, 6 mmol) in PhH(100 mL). After 40 h the mixture was washed with citric acid solution, water and dried (Na$_2$SO$_4$). Evaporation afforded the crude product (247 mg). Preparative TLC (silica gel), Et$_2$O:PhH(2:1), gave 116 mg of 3 (46%), 65 mg of a dimer (26%) and 21 mg of polymer. Recrystallization of 3 (CH$_2$Cl$_2$/diisopropyl ether) afforded colorless needles, mp 123 °C.

## **Y A M A M O T O** Allylation

Synthesis of homoallylamines by catalytic asymmetric allylation of imines with allyl silanes using a chiral bis π-allyl palladium complex, also using allylstannanes (see 1st edition).

| 1 | Yamamoto, Y. | *J. Am. Chem. Soc.* | **1996** | *118* | 6641 |
| 2 | Yamamoto, Y. | *J. Am. Chem. Soc.* | **1998** | *120* | 4242 |
| 3 | Yamamoto, Y. | *J. Org. Chem.* | **1999** | *64* | 2615 |
| 4 | Yamamoto, Y. | *Tetrahedron Lett.* | **2000** | *41* | 131 |
| 5 | Marshall, J.A. | *Chemtracts Org. Chem.* | **1998** | *11* | 855 |

**4-Phenyl-4-benzylamino-1-butene (4).**[3] To a solution of benzylidene benzylamine **1** (90.5 mg; 0.5 mmol) and catalyst **3** (0.025 mmol) in n-hexane (2 mL) was added trimethylallylsilane **2** (114 mg; 1 mmol). The resulting mixture was stirred for about 90 min and then TBAF (0.25 mL; 1.0 M solution in THF) and THF (0.25 mL) were added. Two phases appeared and after 31 h stirring at 20°C, the mixture was quenched (water). Usual work up and chromatography (n-hexane:EtOAc 10:1) afforded 81.7 mg of **4** (69%; 80% ee).

## **YAMAMOTO** Chirality Transfer

Chemoselective, regioselective and E-stereoselective $S_N2'$ displacement with 1,3-chirality transfer by organocyanocopper-trifluoroborate.

| 1 | Yamamoto, Y.; Ibuka, T. | *J.Am.Chem.Soc.* | **1989** | *111* | 4864 |
|---|---|---|---|---|---|
| 2 | Danishefsky, S.J. | *J.Org.Chem.* | **1991** | *56* | 5834 |
| 3 | Yamamoto, Y.; Ibuka, T. | *Synlet.* | **1992** | | 769 |
| 4 | Yamamoto, Y.; Ibuka, T. | *J.Org.Chem.* | **1992** | *57* | 1024 |
| 5 | Krause, N. | *Kontakte (Darmstadt)* | **1993** | *1* | 3 |

**Methyl (E,2R,5R)-2-Methyl-5-(tert-butyl-dimethylsiloxy)-3-hexenoate 2.[4]** To a solution of MeLi (LiI) (1.8 mmol) in $Et_2O$ was added 1.0 M $ZnCl_2$ solution in $Et_2O$ (1.8 mL) and THF (1.5 mL) at -78°C. This solution was added to a solution of CuCN (81 mg, 0.9 mmol) in THF under Ar at -78°C. Under stirring at the same temperature was added a solution of **1** (107 mg, 0.3 mmol) in THF (7 mL). After 30 min stirring at -78°C, the mixture was allowed to warm to 0°C. After 1 h the mixture was quenched with aq. $NH_4Cl$ (PH = 8) and stirring for 30 min at 20°C. Extraction with $Et_2O$, washing the organic phase with 2N HCl, 0.5 M $NaHCO_3$ and brine was followed by drying ($MgSO_4$) and evaporation of the solvent to give an oil. Chromatography (silica gel, n-hexane:EtOAc 10:1) gave 83 mg of **2** (97%) as a colorless oil. Kugelrohr distillation of 90°C/1 mm Hg gave **2**. $[\alpha]_D^{26}$ = -30.01° (c 0.559, $CHCl_3$). Diastereoselectivity > 99:1

## YAMAZAKI–CLAUSEN (GEA) Guanine Synthesis

Synthesis of guanine or of 9-substituted guanines using benzoylisothiocyanate.

| 1 | Yamazaki, A. | *J. Org. Chem.* | **1967** | *32* | 1825 |
| 2 | Yamazaki, A. | *J. Heterocycl. Chem.* | **1978** | *15* | 353 |
| 3 | Townsen, L. D. | *J. Org. Chem.* | **1986** | *51* | 1065,1277 |
| 4 | Clausen, E. P. | *J. Org. Chem.* | **1991** | *56* | 2136 |
| 5 | Nagai, T. | *Org. Prep. Proced. Intn.* | **1993** | *25* | 388 |

**1-[(2-Hydroxyethoxy)methyl]-5-[(thiocarbamoyl)-amino]-1-imidazole-4-carboxami-de (3).**[4] Amine **1** (44 g, 182 mmol) and benzoylisothiocyanate **2** (29.7 g, 182 mmol) was refluxed in Me₂CO for 1 h. MeOH (430 mL) and K₂CO₃ (14.9 g, 108 mmol) in water (45 mL) was added, the mixture was heated to reflux 4 h, the pH adjusted to 8 (AcOH) and cooled to 0 °C. Filtration afforded 39.2 g of **3** (85%), mp 181-183 °C.

**9-[(2-Hydroxyethoxy)methyl]guanine (4).** Thiourea **3** (10g, 38.6 mmol) was added to CuSO₄ (7 g,44 mmol) in 6 N NaOH (80 mL) and stirred for 4 h. After filtration, the filtrate was acidified with HOAc and the solution heated to reflux for a few minutes and cooled to 0 °C. Filtration and recrystallization from water (charcoal) afforded 7.8 g of pure of **4** (85%), mp 250 °C dec.

## Y A M A Z A K I Cyanoaniline Synthesis

Synthesis of o-aminoarylnitriles (useful in pyrimidine synthesis) from nitroquinolines, nitro naphthalenes, and m-substituted ($CF_3$, $COCH_3$ and $COC_6H_5$) nitrobenzenes (see 1st edition).

| 1 | Yamazaki, M. | *Heterocycles* | **1974** | 2 | | 589 |
|---|---|---|---|---|---|---|
| 2 | Yamazaki, M. | *Chem. Pharm. Bull.* | **1981** | 29 | | 1286 |
| 3 | Yamazaki, M. | *Chem. Pharm. Bull.* | **1982** | 30 | | 851 |
| 4 | Yamazaki, M. | *Chem. Pharm. Bull.* | **1985** | 33 | | 1360 |
| 5 | Halama, A. | *J. Chem. Soc. Penkin 1* | **1999** | | | 1839,2495 |

**2-Amino-1-naphthalenecarbonitrile (3).**[3] **1** (5.19 g, 30 mmol) was added to stirred **2** (9.99 g, 90 mmol) and KOH (5.04 g, 90 mmol) in DMF(90 mL). After 24 h solvent was evaporated and the residue hydrolyzed with 5% NaOH (60 mL) for 1 h reflux. Work up afforded 3.80 g of **3** (64%), mp 131.5 °C (75%, KCN is added).

## Z I N K E – Z I E G L E R Calixarene Synthesis

Synthesis of calixarenes (a basket-like macrocyclic compound) from phenols (see 1st edition).

| 1 | Zinke, A.; Ziegler, E. | *Chem. Ber.* | **1941** | 74 | 1279 |
|---|---|---|---|---|---|
| 2 | Gutsche, C. D. | *J. Org. Chem.* | **1986** | 51 | 742 |
| 3 | Gutsche, C. D. | *J. Org. Chem.* | **1990** | 55 | 4487 |
| 4 | Gutsche, C. D. | *Acc. Chem. Res.* | **1983** | 16 | 161 |
| 5 | Gutsche, C. D. | *Top. Curr. Chem.* | **1984** | 123 | 1 |
| 6 | Gutsche, C. D. | *Pure Appl. Chem.* | **1988** | 60 | 1607 |

**p-tert-Butylcalix (4) arene (3).**[3] **1** (100 g, 0.665 mol), 37% formalin **2** (62.3 mL, 0.83 mol) and NaOH (1.2 g, 0.03 mol) in minimum of water was heated for 2 h at 110-120 °C. Add $CHCl_3$, neutralize with HCl, evaporate, heat residue in diphenyl ether (800 mL) under $N_2$, heat to reflux for 2 h. Work up gave 66.5 g of **3** (62%).

## Z A V ' Y A L O V  Pyrrole Synthesis

Pyrrole synthesis from 1,3-dicarbonyl compounds (or enamino ketones) and α-amino acids via cyclization of enamino acid intermediates.

| 1 | Zav'yalov, S.I. | *Izv.Akad.Nauk. SSSR, Ser Khim. Eng.* | **1973** | | 2505 |
|---|---|---|---|---|---|
| 2 | Bach, N.J. | *J. Med. Chem.* | **1980** | 23 | 481 |
| 3 | Berney, D. | *Helv. Chim. Acta* | **1982** | 65 | 1694 |
| 4 | Ceccheti, V. | *J. Heterocycl. Chem.* | **1982** | 19 | 1045 |
| 5 | Cymerman-Craig, J. | *Synth. Commun.* | **1991** | 21 | 1971 |
| 6 | Heron, B.M. | *J. Chem. Soc. Chem. Commun.* | **1999** | | 289 |
| 7 | Andrew, R.J. | *Tetrahedron* | **2000** | 56 | 7267 |

**Isoindole 4.**[5] To **1** (572.66 mg; 2.74 mmol) in anh. EtOH was added tetramethylammonium glycinate **2** (472.33 mg; 3.17 mmol) and the mixture was refluxed under Ar for 2 h. Evaporation under vacuum afforded **3** (yellow powder). Ac$_2$O (7.3 mL) was added to **3** and the mixture was heated under Ar for 2 h at 150-155°C (bath temp). Ac$_2$O was removed in vacuum and the residue was treated with water and extracted with CH$_2$Cl$_2$. The extract was washed with 5% NaHCO$_3$, water, brine and dried (MgSO$_4$). Solvent evaporation and chromatography (silica gel, CHCl$_3$:hexane 1:1) afforded 35 mg of **4** (58%).

## Z E I S E L – P R E Y Ether Cleavage

Acid catalyzed cleavage of aromatic methyl or ethyl ethers. Quantitative methoxy group determination (via AgI). Also ether cleavage with trimethylsilyl iodide.[6] (see 1st edition).

| | | | | | |
|---|---|---|---|---|---|
| 1 | Zeisel, S. | *Monatsh.* | **1885** | 6 | 406 |
| 2 | Belcher, A. | *J. Chem. Soc.* | **1957** | | 4484 |
| 3 | Prey, V. | *Chem. Ber.* | **1941** | 74 | 350 |
| 4 | Burwell, R. L. | *Chem. Rev.* | **1954** | 54 | 635 |
| 5 | Ganem, B. | *J. Org. Chem.* | **1974** | 39 | 3728 |
| 6 | Jung, M.E. | *J. Am. Chem. Soc.* | **1977** | 99 | 968 |

## Z E I S S Ketone Transfer

A method of moving a ring carbonyl function to its alpha carbon.

| | | | | | |
|---|---|---|---|---|---|
| 1 | Zeiss, H. H. | *J. Am. Chem. Soc.* | **1953** | 75 | 5935 |
| 2 | Burgstahler, A. W. | *J. Org. Chem.* | **1969** | 34 | 1562 |
| 3 | Abad, A. | *Org. Prep. Proced. Intn.* | **1991** | 23 | 323 |

**Ketone.** [1] **1** (4.9 g, 19.1 mmol) and **2** (2.52 g, 23.8 mmol) in 95% EtOH (38 mL) was treated with NaOH (1.15 g, 28.7 mmol) in water (1.15 mL). After 8 days, Et$_2$O extraction gave 2.64 g of **3** (40%). **3** (1.86 g, 5.4 mmol) in xylene and 1 N Al (OiPr)$_3$ in iPrOH (15 mL) was refluxed for 100 h. Work up and filtration (Al$_2$O$_3$) gave 0.79 g of **4** (45%). **4** (1.45 g, 4.4 mmol) in EtOAc was ozonized at –60 °C to give 0.29 g of **5** (26%).

## Z I E G L E R Macrocyclic Synthesis

Synthesis of macrocyclic ketones form dinitriles using high dilution (see 1st edition).

| 1 | Ziegler, K. | *Chem. Ber.* | **1934** | 67 | 139 |
|---|---|---|---|---|---|
| 2 | Ziegler, K. | *Liebigs Ann.* | **1937** | 528 | 143 |
| 3 | Newman, M. S. | *J. Org. Chem.* | **1975** | 40 | 2867 |

## Z I M M E R Rearrangement

Rearrangement of α-methylenelactones, lactams and thiolactones to heterocyclic compounds.

| 1 | Zimmer, H. Z. | *J. Org. Chem.* | **1959** | 24 | 28 |
|---|---|---|---|---|---|
| 2 | Zimmer, H. Z. | *Angew. Chem.* | **1961** | 73 | 149 |
| 3 | Zimmer, H. Z. | *J. Heterocycl. Chem.* | **1966** | 3 | 235 |
| 4 | Warnhoff, H. | *Synthesis* | **1972** | | 168 |

**3-Benzamido-6-methylcoumarin (2).**[4] **1** (27.9 mg, 0.1 mmol) in 95% EtOH (1000 mL) was irradiated with a 75-watt General Electric lamp (FG 1096 AX) at 20 °C (cooling). After 2-3 h, evaporation of the solvent afforded **2** as yellow crystals, mp 169 °C.

## Z I N C K E – S U H L Cyclohexadienone Synthesis

Synthesis of cyclohexadienones form phenols by Friedel-Crafts alkylation.

| 1 | Zincke, Th., Suhl, R. | *Chem. Ber.* | **1906** | *39* | 4148 |
|---|---|---|---|---|---|
| 2 | Newman, M. S. | *J. Org. Chem.* | **1958** | *23* | 1236 |
| 3 | Newman, M. S. | *J. Am. Chem. Soc.* | **1959** | *81* | 6454 |
| 4 | Newman, M. S. | *J. Ind. Chem. Soc.* | **1974** | *51* | 95 |

## Z I N I N Benzidine (Semidine) Rearrangement

Acid catalyzed rearrangement of hydrazobenzenes to benzidines and semidines (see 1st edition).

| 1 | Zinin, N. | *J. Prakt. Chem.* | **1845** | *36* | 93 |
|---|---|---|---|---|---|
| 2 | Jakobsohn, P. | *Chem. Ber.* | **1893** | *26* | 688 |
| 3 | Snyder, H. R. | *J. Am. Chem. Soc.* | **1949** | *71* | 289 |
| 4 | Hammond, G. S. | *J. Am. Chem. Soc.* | **1950** | *72* | 20 |
| 5 | Ingold, C. K. | *J. Chem. Soc.* | **1957** | | 1906 |
| 6 | Shine, H. J. | *J. Org. Chem.* | **1963** | *38* | 1232 |

## Z I N N E R  Hydroxylamine Synthesis

Synthesis of O-substituted hydroxylamines from oximes or hydroxamates.

| # | | | | | |
|---|---|---|---|---|---|
| 1 | Zinner, G. | *Angew. Chem.* | **1957** | *69* | 204; 480 |
| 2 | Zinner, G. | *Chem. Ber.* | **1958** | *91* | 302 |
| 3 | Zinner, G. | *Arch. Pharm. (Weinheim)* | **1960** | *293* | 42 |
| 4 | Zinner, G. | *Arch. Pharm. (Weinheim)* | **1970** | *303* | 317 |
| 5 | Zinner, G. | *Synthesis* | **1973** | | 165 |
| 6 | Tamura, Y. | *J. Org. Chem.* | **1973** | *38* | 1239 |

**Hydroxamate (3).**[6] To **2** (34 g, 0.33 mol) and Et₃N (33 g, 0.33 mol) in DMF (90 mL) at 0 °C was added in portions mesitylenesulfonyl chloride **1** (72 g, 0.33 mol). Work up afforded 83 g of **3** (86%), mp 57-58 °C.

**O-Mesitylenesulfonylhydroxylamine (4).** To **3** (75 g, 0.26 mol) in dioxane (50 mL) at 0 °C was added 70% HClO₄ (30 mL) and after 10 min the mixture was poured into water to give 64 g of wet **4**, mp 93-94 °C (from Et₂O-petroleum ether).

# Names Index

ABRAMOV Phosphonylation, 1
ADLER Phenol Oxidation, 2
Akermann, 108
Alder, 89
ALDER (Ene), 3
ALDER–RICKERT Cycloaddition, 4
Allan, 309
ALLEN–MILLAR–TRIPPETT Phosphonium
Rearrangement, 5
ALPER Carbonylation, 6
AMADORI Glucosamine Rearrangement, 7
ANGELI–RIMINI Hydroxamic Acid Synthesis, 7
APPEL Halogenation Reagent, 8
Arbuzov, 243
ARENS–VAN DORP Cinnamaldehyde Synthesis, 11
ARNDT–EISTERT Homologation, 9
ASINGER Thiazoline Synthesis, 10
ATHERTON–TODD Phosphoramidate Synthesis, 11
AUWERS Flavone Synthesis, 12
AUWERS–INHOFFEN Dienone-Phenol
Rearrangement, 12

Bachmann, 134
Backlund, 298
BAER–FISCHER Amino Sugar Synthesis, 13
Baeyer Diarylmethane Synthesis, 1–11
Baeyer–Drewson Indoxyl Synthesis, 1–12
Baeyer Oxindole Synthesis, 1–11
BAEYER Pyridine Synthesis, 13
BAEYER–VILLIGER Ketone Oxidation, 14
Baeyer–Villiger Tritylation, 1–13
BAILEY Crisscross Cycloaddition, 15
BAKER–VENKATAMAN Flavone Synthesis, 16
BALABAN–NENITZESCU–PRAIL Pyrylium Salt
Synthesis, 17
Bally–Scholl Benzanthrene Synthesis, 1–15
Balson Alkylation, 1–15
BAMBERGER Benzotriazine Synthesis, 17
BAMBERGER Imidazole Cleavage, 18
BAMFORD–STEVENS–CAGLIOTI–SHAPIRO
Olefination, 19
BARBIER Reaction, 20
BARBIER–WIELAND Degradation, 21
BARLUENGA Iodination Reagent, 22
Bartoli, 313
Barton, 326
BARTON Deamination, 23
BARTON Decarboxylation, 25
BARTON–KELLOG Olefination, 25
BARTON–McCOMBIE Alcohol Deoxygenation, 26

BARTON Nitrite Photolysis, 23
BARTON Phenylation of Phenols, Enols, 24
Bart–Scheler, 1–21
Batcho, 217
Baudisch, 1–27
Bauer, 146
Baylis, 253
Beak, 253
Bechamp, 1–29
Becker, 241
BECKMANN Rearrangement or Fragmentation, 27
Beech, 39
Belleau, 120
Benary, 1–31, 103
BENARY Conjugated Aldehyde Synthesis, 26
Bender, 149
Ben-Ishai, 407
Benkeser, 32
BERCHTOLD Enamine Homologation, 28
BERGMAN Cycloaromatization, 29
Bernet, 390
BERNTHSEN Acridine Synthesis, 30
BESTMANN Cumulene Ylides, 31
BIGINELLI Pyrimidone Synthesis, 30
BIRCH–HÜCKEL–BENKESER Reduction, 32
Bischler, 1–35
Bischler–Möhlau, 1–35
BISCHLER–NAPIERALSKI Isoquinoline Synthesis,
32
Blaise, 301
Blanc, 40
BLANC–QUELLET Chloroalkylation, 33
BLICKE–PACHTER Pteridine Synthesis, 34
BLOMQUIST Macrocycles Synthesis, 34
BLUM Aziridine Synthesis, 35
BODROUX–CHICHIBABIN Aldehyde Synthesis, 36
BOGER–CARBONI–LINDSEY Heterocycle
Synthesis, 36
BOGER Thermal Cycloaddition, 37
Boord, 1–41
BORCH Reduction, 38
Borman, 398
Borodin, 172
Borsche, 110
BORSCHE–BEECH Aromatic Aldehyde Synthesis, 39
Bosset, 259
BOUVEAULT Aldehyde Synthesis, 38
BOUVEAULT–BLANC Reduction, 40
BOUVEAULT–HANSLEY–PRELOG–STOLL
Acyloin Condensation, 40

Bouveault–Locquin, 1–45
BOYLAND–SIMS Hydroxyaniline Synthesis, 41
BRANDI–GUARNA Rearrangement, 42
Braun, 311
BRAUN (von) Amine Degradation, 43
Braverman, 248
Bredereck, 1–48
BROOK Silaketone Rearrangement, 44
BROWN Acetylene Zipper Reaction, 45
BROWN Hydroboration, 46–47
BROWN Stereoselective Reduction, 48
Bruylants, 179
BRUYLANTS Amination, 41
Bucherer–Le Petit, 1–54
Büchi, 280
BUCHNER–CURTIUS–SCHLOTTERBECK
  Homologation, 49
BUCHWALD–HARTWIG Aryl Halide Amination, 51
BUCHWALD Heterocyclization, 50
BURGESS Alcohol Dehydration, 51
BURTON Trifluoromethylation, 49
Buttenberg, 114

CADOGAN–CAMERON WOOD Cyclization, 52
Caglioti, 19
Cameron Wood, 52
Campbell, 164
CANNIZZARO Oxidation Reduction, 53
Carboni, 36
CARGILL Rearrangement, 54
CARROLL Rearrangement of Allyl Acetoacetic Esters,
  55
Castro, 353
CHAN Reduction of Acetylenes, 56
CHAPMAN Rearrangement, 56
CHATGILIALOGLU Reducing Agent, 57
Chen, 382
Chichibabin, 36
CHICHIBABIN Amination, 58
CHICHIBABIN Pyridine Synthesis, 58
Chodkiewcz, 136
CIAMICIAN–DENNSTEDT Cyclopropanation, 59
CIAMICIAN Photocoupling, 59
CLAISEN–GEUTER–DIECKMANN Ester
  Condensation, 60
CLAISEN–IRELAND Rearrangement, 61
Clark, 103
Clausen, 418
CLAUSON–KLAUS Pyrrole Synthesis, 62
CLAY–KINNEAR–PERREN Phosphonyl Chloride
  Synthesis, 62
CLEMMENSEN, 63
CLIVE–REICH–SHARPLESS Olefination, 64
CLOKE–WILSON Cyclopropylketone Rearrangement,
  65
Coates, 335

COLLMAN Carbonylation Reagent, 66
Colonna, 180
COLVIN Alkyne Synthesis, 67
COMBES Quinoline Synthesis, 65
COMINS Triflating Reagent, 67
CONIA Cyclization, 68
COOPER–FINKBEINER Hydromagnesiation, 69
COPE–MAMLOC–WOLFENSTEIN Olefin Synthesis,
  70
COPE Rearrangement, 70
COREY Enantioselective Borane Reduction, 72
COREY–FUCHS Alkyne Synthesis, 72
COREY Homologotative Epoxidation, 71
COREY–KIM Oxidizing Reagent, 68
COREY NICOLAU GERLACH Macrolactonization, 73
COREY Oxidizing Reagent, 71
COREY–SEEBACH Dithiane Reagent, 74
COREY–WINTER–EASTWOOD Olefination, 75
CORNFORTH Oxazole Rearrangement, 75
CRABBÉ Allene Synthesis, 76
Crafts, 117
CRIEGEE Glycol Oxidation, 76
CRIEGEE Rearrangement, 77
Cristescu, 266
Cristol, 172
Curtius, 49
CURTIUS Rearrangement, 78

DAKIN Phenol Oxidation, 79
DAKIN–WEST Ketone Synthesis, 79
DANHEISER Annulation, 78
DANISHEFSKY Dienes, 80
Darapsky, 1–87
DARZENS Epoxide Synthesis, 81
Darzens–Nenitzeseu, 1–87
DAVID–MUKAIYAMA–UENO Selective Diol
  Oxidation, 84
DAVIDSON Oxazole Synthesis, 85
DAVID–THIEFFRY Monophenylation of Diols, 84
DAVIES Asymmetric Synthesis, 82
DAVIES Oxidizing Reagent, 83
Decker, 107
DE KIMPE Amidine Synthesis, 86
Delepine, 1–89
DELEPINE Amine Synthesis, 88
DEMANJANOV Rearrangement, 88
DE MAYO Photocycloaddition, 87
Denige, 208
Dennstedt, 59
DESS–MARTIN Oxidizing Reagent, 87
Dieckmann, 60
DIELS–ALDER Cyclohexene Synthesis, 89
DIMROTH Rearrangement, 85
DIMROTH Triazole Synthesis, 90
DJERASSI–RYLANDER Oxidation, 90
Doebner, 199

DOEBNER–MILLER Quinoline Synthesis, 91
DOERING–LA FLAMME Allene Synthesis, 91
DONDONI Homologation, 92
O'DONNELL Amino Acid Synthesis, 271
DÖTZ Hydroquinone Synthesis, 93
DOWD Ring Expansion, 94
Drechsel, 110
DUFF Aldehyde Synthesis, 95
Duisberg, 282
DUTHALER–HAFNER Enantioselective Allylation, 96
Dutt–Wormal, 1–102

Eastwood, 75
ECKERT Hydrogenation Catalyst, 97
EHRLICH–SACHS Aldehyde Synthesis, 98
Einhorn–Brunner, 1–104
Eistert, 9
Elbs, 1–105, 106
ELBS oxidation, 98
Emmert, 1–107
Emmons, 171
ENDERS Chiral Reagent, 99
Endo, 335
Engel, 159
Erdman, 392
Erlenmeyer–Plochl–Bergmann, 1–109
Eschenmoser, 139
ESCHENMOSER MEERWEIN Allylic Acetamidation, 102
ESCHENMOSER Methylenation Reagent, 100
ESCHENMOSER Sulfide Contraction, 101
ESCHWEILER–CLARK Amine ethylation, 103
Etard, 1–112
Evans, 248
EVANS Chiral Auxiliary, 104

FAWORSKI–WALLACH Rearrangement, 105
Fedvadjan, 385
FEIST–BENARY Furan Synthesis, 103
FELDMAN Vinylcyclopentene Synthesis, 106
FELKIN Cyclization, 107
Fenton, 315
FERRARIO–AKERMANN Thiocyclization, 108
FERRIER Carbohydrate Rearrangement, 109
FERRIER Chiral Cyclohexanone, 108
Feyst–Benary, 1–115
FINEGAN Tetrazole Synthesis, 110
Finkbeiner, 69
Finkelstein–Gryszkiewicz–Trochimowski–McCombie, 1–121
Firth, 172
Fischer, 13, 197, 305
FISCHER–BORSCHE–DRESCHEL Indole Synthesis, 110
FISCHER Carbene Complexes, 111
Fischer–Hepp, 1–124

Fischer Oxazole, 1–122
Fittig, 1–125
Fleming, 371
FLEMING–MACH Anthracene Synthesis, 112
Flood, 1–124
Forster–Decker, 1–127
FORSTER–DECKER Amine Synthesis, 107
FORSTER Diazo Synthesis, 114
Fraise, 255
Franchimond, 1–128
FRANKEL–SHIBASAKI 1,5-Isomerization, 113
FREEMAN Lithium Reagent, 115
FREUDENBERG–SCHÖNBERG Thiophenol Synthesis, 116
Freund–Gustavson, 1–129
Freytag, 166
FRIEDEL–CRAFTS Alkylation Acylation, 117
FRIEDLÄNDER Quinoline Synthesis, 118
FRIES Phenol Ester Rearrangement, 119
Fritsch, 292
FRITSCH–BUTTENBERG–WIECHEL Acetylene Synthesis, 114
Fuchs, 72
FUJIMOTO–BELLEAU Cyclohexanone Synthesis, 120
Fujiwara, 155–156
FUJIWARA Arylation Carboxylation, 120
FUJIWARA Lantanide (Yb) Reaction, 121

Gabriel, 310
GABRIEL Amine Synthesis, 122
Gabriel–Colman, 1–140
GABRIEL–HEINE Aziridine Isomerization, 123
Gams, 289
Garegg, 196
GARIGIPATI Amidine Synthesis, 124
GASSMAN Oxindole Synthesis, 125
GASTALDI Pyrazine Synthesis, 125
Gatterman, 320
GATTERMAN–KOCH Carbonylation, 126
Gennari, 357
Gerlach, 73
Geuter, 60
GEWALD Aminoheterocycles Synthesis, 127
Ghera, 151
GIESE Free Radical Synthesis, 128
Gilbert, 332
GILMAN–LIPSCHUTZ–POSNER Organocuprate Reagent, 129
GILMAN–VAN ESS Ketone Synthesis, 130
GINGRAS Reagent, 131
Girard–Sandulescu, 1–146
GLASER–SONDHEIMER–CHOKIEWICZ Acetylene Coupling, 132–133
Goldberg, 386
GOLD Reagent, 134

GOMBERG–BACHMANN–GRAEBE–ULLMANN
  Arylation, 134
Graebe, 134
GRAHAM Diazirine Synthesis, 131
GRANÄCKER Homologation, 135
GRIECO Organoselenides, 136
GRIECO Reagent, 137
GRIESS Deamination, 137
Griffith, 219
GRIGNIARD Reagent, 138
GROB–ESCHENMOSER Fragmentation, 139
Grosheintz, 305
GROVENSTEIN–ZIMMERMANN Carbanion
  Rearrangement, 140
Grubbs, 372
GRUBBS Olefin Metathesis, 141
GUARESKI–THORPE Pyridone Synthesis, 142
Guarna, 42
Guette, 142
GUY–LEMAIRE–GUETTE Reagent, 142


Haack, 391
Haaf, 201
HADDADIN–ISSIDORIDES Quinoxaline Synthesis,
  143
Hafner, 96
HAFNER Azulene Synthesis, 144
HAJOSH–PARRISH Enantioselective Aldol
  Cyclization, 145
HALLER–BAUER Ketone Cleavage, 146
Hansdorf, 413
Hansley, 40
HANTSCH Pyridine Synthesis, 148
HANTSCH Thiazole Synthesis, 147
Hartwig, 51
HASS–BENDER–Carbonyl Synthesis, 149
Hassner, 352
HASSNER Aziridine–Azirine Synthesis, 150
HASSNER–GHERA–LITTLE MIRC Ring Closure,
  151
HASSNER–RUBOTTOM–Hydroxylation, 149
Hauser, 346
HAUSER–BEAK Ortho Lithiation, 152
HAUSER–KRAUS Annulation, 153
HAYASHY Rearrangement, 155
HAYASHY UOZUMI Asymmetric Functionalization,
  154
Heathcock, 163
HECK–FUJIWARA Coupling, 155–156
Heine, 123
HELL–VOLHARDT–ZELINSKI Bromination, 157
HENBEST Iridium Hydride Reagent, 157
HENKEL–RAECKE Carboxylic Acid Rearrangement,
  158
HENRY Nitro Aldol Condensation, 158

HERBST–ENGEL–KNOOP–OESTERLING
  Aminoacid Synthesis, 159
HERZ Benzothiazole Synthesis, 159
HESSE–SCHMIDT "Zip" Reaction, 160
Hickinbottom, 166
HILBERT–JOHNSON Nucleoside Synthesis, 161
Hillman, 253
HINSBERG–STOLLE Indole-Oxindole Synthesis, 162
HINSBERG Thiophene Synthesis, 161
HIYAMA Aminacrylate Synthesis, 162
HIYAMA–HEATHCOCK Stereoselective Allylation,
  163
HOCH–CAMPBELL Aziridine Synthesis, 164
Hoesch, 144
HOFFMAN–YAMAMOTO Stereoselective Allylation,
  167
HOFMANN Amide Degradation, 164
HOFMANN Elimination, 165
HOFMANN Isonitrile Synthesis, 165
HOFMANN–LOEFFLER–FREYTAG Pyrrolidine
  Synthesis, 166
HOFMANN–MARTIUS–REILLY–HICKINBOTTOM
  Aniline Rearrangement, 166
HOLLEMANN Pinacol Synthesis, 168
HONZL–RUDINGER Peptide Synthesis, 168
HOOKER Quinone Oxidation-Rearrangement, 169
HOPPE Enantioselective Homoaldol Reaction, 170
Horner, 184, 387
HORNER–WADSWORTH–EMMONS Olefination,
  171
Hosomi, 319
HOUBEN–HOESCH, 144
Huang–Minlon, 413
Hubert, 218, 289
Huckel, 32
HUISGEN Tetrazole Rearrangement, 173
Hull, 306
HUNSDIECKER–BORODINE–CRISTOL–KOCHI
  Halogenation, 172

Imamoto, 403
Inhoffen, 12
Ireland, 61
Issidorides, 143
IVANOV Grignard Reagent, 174

JACOBSEN Asymmetric Epoxidation, 175
Janovsky, 238
JAPP–KLINGEMANN Hydrazone Synthesis, 176
JAPP Oxazole Synthesis, 176
JAROUSSE–MAKOSZA Phase Transfer Reaction, 177
JEGER Tetrahydrofuran Synthesis, 177
Johnson, 161, 182, 227, 1–191
JONES–SARETT Oxidizing Reagent, 178
JULIA–BRUYLANTS Cyclopropyl Carbinol
  Rearrangement., 179

JULIA–COLONNA Asymmetric Epoxidation, 180
JULIA–LYTHGOE Olefination, 181
JUNG–OLAH–VORONKOV Ether Cleavage, 179

Kaas, 62
KABE Chromanone Synthesis, 182
KAGAN–HORNER–KNOWLES Asymmetric
  Hydrogenation, 184
KAGAN–MODENA Asymmetric Oxidation, 183
KAGAN–MOLANDER Samarium Reagent, 185
KAISER–JOHNSON–MIDDLETON Dinitrile
  Cyclization, 182
KAKIS–KIKUCHI Oxidative Aryl Rearrangement, 186
KALUZA Isothiocyanate Synthesis, 187
KAMENTANI Amine Oxidation to Nitrile, 187
KATRITSKY Amine Displacement, 188
KATRITZKY Stereoselective Olefination, 189
KAUFMANN Dimerization, 190
KAWASE N-Acyl Rearrangement, 191
KECK Allylation, 192
KEINAN Silane Reagent, 174
Kellogg, 25
Kendall, 233
KENNEDY Oxidative Cyclization, 193
Khand, 280
KHARASH–LIPSHUTZ–POSNER Cuprate Reagent,
  194
KHARASH–SOSNOVSKY Allylic Oxidation, 195
Kikuchi, 186
KILIANI–FISCHER Sugar Homologation, 197
Kim, 68
Kindler, 405
Kinnear, 62
Kishner, 1–204, 413
Klingemann, 176
KNOCHEL Zinc Vinyl Coupling, 198
KNOEVENAGEL–DOEBNER–STOBBE
  Condensation, 199
Knoop, 159
Knorr, 203, 277
KNORR Pyrazole Synthesis, 200
KNORR Quinoline Synthesis, 200
Knowles, 184
KNUYANTS Fluoroalkylation, 199
Kocheshkov, 336
KOCH–HAAF Carboxylation, 201
Kochi, 126, 172
KOCHI Cross Coupling, 202
KOENIGS–KNORR Glycosidation, 203
Kohler, 1–210
KOLBE Electrolysis, 204
KOLBE–SCHMIDT Salicylic Acid Synthesis, 204
KONAKA Nickel Oxidizing Agent, 205
KÖNIG Benzoxazine Synthesis, 205
KORNBLUM Aldehyde Synthesis, 206
KOSER Tosylation, 206

Kostanecki, 309
KRAPCHO Dealkoxycarbonylation, 207
Kraus, 153
KRESZE Amination Reagent, 203
Krief, 303
Kriewitz, 295
KRÖHNKE–ORTOLEVA Keto Pyridinium Salts, 208
KUCHEROV–DENIGE Mercuric Catalyzed Hydration,
  208
KUCHTIN, 233
KUHN–WINTERSTEIN–GARREGG–
  SAMUELSSON Olefin Synthesis, 196
KULINKOVICH Hydroxycyclopropanation, 209
KURSANOV–PARNES Ionic Hydrogenation, 210

Laatsch, 317
Ladenburg, 1–224
La Flamme, 91
LAPWORTH (BENZOIN) Condensation, 211
LAROCK Annulation, 212
Laszlo, 239
LASZLO Clay Catalyst, 213
La Torre, 385
LAWESSON Thiocarbonylation Reagent, 214
LEBEDEV Methoxymetbylation, 215
LEHN Cryptand Synthesis, 216
LEIMGRUBER–BATCHO Indole Synthesis, 217
LEIMSTEDT–TANASESCU Acridone Synthesis, 215
Lemaire, 142
Lemieux, 227
LEUCKART–PICTET–HUBERT Phenantridine
  Synthesis, 218
LEUCKART Thiophenol Synthesis, 217
LEUKARDT–WALLACH Reductive Amination, 218
LEY–GRIFFITH Ru Oxidation Reagent, 219
LIEBEN Hypohalide Oxidation, 220
LIEBIG Benzylic Acid Rearrangement, 220
LIEPA Phenantrene Synthesis, 221
Linsay, 36, 312
Lipschutz, 129, 194
Little, 151
Loefler, 166
LOSSEN Rearrangement, 222
LUCHE Ce Reducing Agent, 223
LUCHE Zinc Allylation, 224
Lythgoe, 181

MAC DONALD Porphyrin Synthesis, 225
MADELUNG Indole Synthesis, 225
Mah, 112
Makosza, 177
MAKOSZA Vicarious Nucleophylic Substitution, 226
MALAPRADE–LEMIEUX–JOHNSON Olefin (diol)
  Cleavage, 227
Mamlock, 70
MANDER Methoxycarbonylation Reagent, 228

MANN Ether Dealkylation, 224
MANNICH Aminomethylation, 229
MARKOVNIKOV Regioselectivity, 230
MARSHALK Aromatic Alkylation, 230
Martin, 279
MARTIN Dehydrating Reagent, 231
Martius, 166
MASCARELLI Fluorene Synthesis, 231
Matteson, 279
MATTESON Boronic Esters, 232
MATTOX–KENDALL Dehydrohalogenation Synthesis,
    233
Mc Combie, 26
McCORMACK–KUCHTIN–RAMIREZ Phosphole
    Synthesis, 233
Mc Coy, 255
Mc FADYEN–STEVENS Ester Reduction, 234
Mc MURRY, 235
Meerwein, 102, 395
MEERWEIN Alkylating Reagent, 234
MEINWALD Rearrangement, 236
MEISENHEIMER–JANOVSKY Complex, 238
MEISENHEIMER N-Oxide Rearrangement, 237
MELDRUMS Acid, 238
MENKE–LASZLO Nitration of Phenols, 239
MENZER Benzopyran Synthesis, 239
MEYERS Asymmetric Synthesis, 240
MEYER–SCHUSTER Rearrangement, 241
MICHAEL Addition, 242
MICHAELIS–ARBUZOV Phosphonate Synthesis, 243
MICHAELIS–BECKER–NYLEN Phosphonylation,
    241
Middleton, 182, 344
MIDLAND Asymmetric Reduction, 244
MIESCHER Degradation, 245
MIGITA–SANO Quinodimethane Synthesis, 245
MILAS Olefin Hydroxylation, 246
Millar, 5
Miller, 91
MILLER–SNYDER Aryl Cyanide Synthesis, 246
MINISCI Radical Aromatic Substitution, 247
MISLOW–BRAVERMAN–EVANS Rearrangement,
    248
MITSUNOBU Displacement, 249–250
Miyaura, 368
Modena, 183
Moffat, 370
Molander, 185
MOORE Cyclobutenone Rearrangement, 251
Moriarty, 389
MORIN Penicillin Rearrangement, 251
MORI–SHIBASAKI Catalytic Nitrogenation, 252
MORITA–BAYLIS–HILLMAN Vinyl Ketone
    Alkylation, 253
Moser, 404
MOSHER'S ACID For Chirality Determination, 254

MOUSSERON–FRAISSE–McCOY Cyclopropanation,
    255
Mukaiyama, 84
MUKAIYAMA Aldolization, 256–257
Müller, 292–347
Murahashi, 1–269

Nametkin, 395
Napieralsky, 32
NAZAROV Cyclopentenone Synthesis, 258
NEBER–BOSSET Oxindole Cinnoline Synthesis, 259
NEBER Rearrangement, 258
NEF Reaction, 259
NEGISHI Cross Coupling, 260
Nenitzescu, 57
NENITZESCU 5-Hydroxyindole Synthesis, 261
NENITZESCU Indole Synthesis, 261
NERDEL Enol Ether Homologation, 262
NESMEJANOW Aromatic Mercury Halides, 263
NICKL Benzofuran Synthesis, 263
Nicolaou, 73
NICOLAOU Oxidation, 264
NIEMENTOWSKI Quinazolone Synthesis, 264
NISHIMURA–CRISTESCU N-Glycosidation, 266
NOYORI Chiral Homogeneous Hydrogenation,
    267–268
NUGENT–RAJANBABU Epoxide Homolysis, 269
Nylen, 241
NYSTED–TAKAI Olefination, 270

Oesterling, 159
OHSHIRO Bromoalkene Reaction, 272
Olah, 179
OLEKSYSZYN Aminophosphonic Acid Synthesis, 265
OLOFSON Reagent, 273
OPPENAUER Oxidation, 273
OPPOLZER Asymmetric Allyl Alcohol Synthesis, 274
OPPOLZER Cyclopentenone Synthesis, 275
Ortoleva, 208
ORTON Haloaniline Rearrangement, 275
Ostromislensky, 1–282
OVERMAN Pyrrolidine Synthesis, 276

PAAL–KNORR Pyrrole Synthesis, 277
Pachter, 34
PADWA Pyrroline Synthesis, 277
Paquette, 298
PARHAM Cyclization, 278
Parnes, 210
PARNES Geminal imethylation, 278
Parrish, 145
PASSERINI Condensation, 279
PASTO–MATTESON Rearrangement, 279
PATERNO–BÜCHI 2+2 Cycloaddition, 280
PAUSON–KHAND Cyclopentenone Annulation, 280
PAYNE Rearrangement, 281

PEARLMAN Hydrogenolysis Catalyst, 281
PECHMANN (von) Diazo-olefin Cycloaddition, 282
PECHMANN (von)–DUISBERG Coumarin Synthesis, 282
PEDERSEN Crown Ether, 283
PEDERSEN Niobium Coupling Reagent, 284
PERKIN Carboxylic Acid (Ester) Synthesis, 285
PERKIN Coumarin Rearrangement, 285
PERKOW Vinyl Phosphate Synthesis, 286
Perren, 62
Petasis, 372
PETERSON Olefination, 287
Pfannenstiel, 347
PFAU–PLATTNER Cyclopropane Synthesis, 288
Pfenninger, 352
PFITZINGER Quinoline Synthesis, 288
Pfitzner, 370
Pfitzner–Moffat, 1–298
Pictet, 218
PICTET–HUBERT–GAMS Isoquinoline Synthesis, 289
PICTET–SPENGLER Isoquinoline Synthesis, 289
Pilloty–Robinson, 1–300
PINNER Imino Ether Synthesis, 290
PIRKLE Resolution, 290
Plattner, 288
POLONOVSKY N-Oxide Rearrangement, 290
POMERANZ–FRITSCH–SCHLITTER–MULLER Synthesis, 292
Ponndorf, 236
Posner, 129, 194
POSNER Trioxane Synthesis, 293
Prail, 17
Prelog, 40
PREVOST–WOODWARD Olefin Hydroxylation, 294
Prey, 421
PRINS–KRIEWITZ Hydroxyinethylation, 295
PSCHOR Arylation, 296
PUDOVIC Reaction, 297
PUMMERER Sulfoxide Rearrangement, 297

Quellette, 33

Raecke, 158
Rajanbabu, 269
RAMBERG–BACKLUND–PAQUETTE Olefin Synthesis, 298
Ramirez, 233
RAPP–STOERMER Benzofuran Synthesis, 299
RATHKE β-Keto Ester Synthesis, 299
REETZ Titanium Alkylation Reagent, 300
REFORMATSKY–BLAISE Zinc Alkylation, 301
REGITZ Diazo Transfer, 302
Reich, 64
REICH–KRIEF Olefination, 303
Reilly, 166
REIMER–TIEMANN Phenol Formylation, 304

REISSERT–GROSHEINTZ–FISCHER Cyanoamine Reaction, 305
REMFRY–HULL Pyrimidine Synthesis, 306
REPPE Acetylene Reaction, 306
RICHTER (von) Aromatic Carboxylation, 307
RICHTER–WIDMAN–STOERMER Cinnoline Synthesis, 304
Rickert, 4
RIECHE Formylation, 307
RILEY Selenium Dioxide Oxidation, 308
Rimini, 7
RITTER Amidation, 308
ROBINSON–GABRIEL Oxazole Synthesis, 310
ROBINSON–ALLAN–KOSTANECKI Chromone Synthesis, 309
ROBINSON Annulation, 309
Robinson Fould, 1–322
ROELEN Olefin Carbonylation, 310
ROSENMUND Arsonylation, 311
ROSENMUND–BRAUN Aromatic Cyanation, 311
ROSENMUND–SATZEW Reduction to Aldehydes, 312
ROSINI–BARTOLI Reductive Nitroarene Alkylation, 313
ROTHEMUND–LINDSEY Porphine Synthesis, 312
Rozen, 344
ROZEN Hypofluorite Reagent, 314
Rubbottom, 149
Rudinger, 168
RUFF–FENTON Degradation, 315
RUPE Rearrangement, 315
RUPPERT Perfluoroalkylation, 316
RUSSIG–LAATSCH Hydroquinone Monoether Formation, 317
Ruzicka Fukushima, 1–329
Rylander, 90

Sachs, 98
SAEGUSA Enone Synthesis, 318
Saitzew, 312
SAKURAI–HOSOMI Allylation, 319
Samuelson, 196
SANDMEYER–GATTERMANN Aromatic Substitution, 320
SANDMEYER Isatin Synthesis, 320
Sano, 245
Saret, 178
SCHEINER Aziridine Synthesis, 321
SCHENCK Allyl Peroxide Synthesis, 322
SCHIEMANN Aromatic Fluorination, 323
Schlitter, 292
Schlotterbeck, 49
Schmidt, 160, 204
SCHMIDT Rearrangement, 323
SCHMITZ Diaziridine Synthesis, 324

SCHÖLKOPF–BARTON–ZARD Pyrrole Synthesis, 326
SCHÖLLKOPF Amino Acid Synthesis, 325
SCHOLL Polyaromatic Synthesis, 324
SCHOLTZ Indolizine Synthesis, 325
Schonberg, 116
Schöpf, 356
Schorning, 413
Schuster, 241
SCHWARTZ Hydrozirconation, 327
SCHWEITZER Allylamine Synthesis, 328
SCHWEITZER Rearrangement, 328
SCHWESINGER Base, 329
Seebach, 74
Semler–Wolf, 1–342
SEYFERTH Acyllithium Reagent, 330
SEYFERTH Dihalocarbene Reanent, 331
SEYFERTH–GILBERT Diazoalkane Reagent, 332
Shapiro, 19
Sharpless, 64
SHARPLESS Asymmetric Dihydroxylation, 334
SHARPLESS Asymmetric Epoxidation, 333
SHERADSKY–COATES–ENDO Rearrangement, 335
SHESTAKOV Hydrazino Acid Synthesis, 338
SHEVERDINA–KOCHESHKOV Amination, 336
Shibasaki, 113, 252
SHIBASAKI Cyclization, 337
SIEGRIST Stilbene Synthesis, 338
SIMCHEN Azaheterocycle Synthesis, 339
SIMMONS–SMITH Cyclopropanation, 340
SIMONIS Benzopyrone Synthesis, 341
Sims, 41
Sisti, 356
SKATEBØL Dihalocyclopropane Rearrangement, 341
SKRAUP Quinoline Synthesis, 342
SMILES Aromatic Rearrangement, 343
Smith, 340
SMITH–MIDDLETON–ROZEN Fluorination, 344
SNIEKUS Carbamate Rearrangement, 345
Snyder, 246
SOMMELET Aldehyde Synthesis, 346
SOMMELET–HAUSER Ammonium Ylide Rearrangement, 346
Sondheimer, 132
SONN–MULLER Aldehyde Synthesis, 347
Sosnovsky, 195
SOULA Phase Transfer Catalyst, 348
SPECKAMP Acyliminium Ring Closure, 349
Spengler, 289
SPENGLER–PFANNENSTIEL Sugar Oxidation, 347
STAAB Reagent, 350
STAUDINGER Azide Reduction, 350
STAUDINGER Cycloaddition, 351
STAUDINGER–PFENNINGER Thiirane Dioxide Synthesis, 352
STEGLICH–HASSNER Direct Esterification, 352

STEPHEN Aldehyde Synthesis, 353
STEPHENS–CASTRO Acetylene Cyclophane Synthesis, 353
STETTER 1,4-Dicarbonyl Synthesis, 354
Stevens, 19, 234
STEVENS Rearrangement, 355
Stieglitz, 1–365
STILES–SISTY Formylation, 356
STILL Cross Coupling, 359
STILLE Carbonyl Synthesis, 358
STILL–GENNARY Z-Olefination, 357
Stobbe, 199
Stoermer, 299, 304
Stoll, 40
Stolle, 162
STORK Enamine Alkylation, 360
STORK–HUNIG Cyanohydrin Alkylation, 362
STORK Radical Cyclization, 361
STORK Reductive Cyclization, 361
Story, 1–373
STRECKER Aminoacid Synthesis, 363
STRYCKER Regioselective Ueduction, 363
SUAREZ Photochemical Iodo Functionalization, 364
Suhl, 423
SUZUKI Coupling, 366
SUZUKI (KYODAI) Nitration, 365
SUZUKI–MIYAURA Coupling, 368
SUZUKI Selective Reduction, 367
SWARTS Fluoroalkane Synthesis, 369
SWERN–PFITZNER–MOFFAT Oxidation, 370
SZARVASY–SCHÖPF Carbomethoxylation, 356

Takai, 270
TAMAO–FLEMING, 371
Tanasescu, 215
TEBBE–GRUBBS–PETASIS Olefination, 372
Teer Meer, 1–381
TEUBER Quinone Synthesis, 373
Thieffry, 84
THILE–WINTER Quinone Acetoxylation, 373
Thorpe, 142
Tieman, 1–383; 304
TIETZE Domino Reaction, 374
TIFFENEAU Aminoalcohol Rearrangement, 375
TIMMIS Pteridine Synthesis, 375
TISCHENCO–CLAISEN Dismutation, 376
TODA Solid State Reaction, 377
Todd, 11
TORGOV Vinyl Coupling, 378
TRAHANOVSKY Ether Oxidation, 379
TRAUBE Purine Synthesis, 379
TRAUBE Reducing Alvent, 380
TREIBS Allylic Oxidation, 381
Trippett, 5
Trost, 383

TROST–CHEN Decarboxylation, 382
TROST Cyclopentanation, 382
Tsuji, 394
TSUJI–TROST Allylation, 383

Ueno, 84
UGI Multicomponent Condensation, 384
Ullmann, 130
ULLMANN–FEDVADJAN Acridine Synthesis, 385
ULLMANN–GOLDBERG Aromatic Substitution, 386
ULLMANN–HORNER Phenazine Synthesis, 387
ULLMANN–TORRE Acridine Synthesis, 385
Uozumi, 154

VAN BOOM Phosphorylating Reagent, 388
Van Dorp, 11
Van Ess, 130
VAN LEUSEN Reagent, 388
VARVOGLIS–MORIARTI Hypervalent Iodine
Reagent, 389
VASELLA–BERNET Chiral Cyclopentane Synthesis
from Sugars, 390
VEDEJ Hydroxylation, 391
Venkataraman, 16
Verley, 236
Viehe, 391
Villiger, 14
VILSMEIER–HAACK–VIEHE Reagent, 391
VOIGHT α-Aminoketone Synthesis, 392
VOLHARD–ERDMANN Thiophene Synthesis, 392
Volhardt, 157
von BRAUN Amine Degradation, 43
von PECHMANN Diazo-olefin Cycloaddition, 282
von PECHMANN–DUISBERG Coumarin Synthesis,
282
von RICHTER Aromatic Carboxylation, 307
VORBRÜGGEN Nucleoside Synthesis, 393
Voronkov, 179

WACKER–TSUJI Olefin Oxidation, 394
Wadsworth, 171
WAGNER–MEERWEIN–NAMETKIN Rearrangement,
395
WAKAMATSU (SYNGAS) Amino Acid Synthesis, 396
Wallach, 105, 218
WALLACH Azoxybenzene Rearrangement, 395
WALLACH Imidazole Synthesis, 397
WASSERMANN–BORMANN Macrocyclic Lactam
Synthesis, 390
WATANABE Heterocyclization, 597
WEERMAN Degradation, 398
WEIDENHAGEN Imidazole Synthesis, 399
WEINREB Ketone Synthesis, 400
WEISS Annulation, 399
WENCKER Aziridine Synthesis, 401
WENDER Homlogous Diels–Alder Reaction, 402

WENZEL–IMAMOTO Reduction, 403
WESSELY–MOSER Rearrangement, 404
West, 79
Westfalen–Lettree, 1–411
WESTPHAL Heterocycle Condensation, 404
Weygand, 411
WHARTON Olefin Synthesis, 401
WHITING Diene Synthesis, 403
WIDEQUIST Cyclopropane Synthesis, 405
Widman, 304
Wiechel, 114
Wieland, 21
WILKINSON Decarbonylation Rh Catalyst, 406
WILLGERODT–KINDLER Rearrangement, 405
WILLIAMS–BEN ISHAI Amino Acid Synthesis, 407
WILLIAMSON Ether Synthesis, 408
Wilson, 65
Winter, 75, 373
Winterstein, 196
WISSNER Hydroxyketone Synthesis, 408
WITTIG Olefination, 409
WITTIG Rearrangement, 410
Wohl–Aue, 1–423
WOHL–WEYGAND Aldose Dehydration, 411
WOHL–ZIEGLER Bromination, 410
Wolfenstein, 70
WOLF–KISHNER–HUANG MINLON Reduction, 413
WOLFRAM–SCHORNING–HANSDORF
Carboxymethylation, 413
WOLF Rearrangement, 412
Woodward, 294
WOODWARD Peptide Synthesis, 414
WÜRTZ Coupling, 414

YAMADA Peptide Coupling, 415
YAMAGUCHI Lactonization Reagent, 415
YAMAKAZI–CLAUSEN (GEA) Gruanine Synthesis,
418
Yamamoto, 167
YAMAMOTO Allylation, 416
YAMAMOTO Chirality Transfer, 417
YAMAZAKI Cyanoaniline Synthesis, 419

Zaard, 326
ZAV'YALOV Pyrrole Synthesis, 420
ZEISEL–PREY Ether Cleavage, 421
ZEISS Ketone Transfer, 421
Zelinsky, 157
Ziegler, 410, 419
ZIEGLER Macrocyclic Synthesis, 422
Zimmermann, 140
ZIMMER Rearrangement, 422
ZINCKE–SUHL Cyclohexadienone Synthesis, 423
ZININ Benzidine (Semidine) Rearrangement, 423
ZINKE–ZIEGLER Calixarene Synthesis, 419
ZINNER Hydroxylamine Synthesis, 424

# REAGENTS INDEX

Acetic acid  7, 16, 62, 199, 225, 237, 265, 395
Acetic anhydride  17, 79,108, 110, 149, 181,193, 238, 245, 291, 297, 309 370, 373, 413, 420
Acetyl hypofluoride  314
Acyloxazolidinone (Evans)  104
AD-mix-α and AD-mix-β (Sharpless)  333
AIBN  23, 106, 192, 361
Aluminum bromide  278
Aluminum chloride 62, 108, 119, 126, 144, 162, 199, 218, 278, 324, 362, 423
Aluminum isopropoxide  236, 421
Aluminum oxide  65, 158, 277
(R)(S)-Alpine Borane  46, 244
(R)(S) 1-Amino-2(methoxymethyl)- pyrrolidine SAMP/RAMP (Enders) 99
Ammonia  10, 58, 114, 226, 290, 324, 325, 346, 399
Ammonium acetate  85, 199, 325
Ammonium carbonate  13
Ammonium formate  176
Antimony pentachloride  369
Arsenic acid  311, 342
Azides (alkyl, aryl, formate)  321,401
2,2'-Azobisisobutyronitrile (AIBN)  23, 106, 192,361

Barium hydroxide  347
Benzenesulfonyl chloride  234
Benzoquinone  394
Benzoyl chloride  411
Benzoyl isothiocyante  418
Benzylamine  188
Benzyl chloride  305
3-Benzyl-5(2-hydroxyethyl)-4-methyl-1,3- thiazolium chloride (Stetter)  354
Benzyltriethylammonium chloride  11
BINAP  267
Bis(dibenzylideneacetone)palladium  359
(R)(S) Bis-(diphenylphosphine-1,1- binaphthyl) (BINAP) (Noyori)  267
3-Isopinocamphenyl-9-borabicyclo (3,3,1)nona-none (Alpine-Borane) (Brown)  46, 244
R,R-N,N'-Bis-(3,5-di-tertbutyl-salicylidene)- 1,3-cyclohexane-diamino- manganese chloride (R,R)- Jacobsen Catalyst (Salene)  175
Bis(pyridine)iodonium(I)tetrafluoroborate 22
2,4-Bis-(4-methoxyphenyl)-1,3-dithia-2,4- diphosphetane-2,4-disulfide (Lawesson)  214

2,4-Bis-(4-methylphenylthio)-1,3-dithia- 2,4-diphosphetane-2,4-disulfide (Heimgartner's Reagent) see Lawesson  214

2,4-Bis-(4-phenoxyphenyl)-1,3-dithia-2,4- di-phosphetane-2,4-disulfide (Belleau's Reagent) see Lawesson  214
Bis(methoxycarbonyl)sulfurilimide  203
Bis(tributyl)tin oxide  84
Bis(trifluoroacetoxy)iodobenzene (Varvoglis)  389
Bis(trifluoroethyl)phosphonoester (Still- Gennari)  357
Bis(trifluoromethylbenzene methanolo)diphenyl sulfur (Sulfuran) (Martin)  231
Bis(trimethylsilyl)amide (HMDS)  1, 151, 277
Bis(trimethylsilyl)hydroxylamine (Sheverdina)  336
Bis(triphenylphosphine)nickel dichloride 107, 382, 386
Bis(triphenylphosphine)nickel dicarbonyl 386
Bis(triphenylphosphine)palladium Dichloride  132, 416
9-Borabicyclo(3.3.1)nonane (9-BBN)  244
Borane-methyl sulfide  46, 274
Boron trifluoride etherate  12, 61, 109, 167, 234, 312, 373
Bromine  172, 186, 279, 315, 371
Bromine azide  230
Bromine cyanide  43
Bromoform  165
N-Bromosuccinimide (NBS)  164, 166, 172, 245, 249, 407, 410
Bromotrifluoromethane  316
Butylborane  48
t-Butyl hydroperoxide  86, 120, 183, 333
t-Butyl hypochloride  125, 324
Butyllithium  114, 118, 130, 152, 162, 225, 248, 278, 325, 330
t-Butylmagnesium chloride  138, 313, 209

Cadmium chloride  158
(±)10-Camphorsulfonic acid (CSA)  276
Calcium hydroxide  315
Carbon dioxide  204
Carbon monoxide  201, 280, 310
Carbon tetrachloride  8, 11, 35, 72
1,1-Carbonyldiimidazole (CDI) (Staab) 350
(Carbomethoxysulfamoyl)triethyl ammonium salt (Burgess)  51

Catalysts: Grubbs, Hermann, Nugent,
    Petasis, Schrock, Tebbe
    see Grubbs 141,
    Mukaiyama 256,
    Stetter 354,
    Tebbe 372
Catechol borane 72
Cerium ammonium nitrate (CAN) 379
Cerium chloride 223
Cerium fluoride 346
Claifen 27, 239
Chiral auxiliary 82, 99, 104
Chloranil 312
Chloro bis( $\eta^5$-2, 4-cyclopentadiene-1-yl)
    dimethylaluminum methylene
    titanium (Tebbe) 372
Chloro bis( $\eta^5$-cyclopentadienyl)hydro-
    zirconium (Schwartz) 50, 327
Chloroform 59, 304
m-Chloroperbenzoic acid 14, 149, 188,
    237
N-Chlorosuccinimide 68, 275, 298, 377,
    410
N-(5-Chloropyridyl)triflimide (Comin's
    Reagent) 67
Chlorosulfonic acid 110
Chromium (II) and (III) chloride 380
Chromium tricarbonyl naphthalene 93, 113
Chromium trioxide 21, 168, 245, 178, 245
Chromium dimethylpyrazole (Corey) 71
Chromyl chloride 163
Cinconidine 271
Cobalt chloride 367
Cobalt octacarbonyl 6, 280, 310, 396
Comin's Reagent 67
Copper 263, 296, 311, 386
Copper acetate 399
Copper chloride 132, 187, 190, 195, 353,
    394
Copper cyanide 311
Copper iodide 132, 313
Copper thiophenol (Crabbe) 76
Crown ether 272, 254, 283
Cryptands 216
Cumulene ylides 31
Cyclooctadiene (COD) 267
Cyclopentadienyl zirconium 50, 327
Cyclooctapentadienyl titanium(IV)-
    trichloride-(dichloride) 69, 96,
    168, 269

(DHQ)$_2$PHAL; (DHQD)$_2$ and (DHQ)$_2$PYR;
    (DHQD)$_2$PYR (Sharpless ligands)
    334
DABCO 86, 253
DAST 344
Davis 83
DBU 108, 164, 180, 181, 195, 242, 302,
    326, 383
DEAD 249, 268

Diacetoxy iodobenzene (DIB) 364
Diaryl phosphite (arsenite) Mann 224
1,4-Diazabicyclo(2.2.2)octane (DABCO)
    86, 253
1,8-Diazabicyclo(5.4.0)undec-7-ene (DBU)
    108, 164, 180, 181, 195, 242, 302,
    326, 383
Diazoacetic acid methyl ester 282, 288
Diazomethane 9, 49
Dibromodifluoromethane 49
trans-Di(μ-acetato)bis[di-o-tolyl phosphino)
    benzyl] dipalladium (II)
    (Palladacycle)-(Heck-Fujiwara)
    159, 366
2,3-Dichloro-5,6-dicyano-1,4-benzo-
    quinone (DDQ) 277
Dichloromethylene dimethylammonium
    chloride (Viehe) 391
Dichloromethyl methyl ester (Rieche) 307
N,N-Dicyclohexylcarbodiimide (DDC) 139,
    277, 352, 370
DDQ 277
Diethylaminosulfur trifluoride (DAST) 344
Diethyl azodicarboxylate (DEAD) 249, 268
Diethyl phosphite 272
Diethyl phosphenyl cyanide 415
(±) Diethyl tartarate 183, 184, 333
Diiodosilane (Keinan) 174
Dimethylamino-2-azaprop-2-ene-1-ylidene
    dimethylammonium chloride
    (Gold) 134
Dimethylamino-1-methoxyethene
    (Eschenmoser-Meerwein) 102
Dimethyl(diazomethyl)phosphonates
    (Seifert-Gilbert) 332
p-Dimethylaminopyridine (DMAP) 352
Dimethylformamide 38, 207, 226, 265, 391
Dimethyl(methylene)ammonium iodide
    (Eschenmoser) 100
(4S,5S)-4-(2,2'-Dimethyl-4-phenyl-1,3-
    dioxan-6-yl)-1-phenyl-4H-1, 2,4-
    triazoline-perchlorate 211
Dimethyl sulfate 13, 107, 241
Dimethyl sulfoxide (DMSO) 206, 207,
    222, 246, 258, 296, 318, 321, 370
Dimethyl titanium dichloride 300
1,3-Dimethyl-3,4,5,6-tetrahydro-2(II)-
    pyrimidone (DMPU) 181
3,5-Dinitrochlorobenzene 144
2,4-Dinitrofluorobenzene 411
2,4-Dinitrophenylhydrazine 233
(S)-2-(Diphenylphosphine)-2'-methoxy-1,1'
    binaphthyl (MOP) 154
Diphenylsilane 210
Diphenyldisulfide 106
1,1'-bis(Diphenylphosphino-ferrocene) 51
Dipotassium iron tetracarbonyl 66
1,3-Dithiane (Corey-Seebach) 74
Lithium 4,4-di-t-butylbiphenylide (LiDBB)
    115

Electrolysis 204
Ethylazido formate 321
Ethyl chloroformate 43, 187
Ethyl iodide 85
Ethylmagnesium bromide(chloride) 41,
    138, 162, 403
Ethyl orthoformate 75, 105
Ethylene diammonium acetate 374
(-)-3-Exo(dimethylaminoisoborneol) 274

Flash Vacuum Pyrolysis 321
Fluorosulfonic acid 126
Formaldehyde 103, 105, 385, 399, 419
Formic acid 103, 105, 201, 218, 315, 349
Formamide 39, 122, 247
Fremy's salt 375
Fragmentation 139

Grignard Reagent 138, 174

Heat 3, 7, 28, 29, 37, 42, 68, 70, 75, 89,
    102, 147, 148, 239, 251, 263, 265,
    335, 378
Hermann catalyst 141
(2,3,4,5,6,6) or (2,3,4,4,5,6)Hexachloro-
    cyclohexa-2,4 or (2,5)diene-1-one
    142
Hexamethyldisiloxane 241
Hexamthylenetetramine 88, 95, 346
HMDS 1, 151, 277
Hydrazine 15, 36, 200, 239, 328, 338,
    401, 413
Hydrobromic acid 33, 179, 182
Hydrochloric acid 12, 28, 30, 120, 229,
    259, 290, 392, 423
Hydroxy(tosyloxi)iodobenzene (Koser) 206
Hydrofluoric acid 65
Hydriodic acid 225, 404, 421
Hydrogene 184, 218, 259, 261, 407
Hydrogene peroxide 14, 46, 70, 77, 169,
    180, 246, 247, 371
Hydroxylamine 135, 222, 246
Hypofluoric acid acetonitrile 314

IBDA 389
IBX 264
Imidazole 196, 238
Immobilized poly-L-leucine (Julia-Colonna)
    180
Indium trichloride 256
Iodopyridine tetrafluoroborate (Barluenga)
    22
Iodine fluoride 344
Iodo obenzene diacetate (IBDA) 389
o-Iodoxybenzoic acid (IBX) 264
Iridium hydride (H₂IrCl₆) 157
Iron dibenzoylmethane Fe(DBM)₃ 197, 202
Iron chloride 377
Iron nitrate 239
Iron oxide 413

Iron pentacarbonyl 65
Iron sulfate 247, 315
Isoprene 422
KAPA 45, 160

Lantanum (III) chloride 297
Lantanum isopropoxide 236, 273
La-Li-(S)Binol (Pudovik) 297
Lantanum-Nickel alloy 403
Lead acetate (II) and (IV) 25, 76, 105,
    172, 177
Lithium 33, 330
Lithium aluminum hydride (LAH) 19, 164
Lithium diisopropylamide (LDA) 151, 153,
    232, 240
Lithium perchlorate 3
Lithium tetramethylpiperidine (LTMP) 410
Lithium trimethoxyaluminum hydride 48,
    403
Magnesium 20, 26, 35, 91
Magnesium amalgam 168
Magnesium monoperoxyphthalate (MPPP)
    14
Manganese dioxide 412
Martin sulfurane reagent 231
Meldrum's acid 238
Mercuric acetate 128, 266, 371
Mercuric chloride 92, 108, 263
Mercuric oxide 172, 208
Mercuric trifluoroacetate 381
Methanesulfonic acid 27, 258, 297
Methanesulfonyl chloride 303
Methyl chloroaluminum amide 124
Methyl hypofluoride 314
N-Methoxy-N-methylamide (Weinreb) 400
Methyl iodide 165
Methyllithium 77
Methylmagnesium iodide 164
Methylphosphonic dichloride 233
Methyl cyanoformate 228
Morpholine 360
MOP 154
Montmorilonite K-10 (Laszlo) 213
MMPP 14
MoOPH 391

Nafion-H (Olah) 117
1-(1-Naphthyl)ethyl isocyanate (Pirkle) 290
Nickel chloride 386
Nickel cyanide 306
Nickel peroxide 205
Nickel tetrakistriphenylphosphine 260
Nickel dichloride triphenylphosphine 107
Nitric acid
Nitromethane(ethane) 13, 411, 414
o-Nitrophenylselenocyanate 136
p-Nitrosodimethylaniline 98
Nitrogen dioxide 365
Nugent reagent 141

Organocuprate  129, 194, 242, 336,
        414, 417
Osmium tetroxide  227, 246, 334
Oxalic acid  398
Oxalyl chloride  370
Oxaborolidine (Corey)  72
Oxodiperoxymolybdenum(pyridine)hexa-
        methylphosphoric triamide
        (MoOPH) (Vedej)  391
Ozone  151, 163, 365, 421

Palladacycle  159, 366
Palladium  406
Palladium acetate  120, 212, 318, 382, 394
Palladium/BaSO₄  11, 312
Palladium/Carbon  259, 261, 422
Palladium chloride  132, 154, 212, 318,
        394
PdCl( η³-C₃H₅)₂  154
Palladium tris(dibenzoylmethane)  275, 383
Palladium hydroxide  281
Palladium tetrakistriphenylphosphine  260,
        377, 383, 416
Pentacarbonyl[methoxy(phenyl)carbene]
        chromium (O) (Dötz)  93
Peracetic acid  236, 237
Perchloric acid  424
(Perfluoroalkyl)trimethylsilane (Ruppert)
        316
Periodinane  87
Petasis Ti catalyst  372
Phase Transfer Catalyst (PTC)  78, 177,
        203, 262, 271, 305, 348
Phenylmagnesium bromide  21, 245
Phenyl(trichloromethyl)mercury (Seyferth)
        331, 341
Phenylselenyl bromide  64, 303
2-(Phenylsulfonyl)-3-phenyloxaziridine
        (Davis)  83
Phosgene  290
Phosphazene base (Schwesinger base)
        329
Phosphite triethyl  382
Phosphorous oxychloride  218, 289, 310,
        391, 397
Phosphorous pentachloride  399, 347, 397
Phosphorous pentoxide  27, 32, 241, 258,
        341, 370
Phosphorous pentasulfide  392
Phosphorous tetraiodide  196
Phosphorous tribromide/trichloride  157
Photochemical reaction  23, 59, 65, 87,
        106, 119, 172, 280, 412, 422
Phthalocyanine ligand (V, Mn, Co, Ni, Pd)
        97
Platinum oxide  159
Polyphosphoric acid (PPA)  222
Potassium acetate  55
Potassium-3-aminopropylamine (KAPA)
        45, 160

Potassium hexacyanoferrate  334
Potassium t-butoxide  81, 161
Potassium cyanide  41, 197, 211, 307, 354
Potassium hydroxide  5, 34, 53, 118, 220,
        281, 285, 288, 338
Potassium isocyanate  338
Potassium nitrosodisulfonate (Fremy's
        Salt (373)
Potassium permanganate  58, 169
Potassium persulfate  41, 98
Pressure  184, 204, 218, 244, 384
Proline  145
Pyridine  108, 421
Pyridinethiol  73

Pyridine p-toluenesulfonate (PPTS)
        Grieco)  137
Pyridinium chlorochromate (PCC) (Corey)
        71

Quinidine  262
Quinone  318

Rhenium heptoxide  193
Rhodium chloride dicyclooctadiene  184
Rhodium diacetate  49
Rhodium dichlorodicarbonyl  402
Rhodium tris triphenylphosphine carbonyl
        6
Rhodium tetracarbonyl (Alper)  6
Ruthenium trichloride  397
Ruthenium (IV) oxide  90

Salene  175
Samarium  340
Samarium iodide  181, 185
Schrock catalyst  141
Schwartz cyclopentadienyl zirconium
        50, 327
Schwesinger base  329
Selenium dioxide  308
Silver acetate  294
Silver benzoate  294
Silver oxide  165, 186, 238, 318
Silver salts  64, 108, 186
SMEAH  56
Sodium amide  112, 146, 343, 355
Sodium azide  35, 78, 150, 323, 375
Sodium bis(2-methoxyethoxy)aluminum
        Hydride (SMEAH)  56
Sodium borohydride  92, 128, 197, 218,
        223, 240, 367, 377, 398
Sodium cyanide  34
Sodium diformylamide  122
Sodium cyanoborohydride  38, 107, 398
Sodium hydride  171, 197, 199, 255, 258,
        292, 343
Sodium hypochlorite  114, 131, 164, 220,
        338, 398
Sodium methoxide  24, 56, 60, 90, 226,

285, 375, 379
Sodium nitrite 39, 88, 125, 130, 168, 215,
        217, 231, 263, 296, 304, 320, 323,
        375
Sodium periodate 2, 13, 90, 193, 227
Sodium persulfate 125, 230
Sodium triacetoxyborohydride 38
Stibium pentafluoride 126, 369
Sulfur 10, 108, 127
Sulfurane 231
Sulfur dichloride 159
Sparteine 170
N-Sulfonylhydroxylamine 7
2-Sulfonyl oxaziridine (Davis) 83

Sulfuric acid 16, 17, 153, 176, 200,
        215, 241, 258, 282, 308, 395, 401

Tebbe Ti-Al reagent 372
Tetrabutylammonium difluorotriphenyl-
        Stanate (Gingras) 131
Tetrafluoroboric acid 323
Tetrakis(diethylamino)titanium (Reetz) 300
Tetrakis(triphenylphosphine)nickel(O) 243,
        260, 368
Tetrakis(triphenylphosphine)palladium(O)
        252, 260, 358, 359, 368, 383
Tetramethyl ethylenediamine (TMEDA)
        46, 270
Tetramethylpiperidide lithium (LTMP) 112
Tetramethylsilane 278
Tetramethyl-t-butylguanidine 326
Tetraphenylporphyrine (Schenk) 322
Tetrapropylammonium perruthenate (Ley-
        Griffith) 219
1,2,4,5-Tetrazine (Boger-Carboni-Lindsey)
        36
Thalium tris(trifluoroacetate) (Larock) 211
Thalium ethylate 408
Thiazoline (Rhodanine) 135
Thionyl chloride 34
Thiophenol 248
Thiophosgene 75
2-Thioxo-4-thiazolidone 135
Tin chloride (II/IV) 218, 289, 324, 353, 393
Tin hydroxide 137
Titanium alkoxide 209, 218, 333, 376
Titanium tetrachloride 33, 78, 168, 189,
        235, 270, 307, 319
Titanium dichloride dimethyl 300
Titanium complex (3THF, Mg$_2$Cl$_2$, TiNCO)
        (Shibasaki) 252, 337
p-Toluenesulfonic acid 12, 54, 251, 261,
        292, 405, 407
p-Toluenesulfonyl azide 19, 302
p-Toluenesulfonyl chloride 292
p-Toluenesulfonyk methylisocyanate
        (TosMIC) (Van Leusen) 338
TMEDA 46. 270
TosMIC 388

Triacetoxyborohydride 38, 107
Tributylstannane 23, 26, 94, 361
Tributyltin chloride 69, 245
2,4,6-Trichlorobenzoyl chloride 415
Triethylaluminum 270, 306
Triethylamine 34, 123, 143, 187, 205, 424
Triethyloxonium tetrafluoroborate
        (Meerweiin 234
Trifluoroacetic acid/anhydride 120, 191,
        221, 247, 279, 297
(Trifluoromethyl)trimethylsilane (Ruppert)
        316
Triethylsilane 210, 406
Triethylsilylhydrotrioxide 293
Trimethylaluminum 270
Trimethylsilyl acetate 295
Trimethylsilyl chloride 4, 92, 266, 318
Trimethylsilyl cyanide 362
Trimethylsilyl diazomethane 67
Trimethylsilyl thiazole (Dondoni) 92
Trimethylsilyl triflate 393
Triphenylbismuth diacetate 84
Triphenylbismuth carbonate (dichloride) 24
Triphenylphosphine 8, 25, 35, 72, 101,
        196, 249, 275, 322, 409
Triphenylphosphine copperhydride
        (Stryker) 363
Triphenyltin hydride 128
1,1,1-Tris(acetyloxy)-1,1-dihydro-1,2-
        benzo-iodoxol-3(1H)-one
        (Periodinane) (Dess-Martin) 87
Tris(trimethylsilyloxy)ethylene (Wissner)
        408
Tris(triphenylphosphine)chlororhodium
        (Wilkinson) 406

Urea 231
Ultrasound 301

Vanadium oxytrifluoride (Liepa) 221
Viehe imminium reagent 391
Vinyl chloroformate 273

Wilkinson 406

Xanthates 217

Yterbium 121
Yterbium triflate (Mukaiyama) 256

Zinc 135. 198, 217, 224, 235, 270
        351, 377, 390
Zinc amalgam 63, 301
Zinc copper couple 235, 340, 414
Zinc chloride 30, 33, 80, 117, 166, 232
Zinc cyanide 126

# REACTIONS INDEX

Acetoxylation 373
Acetylene coupling 132-133, 306, 353
Acylation 117, 108
Acyl anion 74
Addition 121, 129, 174, 194, 354
Addition regioselective 150, 230
Addition stereoselective 150, 232, 242, 256, 300
Aldolization 13, 158
Aldolization enantioselective 104, 145, 170, 256
Allene synthesis 76, 91
Alkylation 103, 117, 152, 230, 234, 253, 300, 301, 360, 362
Alkylation germinal ("C") 278
Alkylation mono ("N") 107
Alkylation ("O") 234, 317
Allylation 192, 224, 319, 328, 383, 416
Allylation enantioselective 96, 274
Allylation stereoselective 163, 167, 319
Allylic acetamidation 102
Alkylation 67, 72, 114
Amidation 308
Amidine synthesis 86, 124, 290
Amination 13, 41, 51, 58, 88, 107, 122, 188, 203, 336
Aminoacid synthesis 159, 271, 325, 338, 363, 396, 407
Aminomethylation 134, 229
Annulation 32, 39, 58, 78, 85, 90, 120, 144, 153, 161, 182, 205, 212, 215, 231, 239, 263, 265, 276, 278, 309, 280
Annulation regiocontrolled 153
Antracene synthesis 112
Aromatization 221, 423
Arylation 120, 130, 296, 386
Arsonylation 311
Asymmetric synthesis 82, 99, 154, 232, 240, 274
Azaheterocyclization 110, 118, 162
Aziridine synthesis 35, 123, 131, 150, 164, 321, 324, 401

Bromination 142, 157, 410

Calixarene synthesis 419
Carbonylation 6, 66, 149, 259, 310
Carboxylation 120, 201, 203, 204, 208, 285, 307, 354, 358
Carboxymethylation 413
Cascade reaction 374
Chiral catalyst 213
Chiral induction 3
Chiral reagents 104, 254, 417
Chiral synthesis 390, 417
Chloromethylation 32
Cleavage 90, 137, 144, 146, 179, 227, 421

Condensation 11, 12, 26, 40, 60, 182, 199, 211, 213, 279, 282, 299, 324, 352, 358, 375, 378, 385, 387, 392, 393, 400, 404, 418
Complex 238
Conversion 259, 318
Coupling 20, 152, 155-156, 185, 198, 202, 235, 284, 366, 368, 378, 414
Criss-cross reaction 15
Cross coupling 197, 204, 260, 359, 366
Crown ether 283
Cryptands 216
Cyanation 311, 419
Cyclization 10, 12, 13, 16, 17, 30, 32, 34, 38, 52, 62, 106, 108, 110, 148, 182, 231, 258, 262, 275, 278, 288, 292, 337, 382, 397, 402, 420
Cyclization chiral 108
Cyclization dipolar (2+2; 1+3; 4+2) 31, 87, 280, 282, 332, 351
Cyclization enantioselectivity 145
Cyclization photochemical 87
Cyclization oxidative 193
Cyclization regioselective 93
Cyclization stereoselective 107, 390
Cyclization thermal (4+2) 37, 68
Cycloaddition (2+2; 3+1; 4+2; 5+2; 6+2) 4, 15, 31, 89, 280, 282, 351
Cycloaromatization 4, 29
Cyclocondensation 60, 339, 353
Cyclopropanation 59, 111, 255, 288, 340, 405

Dealkoxycarbonylation 207
Dealkylation 224, 273
Deamination 23, 137
Decarbonylation 406
Decarboxylation 25, 79, 382
Degradation 21, 43, 78, 164, 245, 315, 398, 411
Dehydration 8, 51, 231
Dehydrohalogenation 233
Diazo reaction 114, 302, 332
Diene synthesis 89, 104, 402, 403
Diene synthesis, regio and stereo controlled 4, 80
Dihalocarbene 331
Dihydroxylation asymmetric 334
Dimerization 168, 190
Dismutation 376
Displacement 249
Domino reaction 374

Electrochemistry 204
Elimination 165
Ene reaction 3, 107
Epoxidation 71, 81
Epoxidation asymmetric 175, 180, 333

Esterification 299, 352
Etherification 408

Fluorination 131, 314, 323, 344, 369
Fluoroalkylation 199, 369
Formylation 36, 38, 39, 95, 98, 130, 206,
    304, 307, 346, 347, 353, 356, 391
Fragmentation 27, 139
Free radical reaction 128
Functionalization 23, 364, 408
Functionalization asymmetric 154
Functional group transformation 318

Glycosidation 203, 266

Haloalkylation 33
Halogenation 8, 172
Halogenation regioselective 142
Heteroannulation 10, 34, 36, 39, 50, 52,
    58, 62, 85, 90, 276, 299, 312
Heterocyclization 17, 35, 91, 103, 125,
    127, 131, 142, 143, 147, 148, 159,
    161, 162, 164, 166, 167, 200, 205,
    215, 217, 221, 225, 261, 277, 282,
    288, 289, 292, 299, 304, 306, 309,
    310, 320, 325, 326, 339, 341, 342,
    375, 379, 392, 397, 399, 404, 418,
    420
Homolysis 269
Homologation 9, 28, 49, 50, 71, 135, 136,
    197, 262
Hydrazone synthesis 176, 338
Hydroboration 46
Hydration 269
Hydroxylation 41, 149, 246
Hydrogenation stereoselective 267
Hydrogenation "adjustable" 97
Hydromagnesation 69
Hydroquinone synthesis 93
Hydrozirconation 327
Hydrogenation asymmetric 184, 267
Hydrogenation ionic 210
Hydrogenolysis 281
Hydrolysis 174
Hydroxycyclopropanation 209
Hydroxylation 149, 246, 294, 391, 408
Hydroxylation stereospecific 371
Hydroxymethylation 295

Iodination 22, 364, 389
Isomerization 113, 123

Ketone synthesis 130

Lactam macrocyclization 34, 73, 398, 422
Lactonization 73, 415, 398
Lithiation 73, 415, 398

Metathesis 141
Methylenation 100

Methoxycarbonylation 228
Methoxymethylation 215
Migration (alkyl) 85
Michael Initiated Ring Closure (MIRC) 151
Multi step reaction 246, 259, 269, 290, 293
    305, 317
Multi component reaction 384

Nitration 213, 239, 365
Nitration regioselective 142
Nitroaldol condensation 158
Nitrogenation catalytic 252
Nitrosation 142
Nucleoside synthesis 161, 393

Olefination 19, 25, 64, 70, 75, 171, 181,
    270, 298, 357, 372, 383, 401, 409
Olefination stereoselective 75, 171, 189,
    287, 303
Organometallics 20, 138, 174, 263
Oxidation 2, 14, 41, 69, 71, 76, 83, 87, 90,
    98, 169, 178, 187, 195, 205, 219,
    220, 264, 273, 308, 322, 347, 370,
    373, 379, 381, 394
Oxidation symmetric 183
Oxidative aryl rearrangement 186
Oxidative cyclization 193
Oxidation selective 84

Peptide synthesis 168, 414, 415
Perfluoroalkylation 316
Phase transfer catalyses 177, 348
Phenylation 24, 84, 131
Photocoupling 59
Photolysis 23
Phosphazene base 329
Phosphonation 233, 243, 265, 286, 297
Phosphonation stereoselective 1, 11
Phosphonylation 1, 62, 241, 286, 388
Porphyrin synthesis 312
Protection 137

Quinone synthesis 373

Reagents 273, 284, 350, 357, 388, 389,
    391
Rearrangement 5, 7, 12, 16, 27, 42, 44,
    54, 56, 61, 65, 75, 77, 85, 88, 105,
    109, 116, 153, 158, 169, 173, 179,
    191, 222, 236, 248, 251, 258, 323,
    328, 341, 422
Rearrangement benzylic 220
Rearrangement catalytic 12, 17, 18, 54,
    55, 77, 119, 186, 241, 281, 285,
    315, 341, 343, 355, 375, 395, 404,
    423
Rearrangement hetero (3,3, and 3,5) 335
Rearrangement intramolecular 153, 279,
    285, 297, 345, 346, 395
Rearrangement photochemical 54, 119

Rearrangement sigmatropic  3, 70, 140,
        248, 410
Rearrangement stereoselective  281
Rearrangement thermal  55, 158, 166,
        251, 328, 335, 405
Redox reaction  53
Reduction  32, 40, 57, 63, 223, 234, 236,
        272, 312, 350, 413
Reduction enantioselective  72, 244
Reduction regioselective  272, 363
Reductive amination  38, 103, 218
Reductive cyclization  361
Reductive iodination  174
Reductive nitroarene alkylation  313
Regiospecific synthesis  230
Resolution optical  290
Ring closure  151, 349
Ring contration  412
Ring enlargement  5, 12, 14, 94, 375

Solid state reaction 377
Stilbene synthesis 338
Substitution  129, 247, 320, 386
Sulfide contraction  101

Tandem reaction  374
Thiacarbonylation  214
Thiocyclization  108
Trifluoromethylation  49, 67

Vicarious nucleophilic substitution  226
Vinyl cyclopentadiene synthesis  106

Ylide rearrangement  346

Zipper reaction 45, 160

# Functional Group Transformations Index

| From: To: → | Alkanes | Cycloalkanes | Alkenes | Alkynes | Aryls | Halogen compounds | Alcohols | Phenols | Ethers, Quinones | B, S and Si compounds | P and Bi compounds | Nitro, N Azo, A Hydraz |
|---|---|---|---|---|---|---|---|---|---|---|---|---|
| Alkanes | 9 | 278, 280, 382, 422 | 64, 100, 139, 303, 318, 338 | | | 157, 369 | 381 | | | | | 23 |
| Cycloalkanes | | 9, 28, 64, 94, 106, 160, 262, 341, 395 | 179 | | 251 | 157 | 154, 195 | | | | | |
| Alkenes | 184, 210 | 3, 37, 59, 68, 78, 80, 87, 89, 107, 111, 113, 185, 203, 255, 258, 288, 331 | 26, 70, 91, 140, 141, 262, 359 | 144 | 4, 12, 112, 197, 221 | 22, 230 | 46, 48, 69, 208, 246, 294, 334 | | | | | |
| Alkynes | | 275, 280, 306, 381, 402 | 56, 76, 260, 274, 315, 363 | 45, 132, 353 | 29 | | | | | | | |
| Aryls | | 32, 423 | 155 | | 49, 56, 93, 130, 153, 155, 190, 230, 296, 324, 343, 366, 368, 386 | 33, 142, 169 | 98 | | | | 108 | 142, |
| Halogen compounds | 57, 192, 272, 380, 410 | 151, 272 | 272, 314 | | 314 | 275 | | | | | | |
| Alcohols | 26 | 26 | 51, 75, 196, 383 | | 117 | 8 | | 408, 240 | 84, 137, 213, 234 | | | 249 |
| Phenols | | | 383 | | | | | 24, 373 | 2, 317, 373 | 110 | | |
| Ethers, Quinones | | | 269 | | | | 77, 179, 224, 281, 421 | 61, 169, 224, 421 | | 5, 183, 297 | | |
| B, S and Si compounds | 101 | | 181, 287, 298 | | | | 249, 371 | | | | 44 | |
| P and Bi compounds | | | | | | | | | | 217 | 62, 241, 243, 265, 286, 297 | |
| Nitro, Nitroso, Azo, Azoxy, Hydrazo, Azides | | | 25 | | | | | | | | | 114, 395 |
| Amines | 23 | | 70 | | 137, 231 | 320, 323 | 88 | 17, 41 | | | | |
| Organometallic compounds | | | 377 | | 115 | | | | | | | |
| Transition metal compounds | | 242 | | | 198 | | 391, 403 | | | | | |
| Aldehydes | 63 | | 19, 199, 235, 245, 270, 397, 409 | 71 | | 174 | 20, 53, 59, 96, 104, 158, 163, 223, 236, 287 | 79 | | | 1, 11 | |
| Ketones | 63, 413 | 146, 309, 399, 405 | 19, 71, 171, 199, 233, 270, 401 | 67, 332 | | 149, 344 | 20, 145, 157, 168, 224, 244, 377 | | 405 | | | |
| Acids, Anhydrides, Esters | 23 | | | | | 344 | 40 | | | | | 176 |
| Amides, Amidines, Nitriles | | | | | | 25 | | | | | | |
| Hydroxy-aldehydes or -ketones, Sugars, Hydroxy acids | | 108 | | | | | | | | | | |
| Amino acids, Peptides | | | | | | | | | | | | |
| Miscellaneous, including heterocycles | | | 144 | | | | | | | | | |

# Functional Group Transformations Index

| Organometallic compounds | Aldehydes | Ketones | Acids, Anhydrides, Esters | Amides, Amidines, Nitriles | Hydroxy-aldehydes or -ketones, Sugars, Hydroxy acids | Amino acids, Peptides | Heterocycles | | | | Nucleosides | Miscellaneous, including heterocycles |
|---|---|---|---|---|---|---|---|---|---|---|---|---|
| | | | | | | | 3,4,5 ring 1 heteroatom | 3,4,5 ring 2 or more heteroatoms | 6,7 and large ring 1 heteroatom | Other heterocycles | | |
| | 98 | 309, 38 | 311.472 | 120 | | | | | 42 | | | |
| | | 308, 406 | 120 | | | | | 65 | | | | 362 |
| | 208, 236 | 6, 186, 394 | 201 | | | 50, 150, 175, 180, 277, 321, 322, 333 | 212 | | 17, 212 | 22 | | 362 |
| 152 | | | 391, 394 | | | 50 | | | | | | |
| | 38, 95, 126, 307 | | 120, 204, 247, 307, 311, 413 | 110 | | | | | 218 | 18, 226, 367 | | 226, 345, 419 |
| 12 | 36, 66, 206, 346 | 316 | | | | 103 | | 147, 252 | | | | |
| | 71, 76, 84, 87, 178, 219, 227, 273, 370 | 68, 71, 72, 84, 205, 375 | | | | 51, 177, 193, 231 | | 85 | | 216 | | 73 |
| | 309 | | | | | 263, 299 | | | 239, 282, 309, 341 | | | 416 |
| | | 379 | | | | 62 | | | 278 | 205 | | 417 |
| | | | | 128 | | | | | | 74 | | |
| | | | | | 388 | | | 31 | | 31, 304 | | |
| 23 | 146 | 146, 258, 400 | 412 | 356 | | 52, 217, 313, 326 | 52, 90, 110, 267 | 52, 215 | | 328 | | 32 |
| 07, 212, 356 | 39 | 375 | | 187 | | 125, 182, 164, 261, 276, 320, 401 | 134, 159 | 91, 118, 289, 292, 342, 397 | | 30, 65, 259, 265 | | 83 |
| 138 | 356 | | 254, 330 | 128 | | | 15 | | | | | 327 |
| | 92, 99, 350 | 358 | 7, 105 | 105 | | 396 | | 337 | | 121, 301 | | 129, 131, 194, 281, 300 |
| | 270 | 92, 99, 364, 362 | 53, 75 | 87, 246, 376 | 211 | 62, 365 | | | 56, 142, 146 | 30 | | 324 |
| | 136, 347 | 136 | 14, 27 | 27, 195 | 256 | 81, 127, 161, 200, 277, 420 | 10, 39, 85 | 27, 125, 142, 148, 182, 200 | 32 | | | |
| | 234, 312 | 79, 245 | 117 | 125 | | 212, 259, 420 | 90 | 182, 212, 330 | 265 | | | |
| | 398 | 20, 34, 119, 130, 330 | | 124, 290 | 40 | 127, 131, 326 | 182 | 118, 142, 282, 339 | 339, 343, 379, 392, 404 | | 181, 288, 393 | 73 |
| | | | | | 7, 13, 109, 197, 203, 315, 411 | 159 | 176 | | | | | |
| | | 79 | | | | 163, 271, 414, 415 | | | | | | |
| | | 240 | 83, 245, 285, 288 | | | 325, 407 | 35 | 75, 123, 173, 191, 285 | 12, 13, 85, 143, 166, 182, 188, 291, 305 | 36, 162, 225, 231, 312, 339, 349, 385, 387, 398 | | 170, 232, 240, 253, 267, 283, 319, 348 |

445

Printed and bound by CPI Group (UK) Ltd, Croydon, CR0 4YY

03/10/2024

01040413-0016